Healthy Ageing and Longevity

Volume 11

Series Editor

Suresh I.S. Rattan, Department of Molecular Biology and Genetics, Aarhus University, Aarhus, Denmark

Rapidly changing demographics worldwide towards increased proportion of the elderly in the population and increased life-expectancy have brought the issues, such as "why we grow old", "how we grow old", "how long can we live", "how to maintain health", "how to prevent and treat diseases in old age", "what are the future perspectives for healthy ageing and longevity" and so on, in the centre stage of scientific, social, political, and economic arena. Although the descriptive aspects of ageing are now well established at the level of species, populations, individuals, and within an individual at the tissue, cell and molecular levels, the implications of such detailed understanding with respect to the aim of achieving healthy ageing and longevity are ever-changing and challenging issues. This continuing success of gerontology, and especially of biogerontology, is attracting the attention of both the well established academicians and the younger generation of students and researchers in biology, medicine, bioinformatics, bioeconomy, sports science, and nutritional sciences, along with sociologists, psychologists, politicians, public health experts, and health-care industry including cosmeceutical-, food-, and lifestyle-industry. Books in this series will cover the topics related to the issues of healthy ageing and longevity. This series will provide not only the exhaustive reviews of the established body of knowledge, but also will give a critical evaluation of the ongoing research and development with respect to theoretical and evidence-based practical and ethical aspects of interventions towards maintaining, recovering and enhancing health and longevity.

More information about this series at http://www.springer.com/series/13277

Daniel Muñoz-Espin · Marco Demaria
Editors

Senolytics in Disease, Ageing and Longevity

Springer

Editors
Daniel Muñoz-Espin
CRUK Cambridge Centre Early Detection Programme, Department of Oncology
Hutchison/MRC Research Centre
University of Cambridge
Cambridge, UK

Marco Demaria
European Research Institute for the Biology of Aging (ERIBA)
University Medical Center Groningen
Groningen, The Netherlands

ISSN 2199-9007 ISSN 2199-9015 (electronic)
Healthy Ageing and Longevity
ISBN 978-3-030-44902-5 ISBN 978-3-030-44903-2 (eBook)
https://doi.org/10.1007/978-3-030-44903-2

This Springer imprint is published by the registered company Springer Nature Switzerland AG
The registered company address is: Gewerbestrasse 11, 6330 Cham, Switzerland

Preface

The population continues to age at an increasing rate in virtually every country worldwide. According to data from the 2019 Revision of World Population Prospects (United Nations), by 2050 one in six people will be over age 65, with peaks of one in four people in Europe and Northern America. Remarkably, the number of individuals aged 80 years or over is projected to triple, from 143 million in 2019 to 426 million in 2050. This situation is expected to increase dramatically the incidence of multiple age-related pathologies, and to have an enormous impact on public health.

Cellular senescence, a response to damage and stress characterized by a stable cell cycle arrest and a complex secretory phenotype (SASP), was conceived, from the initial experiments of Leonard Hayflick and Paul Moorhead (1961), as an underlying cause of ageing. However, until recently, a formal *in vivo* demonstration of cellular senescence as a hallmark of ageing has remained elusive. Only in the last 10 years, the development of genetically-engineered mouse models allowed to conclude that the selective elimination of senescent cells attenuates a number of age-related dysfunctions and promotes healthspan and lifespan. This notion boosted the development of senotherapies—therapies aimed at interfering with senescent cells. Among those, the most investigated senotherapies are based on compounds that can either selectively kill senescent cells, also called senolytics, or inhibit the SASP, also called senostatic or senomorphics. Preclinical validation of various senolytics confirmed that the eradication of senescent cells delays, and in some cases reverts, a number of age-related disorders. These pathologies include, among others, cardiovascular diseases, neurological disorders, type 2 diabetes, inflammatory diseases, fibrosis, geriatric syndromes, musculoskeletal impairments, and cancer.

Cellular senescence is a very heterogeneous process that depends on a number of variables including the trigger or type of stress, the particular cell type and the tissue of origin. For this reason, a universal marker able to identify senescent cells unequivocally and to serve as a target for interventions seems to be lacking. As a consequence, one main strategy to identify drugs with senolytic activity has been the use of large-scale screening methods based on siRNA and small compounds libraries in the context of cytotoxicity readouts. This is the case of the combination of the tyrosine kinase inhibitor, dasatinib and the flavonoid, quercetin, which promotes senolysis in particular cell types. Also, as senescent cells upregulate pro-survival pathways to become resistant to cell death inducers, it has been shown that inhibitors of the Bcl-2 family of antiapoptotic proteins (e.g. ABT-263 and ABT-737) are potent and more specific senolytics. Similarly, other strategies include the use of molecules interfering with the interaction between p53 and FOXO4 (FOXO4-DRI peptides) or between p53 and MDM2 (UBX0101), facilitating p53-mediated apoptosis. The identification of vulnerabilities in senescent cells has been occasionally based on some markers and features of senescent cells, such as increased lysosomal senescence-associated β-galactosidase (SAβ-gal) activity, and this includes chaperone HSP90 inhibitors (e.g. 17-AAG and 17-DMAG) and drug delivery systems (e.g. nanocarriers coated with galacto-oligosaccharides, GalNPs). Finally, screening of compound libraries has allowed the identification of natural compounds with senolytic activity, including the flavonoid, fisetin, piperlongumine and more recently, cardiac glycosides.

Unfortunately, this arsenal of senolytics is not exempt from side effects and toxicities, which limits their clinical potential. For example, it is well known that ABT-263 (navitoclax) causes neutropenia and thrombocytopenia. Therefore, despite the successful preclinical validation of a number of formulations with senolytic activity in different animal disease models, drug-associated toxicities have compromised validation in early phase clinical trials. The development of more selective second-generation senolytics requires a more accurate understanding of the molecular properties of senescent cells and the identification of senescence biomarkers. Despite these aforementioned limitations, some senolytic agents have already advanced to trials testing. This includes UBX0101 for the treatment of osteoarthritis (now in phase II), the cocktail dasatinib and quercetin for the management of idiopathic pulmonary fibrosis and chronic kidney disease, and ABT-263 in combination with senescence-inducing chemotherapies for the treatment of advanced solid tumors.

The SASP is considered to be the main senescence-associated component responsible for tissue degeneration and inflammation, and a major driver of age-related disorders, particularly when senescent cells accumulate and persist in organs. Therefore, in addition to senolytics, an innovative and exciting therapeutic strategy is the development of compounds capable of manipulating the detrimental effects of the senescent cells via the inhibition of SASP components. Among these senomorphic or senostatic compounds, inhibitors of NFκB (e.g. metformin, resveratrol, simvastatin), JAK/STAT (e.g. ruxolitinib) and mTOR (e.g. rapamycin) have been successfully tested at the preclinical level. This strategy is however still in its infancy, and may also exert 'on-target' side effects such as attenuation of tumor immune surveillance and retardation of tissue repair and remodeling: processes where short-term senescence and SASP seem to play beneficial functions.

Finally, it is worth mentioning that there is an emerging interest within the scientific community to develop tools to detect and monitor senescent cells *in vivo*. This would allow clinicians to monitor longitudinally the burden of senescent cells in multiple age-related disorders and also, the assessment in cancer patients in response to radiotherapy and chemotherapy. To date, a collection of senoprobes has been developed, mainly consisting of fluorescent SAβ-gal-dependent probes and also in diagnostic nanoparticles. However, the translation of senoprobes to the clinic may require the adaptation of more conventional bioimaging systems in humans, like Magnetic Resonance Imaging (MRI) and Positron Emission Tomography (PET), by developing contrast agents and radionuclides, respectively. The field of senoprobes is very much at its beginning but it is expected to experience an extraordinary expansion in the next few years provided that more selective biomarkers of cellular senescence are identified.

Preclinical studies have convincingly concluded that the eradication of senescent cells can alleviate multiple age-related pathologies in animal models. Despite these exciting results, many challenges still remain: (i) better insights in the intrinsic mechanisms of cellular senescence and the identification of more specific senescence-associated biomarkers, (ii) the development of more selective senolytics and senoprobes, and (iii) a deeper understanding on the required doses, biodistribution and timeline of elimination of potential senotherapeutics. Despite these limitations, some first-generation senolytics are already in clinical trials and the scientific community awaits the translation of successful preclinical studies to promising, confirmed and proven principles. We are now entering an exciting era in which we are uniquely positioned to translate anti-senescence therapies to medical applications, a strategy that may have an important impact on multiple human age-related disorders and, ultimately, the promotion of healthy ageing.

Cambridge, UK — Daniel Muñoz-Espin
Groningen, The Netherlands — Marco Demaria

Contents

Part I
Origins and Development of Senotherapies

Chapter 1
Senolytic Drug Development

Yonghan He, Guangrong Zheng, and Daohong Zhou

Abstract Cellular senescence is a hallmark of aging because senescent cells (SnCs) accumulate with age and play a causative role in many age-related diseases. Selectively eliminating SnCs has been emerging as a new strategy for treating age-related diseases and extending healthspan. Small molecules that targeting different SnC anti-apoptotic pathways (SCAPs) to selectively kill SnCs are termed senolytics. Up to date, several classes of senolytic agents, including naturally occurring compounds and their derivatives, and targeted therapeutics, have been identified. Here we discuss the biological significances of cellular senescence in aging, and summarize some of the known naturally occurring and targeted senolytic agents and their targets. As most of the known naturally occurring compounds or targeted senolytics have limitations to be developed as therapeutics for human applications, development of more specific and potent senolytic agents that can reduce the on-target and/or off-target toxicity of senolytics, is urgently needed to improve healthy aging in humans.

Keywords Cellular senescence · Senolytics · Senescence-associated secretory phenotype · Aging · Age-related diseases

1.1 Cellular Senescence

Cells become permanently growth arrested after extensive replication or as a result of exposure to stress, which prevents propagation of genetically unstable and damaged cells, and promotes their removal by the immune system (Childs et al. 2015). Therefore, cellular senescence normally functions as a vital tumor suppressive mechanism and also plays an important role in tissue damage repair. However, if the

Y. He · D. Zhou (✉)
Department of Pharmacodynamics, College of Pharmacy, University of Florida, 1345 Center Drive, Gainesville 32610, FL, USA
e-mail: zhoudaohong@cop.ufl.edu

G. Zheng
Department of Medicinal Chemistry, College of Pharmacy, University of Florida, 1345 Center Drive, Gainesville 32610, FL, USA

D. Muñoz-Espin and M. Demaria (eds.), *Senolytics in Disease, Ageing and Longevity*, Healthy Ageing and Longevity 11,
https://doi.org/10.1007/978-3-030-44903-2_1

increase in senescent cell (SnC) production persists beyond the immune clearance capacity or the immune system is compromised and cannot efficiently remove SnCs, SnCs can accumulate. Under such circumstances, SnCs can play a causative role in aging and age-related diseases by inducing chronic oxidative stress and inflammation via increasing the production of reactive oxygen species (ROS) and secretion of a plethora of inflammatory mediators (e.g., cytokines and chemokines), growth factors, and extracellular proteases—termed the senescence-associated secretory phenotype (SASP) (Campisi 2013). Therefore, it has been suggested that inhibiting the induction of senescence might be detrimental, but promoting SnC clearance is beneficial. This suggestion is supported by the findings: (1) SnCs accumulate with aging, particularly at the sites of age-related pathologies (Childs et al. 2015); (2) SnCs can contribute to 'inflammaging' and other age-related pathologies in part via expression of SASP (Tchkonia et al. 2013); and (3) clearance of SnCs using a transgenic approach delays the onset of several age-related diseases and disorders including cancer in naturally aged mice and prolongs their lifespan (Baker et al. 2016). These findings demonstrate that SnCs are novel therapeutic targets of aging and age-related diseases.

1.2 SnCs are Emerging Therapeutic Targets

Although SnCs have been associated with various biological and pathological processes with aging, the causal relationship between SnCs and age-related diseases and disorders remained unclear until the year of 2011 (Baker et al. 2011). Baker et al. designed a transgenic strategy for the clearance of SnCs in progeroid mice and demonstrated that SnC removal can significantly delay the onset of several age-related pathologies and prolong the healthspan. The study provides solid evidence supporting that SnCs are causally implicated in generating age-related phenotypes (Baker et al. 2011). Using the same genetic approach, they revealed that clearance of SnCs not only delayed the onset of tumorigenesis and attenuated age-related deterioration of several organs, but also significantly extended the lifespan of normal mice (Baker et al. 2016). Since then, an increasing body of evidence has accumulated to demonstrate that SnCs play a causative role in a variety of diseases using mouse models, including atherosclerosis (Childs et al. 2016), osteoarthritis (Jeon et al. 2017), Parkinson's disease (Chinta et al. 2018), Alzheimer's disease (Zhang et al. 2019), diabetes (Palmer et al. 2015; Aguayo-Mazzucato et al. 2019), cancer (Takasugi et al. 2017; Demaria et al. 2017), pulmonary fibrosis (Schafer et al. 2017; He et al. 2019) and many other diseases (Childs et al. 2015; Tchkonia et al. 2013; Kirkland and Tchkonia 2017; Childs et al. 2017; Niedernhofer and Robbins 2018). Therefore, SnCs have been emerged as therapeutic targets for many of these age-related diseases.

1.3 Senolytics

SnCs can endure sustained DNA damage, oxidative stress, proteotoxicity and other stressors because they are protected from induction of apoptosis by various SnC anti-apoptotic pathways (SCAPs) (Kirkland and Tchkonia 2017; Childs et al. 2017; Niedernhofer and Robbins 2018). However, these SCAPs can also function as the Achilles' heel of SnCs. Molecularly targeted inhibition of SCAPs with a small molecule can selectively kill SnCs. These small molecules are termed senolytics, whereas those that can suppress SASP, named senomorphics. Both senolytics and senomorphics have the potential to prevent and treat age-related diseases and to extend healthspan (Kirkland and Tchkonia 2017; Childs et al. 2017; Niedernhofer and Robbins 2018). However, compared to senomorphics, senolytics may provide greater promise and better benefits as anti-aging therapeutics because permanent elimination of SnCs by senolytics requires less drug exposure, produces less drug toxicity, and leads to a more durable effect than suppression of SASP by senomorphics. Therefore, development of senolytics has become a more attractive strategy to combat aging and age-related diseases. This hypothesis has yet to be tested in future studies.

To date, several classes of senolytic agents have been identified (Fig. 1.1), including (1) naturally occurring compounds and their derivatives such as quercetin (Xu et al. 2018; Zhu et al. 2015; Hickson et al. 2019), fisetin (Yousefzadeh et al. 2018; Zhu et al. 2017), piperlongumine and analogs (Wang et al. 2016; Zhang et al. 2018; Liu et al. 2018), curcumin analogs (Li et al. 2019) and cardiac glycosides (Guerrero et al. 2019; Triana-Martínez et al. 2019); and (2) targeted therapeutics such

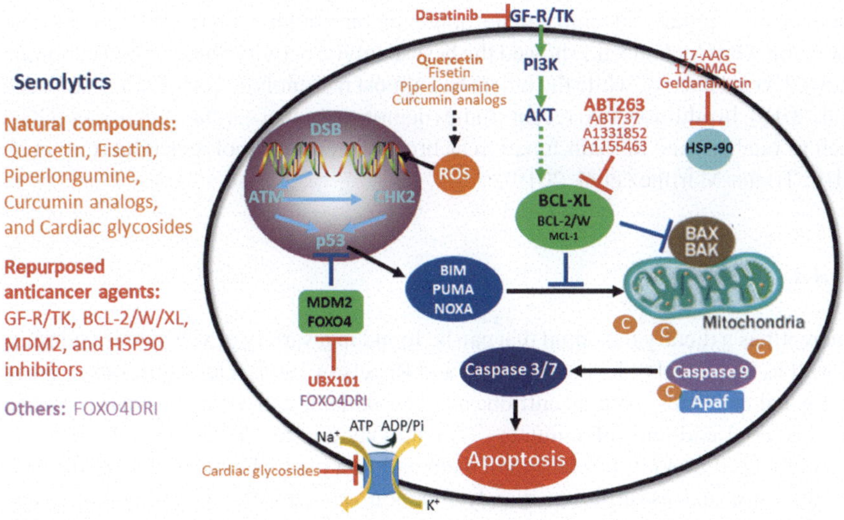

Fig. 1.1 Known senolytics that target different senescent cell (SnC) anti-apoptotic pathways (SCAPs)

as dasatinib (Xu et al. 2018; Zhu et al. 2015; Hickson et al. 2019), a non-specific tyrosine kinase inhibitor; inhibitors of the anti-apoptotic Bcl-2 family proteins (Zhu et al. 2015; Zhu et al. 2017; Chang et al. 2016; Yosef et al. 2016); HSP90 inhibitor and histone deacetylase (Fuhrmann-Stroissnigg et al. 2017); UBX0101 (Jeon et al. 2017), an inhibitor of the MDM2/p53 protein interaction; and a modified FOXO4-p53 interfering peptide (IP) (Baar et al. 2017). The following is a brief description of each of these known senolytics.

1.3.1 Senolytic Natural Compounds

Many natural products have anti-aging effects and are used as traditional medicines and nutritional supplements to prevent or treat various age-related diseases, such as resveratrol (Knutson and Leeuwenburgh 2008), berberine (Xu et al. 2017), rutin (Li et al. 2016; Yang et al. 2012), catechin (Assuncao and Andrade 2015; Bernatoniene and Kopustinskiene 2018), proanthocyanidin (Liu et al. 2018), ginkgo biloba extract (EGb 761) (Sastre et al. 1998), ursolic acid (He et al. 2014; He et al. 2013) and other phytomolecules (Mukherjee et al. 2011). Most of them are antioxidants and exert their anti-aging functions mainly by reducing oxidative damage. However, only a few of them have been identified as senolytics, including quercetin (Xu et al. 2018; Zhu et al. 2015; Hickson et al. 2019), fisetin (Yousefzadeh et al. 2018; Zhu et al. 2017), piperlongumine and analogs (Wang et al. 2016; Zhang et al. 2018; Liu et al. 2018), the curcumin analog EF24 (Li et al. 2019) and goldenrod extracts (Lämmermann et al. 2018), some of which have been validated by a recent study done by Yousefzadeh et al. (2018). They tested a panel of flavonoids, including resveratrol, curcumin, luteolin, fisetin, rutin, epigallocatechin gallate, apigenin, pirfenidone, myricetin, catechin and quercetin. Of which, fisetin showed the best senolytic activity, luteolin and curcumin showed weak activity, while the others had almost no senolytic activity (Yousefzadeh et al. 2018). In addition, two recent studies demonstrate that cardiac glycosides (CGs) such as ouabain and digoxin function as broad-spectrum senolytics (Guerrero et al. 2019; Triana-Martínez et al. 2019).

1.3.1.1 Quercetin

Quercetin is a dietary flavonoid that can be found in a variety of vegetables and fruits as well as in tea and red wine (Formica and Regelson 1995). Quercetin shows broad biological activities, such as anti-obesity, antioxidant, anti-viral, anti-carcinogenic, anti-bacterial and anti-inflammatory (Anand David et al. 2016). Indeed, quercetin is widely used as a nutritional supplement and as a phytochemical remedy for various diseases, such as cardiovascular dysfunction, diabetes/obesity, inflammation and mood disorders. The strong antioxidant activity of quercetin enables it to quench free radicals from forming resonance-stabilized phenoxyl radicals. Nevertheless, the low bioavailability, chemical instability and poor water solubility greatly hinder its applications (Wang et al. 2016). Thus, various strategies have been developed to improve its stability, efficacy and bioavailability.

Levels of oxidative stress increase with age. Considering that quercetin is a potent antioxidant, it has been hypothesized that quercetin may delay aging via reducing oxidative damage. Administration of quercetin can reverse cognitive deficits in aged mice (Singh et al. 2003) and promote longevity in *Saccharomyces cerevisiae* (Belinha et al. 2007), which was attributed to its antioxidant activity. It was not identified as a senolytic until 2015 when Dr. Kirkland's group first discovered that quercetin is a senolytic agent (Zhu et al. 2015). However, its senolytic activity was moderate and cell type specific as it can only kill senescent human endothelial cells but not senescent preadipocytes. Interestingly, when it was combined with dasatinib, they became more effective than either agent alone in killing not only senescent human endothelial cells but also preadipocytes and SnCs from many other tissue origins (Zhu et al. 2015). For example, it was shown that naturally aged, radiation-exposed, and progeroid *Ercc1*$^{-/\Delta}$ mice exhibited a significant reduction in SnC burden after the treatment with the combination of quercetin and dasatinib. More importantly, this combination treatment improved the functions of multiple organs and delayed many age-related pathologies in these mice, particularly extending the healthspan in *Ercc1*$^{-/\Delta}$ mice. Since then, the combination of quercetin and dasatinib have been widely used to treat a variety of age-related diseases in mouse models (Zhang et al. 2019; Schafer et al. 2017; Roos et al. 2016; Ogrodnik et al. 2017; Nath et al. 2018; Musi et al. 2018; Ogrodnik et al. 2019), including atherosclerosis (Roos et al. 2016), pulmonary fibrosis (Schafer et al. 2017), hepatic steatosis (Ogrodnik et al. 2017), chronic kidney disease (Nath et al. 2018), Alzheimer's disease (Zhang et al. 2019; Musi et al. 2018), and obesity (Ogrodnik et al. 2019). Moreover, two clinical studies have been conducted to evaluate the safety of the combination of quercetin and dasatinib in patients with idiopathic pulmonary fibrosis (Justice et al. 2019) and diabetic kidney disease (Hickson et al. 2019). The results from these clinical studies show that quercetin and dasatinib treatment was well tolerated and could reduce SnC burden in these patients. However, quercetin is a polypharmacologic agent and its mechanisms of action have not been well defined nor have their molecular targets been identified and characterized. It remains unclear whether its therapeutic effects are mediated by its senolytic activity, particularly considering that it is not a potent senolytic agent and can only kill SnCs derived from a limited number of tissue origins alone or in combination with dasatinib in vitro (Yousefzadeh et al. 2018; Hwang et al. 2018; Grezella et al. 2018).

1.3.1.2 Fisetin

Fisetin is widely studied flavonoid extracted from various fruits and vegetables such as apples, persimmons, grapes, cucumbers, strawberries and onions (Arai et al. 2000). It is commonly used as a nutritional supplement and has a highly favorable safety profile. In Japan, the average dietary intake of naturally occurring fisetin is approximately

0.4 mg/day (Arai et al. 2000; Kimira et al. 1998), apparently without any adverse effects. Fisetin has numerous beneficial biological effects, including anti-oxidant, anti-tumor, anti-angiogenic, anti-inflammatory, anti-hyperlipidemic and neuroprotective effects (Pal et al. 2016; Khan et al. 2013; Sundarraj et al. 2018). Like many other flavonoids, fisetin acts as an antioxidant that can scavenge free radicals to confer marked antioxidant activity and significant biological effects. Its anti-oxidative activity has been confirmed by both cyclic voltammetry assays and quantum-chemical-based calculations (Marković et al. 2009). Accumulating data suggest fisetin as a potent anti-tumor agent that can inhibit cancer cell proliferation and induce cancer cell apoptosis in a variety of cancer cell lines (Lall et al. 2016). Interestingly, the effects are limited to cancer cells, as normal cells are less sensitive to fisetin treatment (Lall et al. 2016), showing good selectivity against normal and cancer cells.

In 2017, fisetin was first found to selectively cause cell death in SnCs but not in proliferating human umbilical vein endothelial cells (HUVECs) (Zhu et al. 2017). However, it had no senolytic activity on senescent IMR-90 cells or primary human preadipocytes (Zhu et al. 2017), indicating that its senolytic activity is cell-specific. The senolytic activity of fisetin was validated by another study in which a series of flavonoid polyphenols were tested for senolytic activity using SnCs (Yousefzadeh et al. 2018). Among the flavonoids tested, fisetin was the most potent one to induce SnC death. More importantly, treatment of progeroid $Ercc1^{-/\Delta}$ and naturally aged mice with fisetin reduced SnC burden in multiple tissues, which resulted in a significant improvement in tissue homeostasis, reduced age-related pathology and moderately extended median and maximum lifespan of naturally aged mice (Yousefzadeh et al. 2018). Again, it has yet to be determined whether the therapeutic effects are mediated by its senolytic activity as fisetin is also a polypharmacologic agent that has been shown to extend the replicative lifespan of *S. cerevisiae* (Howitz et al. 2003) and the lifespan of D. melanogaster (Wood et al. 2004) in part via activation of sirtuins.

1.3.1.3 Piperlongumine

Piperlongumine is a biologically active extract from Piper species. It is the major alkaloid from long pepper and other important medicinal plants (Bezerra et al. 2013). Piperlongumine has wide pharmacological activities, such as anti-tumor, anti-angiogenic, anti-platelet aggregation, anti-metastatic, anti-nociceptive, anti-depressant, anti-atherosclerotic, anti-diabetic, and anti-bacterial (Bezerra et al. 2013). The anti-cancer activities of piperlongumine have been widely studied. It can kill various cancer types, including leukemia and solid tumors, such as skin, colon, breast, lung, central nervous system (CNS), nasopharyngeal, pancreatic, osseous, renal, bladder and prostate cancers (Bezerra et al. 2013; Piska et al. 2018). Interestingly, piperlongumine shows selective cytotoxicity over cancer cells and only displays weak cytotoxicity to normal cells (Bezerra et al. 2013). For example, it can suppress leukemia cell growth and reduce tumor cell viability by inducing apoptosis, but only has weak cytotoxicity to normal lymphocytes (Bezerra et al. 2007). Mechanistic studies reveal that piperlongumine functions as an antitumor agent via regulating multiple signal transduction pathways, including the mitochondrial apoptosis pathway, receptor tyrosine kinase (Raf-1) and extracellular signal-regulated kinases (ERK1/2) (Bezerra et al. 2013). Additionally, piperlongumine can suppress tumor progression and migration in vivo. The anticancer effect of piperlongumine has been proposed through its inhibition of oxidative stress response enzymes such as GSTp1 and CRB1, resulting in selective induction of ROS production in cancer cells but not in normal cells (Bezerra et al. 2013).

By screening a library of small molecules that target pathways predicted to be important for SnCs survival, Wang et al. identified piperlongumine as a novel lead for the development of senolytic agents (Wang et al. 2016). Piperlongumine selectively kills senescent human WI-38 fibroblasts induced by ionizing radiation, replicative exhaustion, or ectopic expression of the oncogene *Ras*. It induces SnCs apoptosis via activing the caspase cascades as pretreatment with the pan-caspase inhibitor Q-VD-OPh (QVD) can significantly block the apoptosis. Piperlongumine was reported to kill cancer cells by inducing the production of ROS. However, it cannot induce ROS production in SnCs (Wang et al. 2016). Interestingly, piperlongumine synergistically killed SnCs in combination with ABT263, a Bcl-2/Bcl-xL inhibitor. Initial structural modifications to piperlongumine identified a series of analogs with improved potency and/or selectivity in inducing SnC death (Liu et al. 2018). However, the mechanisms by which piperlongumine kills SnCs are largely unknown. Dr. Zhou's lab identified a series of potential molecular targets of piperlongumine using a piperlongumine-based chemical probe to pull-down piperlongumine-binding proteins from live cells. One of them is oxidation resistance 1 (OXR1), an important antioxidant protein that regulates the expression of a variety of antioxidant enzymes. They found that OXR1 was upregulated in senescent WI-38 fibroblasts. Piperlongumine can bound to OXR1 directly and induce its degradation through the ubiquitin-proteasome system in an SnC-specific manner (Zhang et al. 2018). These findings provide new insights into the mechanism by which SnCs are highly resistant to oxidative stress and suggest that OXR1 is a novel senolytic target of piperlongumine that can be further exploited for the development of new senolytic agents. However, whether piperlongumine and its analogs can function as a senolytic agent in vivo has yet to be determined.

1.3.1.4 EF24

Curcumin is a hydrophobic polyphenol derived from the rhizome of the herb *Curcuma longa,* and is a well-defined natural compound with a variety biological activities, such as anti-cancer, anti-oxidation, anti-inflammation, and anti-microbial (Aggarwal and Harikumar 2009; Anand et al. 2008; Gupta et al. 2013; Hatcher et al. 2008; Maheshwari et al. 2006). Curcumin was found to have therapeutic potential and benefits in delaying aging and has been used to prevent and treat certain age-associated diseases (Grill et al. 2018; Takano et al. 2018; Yang et al. 2017). It has been shown to extend lifespan and healthspan in *Drosophila melanogaster* (fruit fly) (Chandrashekara et al. 2014) and *Caenorhabditis* elegans (Liao et al. 2011). However, its low potency and poor bioavailability limit its clinical applications

(Shoba et al. 1998). A series of curcumin analogs have thus been developed in order to improve its bioavailability and therapeutic efficacy, such as EF24 (Adams et al. 2005; He et al. 2018), HO-3867 (Selvendiran et al. 2009), 2-HBA (Dinkova-Kostova et al. 2007) and dimethoxycurcumin (DIMC) (Tamvakopoulos et al. 2007), which were demonstrated to be more active than curcumin in reducing age-dependent deterioration, such as cancer and inflammation.

Curcumin was reported to have weak senolytic activity in a recent study (Yousefzadeh et al. 2018). However, Li et al. identified EF24 as a more potent senolytic agent than other curcumin analogs tested including HO-3867, 2-HBA and DIMC (Li et al. 2019). They revealed that EF24 reduced cell viability not only in ionizing radiation induced SnCs, but also in SnCs induced by extensive replication or ectopic transfection of the *Ras* oncogene. Moreover, EF24 displayed broad-spectrum senolytic activity against different types of SnCs, including human IMR-90 fibroblasts, HUVECs and human renal epithelial cells (Li et al. 2019). EF24 was reported to induce apoptosis in various tumor cells in part by inducing ROS production and endoplasmic reticulum stress. However, EF24 did not induce ROS production in SnCs, indicating that its senolytic activity is ROS-independent. Instead, they found that EF24 could reduce the expression of Bcl-xL and Mcl-1 in SnCs but not in normal cells, probably via proteasomal degradation. The findings provide new insights into the mechanisms by which curcumin analogs function as anti-aging agents, and suggest the potential of EF24 to be a novel senolytic agent for the treatment of age-related diseases.

1.3.1.5 Cardiac Glycosides (CGs)

Two groups recently discovered that CGs including ouabain and digoxin are potent and broad-spectrum senolytics (Guerrero et al. 2019; Triana-Martínez et al. 2019). These CGs could kill a variety of SnCs from different species and tissues origins and induced by different stressors in vitro. They selectively killed SnCs primarily by inhibiting the Na^+/K^+ ATPase on the plasma membrane, which caused the disturbance of intracellular concentrations of Na^+, K^+ and H^+ more profoundly in SnCs than non-SnCs, and subsequently led to the loss of membrane potential and acidification of the cells. In addition, SnCs are more susceptible to this disturbance than non-SnCs because SnCs exhibit partial depolarization of their plasma membrane and have a higher concentration of H^+ than non-SnCs under basal conditions.

CGs also exhibit strong senolytic activity in vivo. For example, it was shown that administration of ouabain or digoxin to mice effectively eliminated oncogene-induced pre-neoplastic SnCs in the liver and pituitary, SnCs induced by radiation in the lungs, and SnCs accumulated in different tissues with aging. The elimination of SnCs in these mice led to decrease in the expression of SASP, suppression of tissue fibrosis and improvement of various physiological functions. Furthermore, CGs were highly cytotoxic to senescent cancer cells induced by various chemotherapeutic agents, resulting in a significant improvement in tumor response to chemotherapy in a lung cancer xenograft mouse model and a patient-derived breast cancer xenograft mouse model. These findings suggest that CGs have the potential to be used as effective treatments for a variety of age-related diseases including cancer.

1.3.2 Targeted Senolytics

Almost all targeted senolytics identified to date are repurposed anticancer agents that target SCAPs. These senolytics are in general more potent than naturally occurring senolytic compounds with the exception of CGs. However, these repurposed senolytics usually possess various on-target and/or off-target toxicities, which may preclude their clinical use as anti-aging agents as older people are more susceptible to adverse drug effects than younger individuals and less tolerant of cancer drug toxicity. Therefore, strategies to reduce on-target and/or off-target toxicity of known targeted senolytics are needed to generate safer targeted senolytics for clinical translation.

1.3.2.1 Dasatinib—A Pan Tyrosine Kinase Inhibitor

Dasatinib was one of the first senolytics discovered by Zhu et al. (2015). Dasatinib is a pan tyrosine kinase inhibitor that is known to promote tumor cell apoptosis via inhibiting a variety of cell survival pathways, including the down-stream pathway of ephrins or ephrin-B (EFNB) family members that are upregulated in SnCs. Dasatinib preferentially reduced the viability in senescent human preadipocytes, but was much less cytotoxic to senescent HUVECs. However, when it was combined with quercetin, they were more potent than either agent alone in killing different types of SnCs. Therefore, the combination of dasatinib and quercetin have been widely used to clear SnCs in various mouse models to treat different age-related diseases and tested in two clinical trials as we discussed earlier in Sect.1.3.1.1. However, the mechanism of action of dasatinib and the specific tyrosine kinase targeted by dasatinib to mediate its senolytic activity have yet to be determined. Identification of the specific senolytic target of dasatinib can lead to the development of more specific and potent senolytic agents that can reduce the off-target toxicity of dasatinib.

1.3.2.2 Inhibitors of the Bcl-2 Family Antiapoptotic Proteins

Resistance to apoptosis is a hallmark of SnCs (Wang 1995; Childs et al. 2014; Sasaki et al. 2001; Hampel et al. 2005; Soto-Gamez et al. 2019). Various SnCs may use different SCAPs to resist apoptosis. The Bcl-2 family proteins, consisting of both antiapoptotic and proapoptotic proteins, play important roles in the regulation of apoptosis (Youle and Strasser 2008). These proteins share sequence homology within conserved regions known as Bcl-2 homology (BH) domains. The Bcl-2 antiapoptotic proteins are multi-BH-domain proteins including Bcl-2, Bcl-xL, Mcl-1, Bcl-w and Bfl1. They can inhibit apoptosis by binding to the multi-BH-domain and BH3-only proapoptotic proteins. Among the Bcl-2 antiapoptotic proteins, Bcl-xL has been primarily implicated in SnC resistance to apoptosis because inhibition of Bcl-xL with a Bcl-xL specific inhibitor (such as A-1331852) or a Bcl-2 and Bcl-xL dual inhibitor (such as ABT263 or ABT737) can potently and selectively induce apoptosis in a variety of SnCs (Zhu et al. 2017; Chang et al. 2016; Yosef et al. 2016; Zhu et al. 2016), whereas inhibition of Bcl-2 and Mcl-1 alone with their specific inhibitors has no or weak effect on SnC survival (Chang et al. 2016; Yosef et al. 2016). However, inhibition of Bcl-2 and Bcl-w may contribute to the cytotoxic effect of the Bcl-xL inhibitors ABT263 and ABT737 on SnCs.

The mechanism by which Bcl-xL inhibition selectively induces apoptosis in SnCs may be attributable to the persistent stress endured by SnCs, which can upregulate the expression of some of the proapoptotic proteins such as Bcl-2 antagonist/killer (BAK) (Chang et al. 2016). To counteract the effect of these proapoptotic proteins for survival, SnCs also express a higher level of antiapoptotic proteins such as Bcl-xL (Chang et al. 2016; Yosef et al. 2016). Therefore, inhibition of Bcl-xL with an inhibitor can release Bcl-2-interacting mediator of cell death (BIM) and other BH3 proteins, which in turn activates BAK and/or Bcl-2-associated X protein (BAX). The activation of BAX and/or BAK at the mitochondrial membrane induces their oligomerization and formation of the macropores that causes mitochondrial outer membrane permeabilization (MOMP). MOMP results in the release of cytochrome C from mitochondria to the cytoplasm, which binds to the apoptotic protease-activating factor 1 (APAF1) to form the apoptosome. The apoptosome then induces a cascade activation of the initiator caspase (caspase 9) and executioner caspases (caspases 3, 6 and 7) to dismantle the cells (Czabotar et al. 2014).

Because ABT263 is one of the most advanced Bcl-2 and Bcl-xL dual inhibitor drug candidate, it has been extensively evaluated as a senolytic agent. Dr. Zhou's and other labs have found that ABT263 can potently kill a variety of SnCs in cell culture with a few exceptions (such as senescent chondrocytes and synovial fibroblasts in the osteoarthritic joint), whereas it has minimal effect on their non-senescent counterparts (Zhu et al. 2015; Chang et al. 2016; Yosef et al. 2016). These findings suggest that ABT263 is a potent and broad-spectrum senolytic agent. This suggestion is supported by the finding that treatment of mice with ABT263 can effectively clear SnCs in various murine tissues. More importantly, clearance of SnCs with ABT263 can rejuvenate aged hematopoietic stem cells (HSCs) and the senescent hematopoietic system in aged mice (Chang et al. 2016) and ameliorate several pathological conditions associated with aging such as atherosclerosis, dementia and pulmonary fibrosis (Childs et al. 2016; Pan et al. 2017; Bussian et al. 2018). However, the on-target toxicity of thrombocytopenia induced by Bcl-xL inhibition prevents the use of ABT263 and other Bcl-xL specific inhibitors in clinic even for cancer patients, because platelets also depend on Bcl-xL for survival (Ashkenazi et al. 2017; Gandhi et al. 2011; Leverson et al. 2015; Souers et al. 2013). Therefore, strategies that can be used to overcome this on-target toxicity will be needed in order to generate a safer and more effective Bcl-xL targeting senolytic agent for clinical translation. Alternatively, a combination therapy with lower doses of different senolytic agents may provide a synergy to more effectively clear SnCs while reducing their on-target and off-target toxicity as seen with the combination of quercetin and dasatinib (Zhu et al. 2015) and ABT263 plus piperlongumine (Wang et al. 2016).

1.3.2.3 HSP90 Inhibitors

HSP90 is a molecular chaperone ubiquitously expressed in cells and tissues. It plays an important role in the regulation of protein stability. It is upregulated in many different types of cancers and required for the stability and function of numerous oncogenic signaling proteins as well as certain anti-apoptotic factors (Solárová et al. 2015). Therefore, several HSP90 inhibitors have been developed as potential anticancer agents. It has been well established that SnCs are under proteotoxic stress (Pluquet et al. 2015) and thus potentially are more dependent on HSP90 for survival than non-SnCs. Indeed, Fuhrmann-Stroissnigg et al. recently reported that HSP90 inhibitors such as 17-DMAG are senolytics (Fuhrmann-Stroissnigg et al. 2017). These HSP90 inhibitors can selectively kill a variety of SnCs from mouse and human. Mechanic study reveals that inhibition of HSP90 with an inhibitor disrupts the interaction of HSP90 with the phosphorylated AKT, leading to the destabilization of the active form of AKT that is important for the induction of cellular senescence and SnC survival. More importantly, it was shown that periodic treatment of $Ercc1^{-/\Delta}$ progeroid mice with 17-DMAG reduced the tissue burden of SnCs and delayed the onset of several age-related phenotypes and pathologies. However, to translate 17-DMAG into clinic for the treatment of age-related diseases, we need to generate analogs of 17-DMAG to improve its pharmacokinetic and pharmacodynamic properties and reduce its side effects (Mellatyar et al. 2018).

1.3.2.4 FOXO4-p53 Interfering Peptide (IP) and MDM2 Inhibitors

p53 is a well-known tumor suppressor that acts as a double-edged sword in regulation of cellular senescence and aging (Wu and Prives 2018; Johmura and Nakanishi 2016). Increases in the levels and activity of p53 occur when cells enter a pre-senescent stage upon activation of the DNA damage response (DDR) pathway, which plays an important role in the initiation of cellular senescence. In addition, direct activation of p53 by MDM2 inhibition with nutlin-3a can also induce senescence in mouse fibroblasts without induction of DNA damage and activation of the DDR pathway (Efeyan et al. 2007). However, in many types of cells, p53 levels reduce to a level that is below the basal levels of p53 in non-SnCs when they become senescent (Kim et al. 2015). The reduction of p53 in SnCs may protect them from apoptosis because p53 is one of the most important apoptosis determinants and can induce apoptosis through both transcription-dependent and -independent mechanisms (Fridman and Lowe 2003). In addition, the reduction of p53 activity was found in various tissues in aged mice (Feng et al. 2007), which may contribute to the accumulation of SnCs and higher prevalence of cancer due to reduced apoptosis during aging. Therefore, restoration of p53 activity has the potential to eliminate SnCs via induction of SnC apoptosis.

This hypothesis was supported by the finding that increases in p53 transcriptional activity via disruption of the interaction between FOXO4 and p53 using a FOXO4 peptide selectively induced apoptosis in SnCs in cell culture and effectively cleared SnCs in mice (Baar et al. 2017). In their study, they found that FOXO4 was elevated to maintain cell viability in SnCs. Subsequently, they designed a peptide called FOXO4-DRI which can disrupt PML/DNA-SCARS, release active p53 in SnCs, and selectively and potently target SnCs for p53-dependent apoptosis. In vivo, FOXO4-DRI counteracted chemotherapy-induced senescence and loss of liver function, as well as loss of renal function in fast-aging mice.

However, there are still some challenges to use peptides as a therapeutics in clinic. Alternatively, p53 can be activated by inhibition of the interaction between MDM2 and p53 to selectively kill SnCs. This is because the levels and activities of p53 are primarily regulated at the level of post-transcription via the MDM2-mediated ubiquitination and proteasome degradation (Kruse and Gu 2009), and inhibition of the interaction between MDM2 and p53 with an inhibitor can increase p53 stability and activity (Moll and Petrenko 2003). Indeed, it was reported recently that UBX0101, an inhibitor of MDM2, could selectively kill SnCs in vitro and effectively clear them in mice with post-traumatic osteoarthritis after local therapy. However, systemic treatment with MDM2 inhibitors can be risky because it causes substantial hematopoietic suppression and gastrointestinal toxicity (Tisato et al. 2017). It has yet to be determined whether these adverse effects are on-target toxicities or off-target side effects. Therefore, MDM2 inhibitors may be only suitable for clearing SnCs to treat age-related diseases such as osteoarthritis via local administration. It will be important to find an alternative strategy to activate p53 without causing significant normal tissue toxicity. This would be more desirable for the development of a better senolytic agent that can be safely used in elderly individuals who are more susceptible to drug adverse effects than young people.

1.4 Conclusions

While natural senolytics may have the advantages of low toxicity, they are usually less potent than targeted senolytics and thus have to be combined with other senolytic agents to be effective in clearing SnCs (Zhu et al. 2015), except CGs (Guerrero et al. 2019; Triana-Martínez et al. 2019). The mechanisms of action of most natural senolytics have not been well defined nor have their molecular targets been identified and characterized, making it very difficult to rationally modify the compounds to improve their senolytic activity. In contrast, almost all the targeted senolytics discovered are repurposed anticancer agents except the FOXO4-p53-IP (Kirkland and Tchkonia 2017; Childs et al. 2017; Niedernhofer and Robbins 2018). These repurposed senolytics usually possess various on-target and/or off-target toxicities, which could preclude their clinical use as anti-aging agents because older people are more susceptible to adverse drug effects than younger individuals and less tolerant to cancer drug toxicity. Therefore, strategies to reduce on-target and/or off-target toxicity of known targeted senolytics are urgently needed.

Acknowledgements This work was supported in part by US National Institutes of Health (NIH) grants R01CA211963 (D.Z.), R01CA219836 (D.Z.), R01AG063801 (D.Z.), and R56AG056372 (G.Z.).

Conflict of Interest G.Z. and D.Z. are inventors of a pending patent application for the development of Bcl-xL targeted senolytic agents. D.Z. is a co–founder and an advisor of Unity that develops senolytic agents to treat various age-related diseases in humans.

References

Adams BK, Cai J, Armstrong J, Herold M, Lu YJ, Sun A et al (2005) EF24, a novel synthetic curcumin analog, induces apoptosis in cancer cells via a redox-dependent mechanism. Anticancer Drugs 16(3):263–275

Aggarwal BB, Harikumar KB (2009) Potential therapeutic effects of curcumin, the anti-inflammatory agent, against neurodegenerative, cardiovascular, pulmonary, metabolic, autoimmune and neoplastic diseases. Int J Biochem Cell Biol 41(1):40–59

Aguayo-Mazzucato C, Andle J, Lee TB, Midha A, Talemal L, Chipashvili V et al (2019) Acceleration of β cell aging determines diabetes and senolysis improves disease outcomes. Cell Metab 30(1):129–142.e4

Anand David AV, Arulmoli R, Parasuraman S (2016) Overviews of biological importance of quercetin: a bioactive flavonoid. Pharmacogn Rev 10(20):84–89

Anand P, Thomas SG, Kunnumakkara AB, Sundaram C, Harikumar KB, Sung B et al (2008) Biological activities of curcumin and its analogues (Congeners) made by man and Mother Nature. Biochem Pharmacol 76(11):1590–1611

Arai Y, Watanabe S, Kimira M, Shimoi K, Mochizuki R, Kinae N (2000) Dietary intakes of flavonols, flavones and isoflavones by Japanese women and the inverse correlation between quercetin intake and plasma LDL cholesterol concentration. J Nutr 130(9):2243–2250

Ashkenazi A, Fairbrother WJ, Leverson JD, Souers AJ (2017) From basic apoptosis discoveries to advanced selective BCL-2 family inhibitors. Nat Rev Drug Discov 16(4):273–284

Assuncao M, Andrade JP (2015) Protective action of green tea catechins in neuronal mitochondria during aging. Front Biosci (Landmark Ed). 20:247–262

Baar MP, Brandt RMC, Putavet DA, Klein JDD, Derks KWJ, Bourgeois BRM et al (2017) Targeted apoptosis of senescent cells restores tissue homeostasis in response to chemotoxicity and aging. Cell 169(1):132–147.e16

Baker DJ, Wijshake T, Tchkonia T, LeBrasseur NK, Childs BG, van de Sluis B et al (2011) Clearance of p16Ink4a-positive senescent cells delays ageing-associated disorders. Nature 479(7372):232–236

Baker DJ, Childs BG, Durik M, Wijers ME, Sieben CJ, Zhong J et al (2016) Naturally occurring p16(Ink4a)-positive cells shorten healthy lifespan. Nature 530(7589):184–189

Belinha I, Amorim MA, Rodrigues P, de Freitas V, Moradas-Ferreira P, Mateus N et al (2007) Quercetin increases oxidative stress resistance and longevity in Saccharomyces cerevisiae. J Agric Food Chem 55(6):2446–2451

Bernatoniene J, Kopustinskiene DM (2018) The role of catechins in cellular responses to oxidative stress. Molecules 23(4)

Bezerra DP, Militão GCG, de Castro FO, Pessoa C, de Moraes MO, Silveira ER et al (2007) Piplartine induces inhibition of leukemia cell proliferation triggering both apoptosis and necrosis pathways. Toxicol in Vitro 21(1):1–8

Bezerra DP, Pessoa C, de Moraes MO, Saker-Neto N, Silveira ER, Costa-Lotufo LV (2013) Overview of the therapeutic potential of piplartine (piperlongumine). Eur J Pharm Sci 48(3):453–463

Bussian TJ, Aziz A, Meyer CF, Swenson BL, van Deursen JM, Baker DJ (2018) Clearance of senescent glial cells prevents tau-dependent pathology and cognitive decline. Nature 562(7728):578–582

Campisi J (2013) Aging, cellular senescence, and cancer. Annu Rev Physiol 75:685–705

Chandrashekara KT, Popli S, Shakarad MN (2014) Curcumin enhances parental reproductive lifespan and progeny viability in Drosophila melanogaster. Age (Dordr) 36(5):9702

Chang J, Wang Y, Shao L, Laberge R-M, Demaria M, Campisi J et al (2016) Clearance of senescent cells by ABT263 rejuvenates aged hematopoietic stem cells in mice. Nat Med 22(1):78–83

Childs BG, Baker DJ, Kirkland JL, Campisi J, van Deursen JM (2014) Senescence and apoptosis: dueling or complementary cell fates? EMBO Rep 15(11):1139–1153

Childs BG, Durik M, Baker DJ, van Deursen JM (2015) Cellular senescence in aging and age-related disease: from mechanisms to therapy. Nat Med 21(12):1424–1435

Childs BG, Baker DJ, Wijshake T, Conover CA, Campisi J, van Deursen JM (2016) Senescent intimal foam cells are deleterious at all stages of atherosclerosis. Science 354(6311):472–477

Childs BG, Gluscevic M, Baker DJ, Laberge R-M, Marquess D, Dananberg J et al (2017) Senescent cells: an emerging target for diseases of ageing. Nat Rev Drug Discov 16(10):718–735

Chinta SJ, Woods G, Demaria M, Rane A, Zou Y, McQuade A, et al (2018) Cellular senescence is induced by the environmental neurotoxin paraquat and contributes to neuropathology linked to Parkinson's disease. Cell Rep 22(4):930–940

Czabotar PE, Lessene G, Strasser A, Adams JM (2014) Control of apoptosis by the BCL-2 protein family: implications for physiology and therapy. Nat Rev Mol Cell Biol 15(1):49–63

Demaria M, O'Leary MN, Chang J, Shao L, Liu S, Alimirah F et al (2017) Cellular senescence promotes adverse effects of chemotherapy and cancer relapse. Cancer Discov 7(2):165–176

Dinkova-Kostova AT, Cory AH, Bozak RE, Hicks RJ, Cory JG (2007) Bis(2-hydroxybenzylidene)acetone, a potent inducer of the phase 2 response, causes apoptosis in mouse leukemia cells through a p53-independent, caspase-mediated pathway. Cancer Lett 245(1–2):341–349

Efeyan A, Ortega-Molina A, Velasco-Miguel S, Herranz D, Vassilev LT, Serrano M (2007) Induction of p53-dependent senescence by the MDM2 antagonist nutlin-3a in mouse cells of fibroblast origin. Cancer Res 67(15):7350–7357

Feng Z, Hu W, Teresky AK, Hernando E, Cordon-Cardo C, Levine AJ (2007) Declining p53 function in the aging process: a possible mechanism for the increased tumor incidence in older populations. Proc Natl Acad Sci USA 104(42):16633–16638

Formica JV, Regelson W (1995) Review of the biology of Quercetin and related bioflavonoids. Food Chem Toxicol 33(12):1061–1080

Fridman JS, Lowe SW (2003) Control of apoptosis by p53. Oncogene 22(56):9030–9040

Fuhrmann-Stroissnigg H, Ling YY, Zhao J, McGowan SJ, Zhu Y, Brooks RW, et al (2017) Identification of HSP90 inhibitors as a novel class of senolytics. Nat Commun 8(1):422

Gandhi L, Camidge DR, Ribeiro de Oliveira M, Bonomi P, Gandara D, Khaira D, et al (2011) Phase I study of Navitoclax (ABT-263), a novel Bcl-2 family inhibitor, in patients with small-cell lung cancer and other solid tumors. J Clin Oncol 29(7):909–916

Grezella C, Fernandez-Rebollo E, Franzen J, Ventura Ferreira MS, Beier F, Wagner W (2018) Effects of senolytic drugs on human mesenchymal stromal cells. Stem Cell Res Ther 9(1):108

Grill AE, Shahani K, Koniar B, Panyam J (2018) Chemopreventive efficacy of curcumin-loaded PLGA microparticles in a transgenic mouse model of HER-2-positive breast cancer. Drug Deliv Transl Res 8(2):329–341

Guerrero A, Herranz N, Sun B, Wagner V, Gallage S, Guiho R et al (2019) Cardiac glycosides are broad-spectrum senolytics. Nat Metab 1(11):1074–1088

Gupta SC, Patchva S, Aggarwal BB (2013) Therapeutic roles of curcumin: lessons learned from clinical trials. AAPS J 15(1):195–218

Hampel B, Wagner M, Teis D, Zwerschke W, Huber LA, Jansen-Dürr P (2005) Apoptosis resistance of senescent human fibroblasts is correlated with the absence of nuclear IGFBP-3. Aging Cell 4(6):325–330

Hatcher H, Planalp R, Cho J, Torti FM, Torti SV (2008) Curcumin: from ancient medicine to current clinical trials. Cell Mol Life Sci 65(11):1631–1652

He Y, Li Y, Zhao T, Wang Y, Sun C (2013) Ursolic acid inhibits adipogenesis in 3T3-L1 adipocytes through LKB1/AMPK pathway. PLoS ONE 8(7):e70135

He Y, Li W, Li Y, Zhang S, Wang Y, Sun C (2014) Ursolic acid increases glucose uptake through the PI3K signaling pathway in adipocytes. PLoS ONE 9(10):e110711

He Y, Li W, Hu G, Sun H, Kong Q (2018) Bioactivities of EF24, a novel curcumin analog: a review. Front Oncol 8:614

He Y, Thummuri D, Zheng G, Okunieff P, Citrin DE, Vujaskovic Z et al (2019) Cellular senescence and radiation-induced pulmonary fibrosis. Transl Res 209:14–21

Hickson LJ, Langhi Prata LGP, Bobart SA, Evans TK, Giorgadze N, Hashmi SK et al (2019) Senolytics decrease senescent cells in humans: preliminary report from a clinical trial of Dasatinib plus Quercetin in individuals with diabetic kidney disease. EBioMedicine 47:446–456

Howitz KT, Bitterman KJ, Cohen HY, Lamming DW, Lavu S, Wood JG et al (2003) Small molecule activators of sirtuins extend Saccharomyces cerevisiae lifespan. Nature 425(6954):191–196

Hwang HV, Tran DT, Rebuffatti MN, Li C-S, Knowlton AA (2018) Investigation of quercetin and hyperoside as senolytics in adult human endothelial cells. PLoS ONE 13(1):e0190374

Jeon OH, Kim C, Laberge R-M, Demaria M, Rathod S, Vasserot AP et al (2017) Local clearance of senescent cells attenuates the development of post-traumatic osteoarthritis and creates a pro-regenerative environment. Nat Med 23(6):775–781

Johmura Y, Nakanishi M (2016) Multiple facets of p53 in senescence induction and maintenance. Cancer Sci 107(11):1550–1555

Justice JN, Nambiar AM, Tchkonia T, LeBrasseur NK, Pascual R, Hashmi SK et al (2019) Senolytics in idiopathic pulmonary fibrosis: results from a first-in-human, open-label, pilot study. EBioMedicine 40:554–563

Khan N, Syed DN, Ahmad N, Mukhtar H (2013) Fisetin: a dietary antioxidant for health promotion. Antioxid Redox Signal 19(2):151–162

Kim RH, Kang MK, Kim T, Yang P, Bae S, Williams DW et al (2015) Regulation of p53 during senescence in normal human keratinocytes. Aging Cell 14(5):838–846

Kimira M, Arai Y, Shimoi K, Watanabe S (1998) Japanese intake of flavonoids and isoflavonoids from foods. J Epidemiol 8(3):168–175

Kirkland JL, Tchkonia T (2017) Cellular senescence: a translational perspective. EBioMedicine 21:21–28

Knutson MD, Leeuwenburgh C (2008) Resveratrol and novel potent activators of SIRT1: effects on aging and age-related diseases. Nutr Rev 66(10):591–596

Kruse J-P, Gu W (2009) Modes of p53 regulation. Cell 137(4):609–622

Lall RK, Adhami VM, Mukhtar H (2016) Dietary flavonoid fisetin for cancer prevention and treatment. Mol Nutr Food Res 60(6):1396–1405

Lämmermann I, Terlecki-Zaniewicz L, Weinmüllner R, Schosserer M, Dellago H, de Matos Branco AD et al (2018) Blocking negative effects of senescence in human skin fibroblasts with a plant extract. NPJ Aging Mech Dis 4:4

Leverson JD, Phillips DC, Mitten MJ, Boghaert ER, Diaz D, Tahir SK, et al (2015) Exploiting selective BCL-2 family inhibitors to dissect cell survival dependencies and define improved strategies for cancer therapy. Sci Transl Med 7(279):279ra40

Li T, Chen S, Feng T, Dong J, Li Y, Li H (2016) Rutin protects against aging-related metabolic dysfunction. Food Funct. 7(2):1147–1154

Li W, He Y, Zhang R, Zheng G, Zhou D (2019) The curcumin analog EF24 is a novel senolytic agent. Aging (Albany NY) 11(2):771–782
Liao VH-C, Yu C-W, Chu Y-J, Li W-H, Hsieh Y-C, Wang T-T (2011) Curcumin-mediated lifespan extension in Caenorhabditis elegans. Mech Ageing Dev 132(10):480–487
Liu X, Wang Y, Zhang X, Gao Z, Zhang S, Shi P, et al (2018) Senolytic activity of piperlongumine analogues: Synthesis and biological evaluation. Bioorg Med Chem 26(14):3925–3938
Liu X, Lin X, Mi Y, Li J, Zhang C (2018b) Grape Seed Proanthocyanidin Extract Prevents Ovarian Aging by Inhibiting Oxidative Stress in the Hens. Oxid Med Cell Longev. 2018:9390810
Maheshwari RK, Singh AK, Gaddipati J, Srimal RC (2006) Multiple biological activities of curcumin: a short review. Life Sci 78(18):2081–2087
Marković ZS, Mentus SV, Dimitrić Marković JM (2009) Electrochemical and density functional theory study on the reactivity of fisetin and its radicals: implications on in vitro antioxidant activity. J Phys Chem A 113(51):14170–14179
Mellatyar H, Talaei S, Pilehvar-Soltanahmadi Y, Barzegar A, Akbarzadeh A, Shahabi A et al (2018) Targeted cancer therapy through 17-DMAG as an Hsp90 inhibitor: overview and current state of the art. Biomed Pharmacother 102:608–617
Moll UM, Petrenko O (2003) The MDM2-p53 interaction. Mol Cancer Res 1(14):1001–1008
Mukherjee PK, Maity N, Nema NK, Sarkar BK (2011) Bioactive compounds from natural resources against skin aging. Phytomedicine 19(1):64–73
Musi N, Valentine JM, Sickora KR, Baeuerle E, Thompson CS, Shen Q et al (2018) Tau protein aggregation is associated with cellular senescence in the brain. Aging Cell 17(6):e12840
Nath KA, O'Brien DR, Croatt AJ, Grande JP, Ackerman AW, Nath MC, et al (2018) The murine dialysis fistula model exhibits a senescence phenotype: pathobiological mechanisms and therapeutic potential. Am J Physiol Renal Physiol 315(5):F1493–1499
Niedernhofer LJ, Robbins PD (2018) Senotherapeutics for healthy ageing. Nat Rev Drug Discov 17(5):377
Ogrodnik M, Miwa S, Tchkonia T, Tiniakos D, Wilson CL, Lahat A et al (2017) Cellular senescence drives age-dependent hepatic steatosis. Nat Commun 13(8):15691
Ogrodnik M, Zhu Y, Langhi LGP, Tchkonia T, Krüger P, Fielder E et al (2019) Obesity-induced cellular senescence drives anxiety and impairs neurogenesis. Cell Metab 29(5):1061–1077.e8
Pal HC, Pearlman RL, Afaq F (2016) Fisetin and its role in chronic diseases. Adv Exp Med Biol 928:213–244
Palmer AK, Tchkonia T, LeBrasseur NK, Chini EN, Xu M, Kirkland JL (2015) Cellular senescence in type 2 diabetes: a therapeutic opportunity. Diabetes 64(7):2289–2298
Pan J, Li D, Xu Y, Zhang J, Wang Y, Chen M, et al (2017) Inhibition of Bcl-2/xl with ABT-263 selectively kills senescent type II pneumocytes and reverses persistent pulmonary fibrosis induced by ionizing radiation in mice. Int J Radiat Oncol Biol Phys 99(2):353–361
Piska K, Gunia-Krzyżak A, Koczurkiewicz P, Wójcik-Pszczoła K, Pękala E (2018) Piperlongumine (piplartine) as a lead compound for anticancer agents-Synthesis and properties of analogues: a mini-review. Eur J Med Chem 156:13–20
Pluquet O, Pourtier A, Abbadie C (2015) The unfolded protein response and cellular senescence. A review in the theme: cellular mechanisms of endoplasmic reticulum stress signaling in health and disease. Am J Physiol Cell Physiol 308(6):C415–425
Roos CM, Zhang B, Palmer AK, Ogrodnik MB, Pirtskhalava T, Thalji NM et al (2016) Chronic senolytic treatment alleviates established vasomotor dysfunction in aged or atherosclerotic mice. Aging Cell 15(5):973–977
Sasaki M, Kumazaki T, Takano H, Nishiyama M, Mitsui Y (2001) Senescent cells are resistant to death despite low Bcl-2 level. Mech Ageing Dev 122(15):1695–1706
Sastre J, Millán A, García de la Asunción J, Plá R, Juan G, Pallardó null, et al (1998) A Ginkgo biloba extract (EGb 761) prevents mitochondrial aging by protecting against oxidative stress. Free Radic Biol Med 24(2):298–304
Schafer MJ, White TA, Iijima K, Haak AJ, Ligresti G, Atkinson EJ et al (2017) Cellular senescence mediates fibrotic pulmonary disease. Nat Commun 23(8):14532

Selvendiran K, Kuppusamy ML, Bratasz A, Tong L, Rivera BK, Rink C et al (2009) Inhibition of vascular smooth-muscle cell proliferation and arterial restenosis by HO-3867, a novel synthetic curcuminoid, through up-regulation of PTEN expression. J Pharmacol Exp Ther 329(3):959–966

Shoba G, Joy D, Joseph T, Majeed M, Rajendran R, Srinivas PS (1998) Influence of piperine on the pharmacokinetics of curcumin in animals and human volunteers. Planta Med 64(4):353–356

Singh A, Naidu PS, Kulkarni SK (2003) Reversal of aging and chronic ethanol-induced cognitive dysfunction by quercetin a bioflavonoid. Free Radic Res 37(11):1245–1252

Solárová Z, Mojžiš J, Solár P (2015) Hsp90 inhibitor as a sensitizer of cancer cells to different therapies (review). Int J Oncol 46(3):907–926

Soto-Gamez A, Quax WJ, Demaria M (2019) Regulation of Survival Networks in Senescent Cells: From Mechanisms to Interventions. J Mol Biol 431(15):2629–2643

Souers AJ, Leverson JD, Boghaert ER, Ackler SL, Catron ND, Chen J et al (2013) ABT-199, a potent and selective BCL-2 inhibitor, achieves antitumor activity while sparing platelets. Nat Med 19(2):202–208

Sundarraj K, Raghunath A, Perumal E (2018) A review on the chemotherapeutic potential of fisetin: in vitro evidences. Biomed Pharmacother 97:928–940

Takano K, Tatebe J, Washizawa N, Morita T (2018) Curcumin inhibits age-related vascular changes in aged mice fed a high-fat diet. Nutrients 10(10)

Takasugi M, Okada R, Takahashi A, Virya Chen D, Watanabe S, Hara E (2017) Small extracellular vesicles secreted from senescent cells promote cancer cell proliferation through EphA2. Nat Commun 8:15729

Tamvakopoulos C, Dimas K, Sofianos ZD, Hatziantoniou S, Han Z, Liu Z-L et al (2007) Metabolism and anticancer activity of the curcumin analogue, dimethoxycurcumin. Clin Cancer Res 13(4):1269–1277

Tchkonia T, Zhu Y, van Deursen J, Campisi J, Kirkland JL (2013) Cellular senescence and the senescent secretory phenotype: therapeutic opportunities. J Clin Invest 123(3):966–972

Tisato V, Voltan R, Gonelli A, Secchiero P, Zauli G (2017) MDM2/X inhibitors under clinical evaluation: perspectives for the management of hematological malignancies and pediatric cancer. J Hematol Oncol 10(1):133

Triana-Martínez F, Picallos-Rabina P, Da Silva-Álvarez S, Pietrocola F, Llanos S, Rodilla V et al (2019) Identification and characterization of Cardiac Glycosides as senolytic compounds. Nat Commun 10(1):4731

Wang E (1995) Senescent human fibroblasts resist programmed cell death, and failure to suppress bcl2 is involved. Cancer Res 55(11):2284–2292

Wang Y, Chang J, Liu X, Zhang X, Zhang S, Zhang X, et al (2016) Discovery of piperlongumine as a potential novel lead for the development of senolytic agents. Aging (Albany NY) 8(11):2915–2926

Wang W, Sun C, Mao L, Ma P, Liu F, Yang J et al (2016b) The biological activities, chemical stability, metabolism and delivery systems of quercetin: a review. Trends Food Sci Technol 56:21–38

Wood JG, Rogina B, Lavu S, Howitz K, Helfand SL, Tatar M et al (2004) Sirtuin activators mimic caloric restriction and delay ageing in metazoans. Nature 430(7000):686–689

Wu D, Prives C (2018) Relevance of the p53-MDM2 axis to aging. Cell Death Differ 25(1):169–179

Xu Z, Feng W, Shen Q, Yu N, Yu K, Wang S et al (2017) Rhizoma Coptidis and berberine as a natural drug to combat aging and aging-related diseases via anti-oxidation and AMPK activation. Aging Dis 6:760–777

Xu M, Pirtskhalava T, Farr JN, Weigand BM, Palmer AK, Weivoda MM et al (2018) Senolytics improve physical function and increase lifespan in old age. Nat Med 24(8):1246–1256

Yang Y-C, Lin H-Y, Su K-Y, Chen C-H, Yu Y-L, Lin C-C et al (2012) Rutin, a flavonoid that is a main component of saussurea involucrata, attenuates the senescence effect in D-galactose aging mouse model. Evid Based Complement Alternat Med. 2012:980276

Yang L, Zheng Z, Qian C, Wu J, Liu Y, Guo S et al (2017) Curcumin-functionalized silk biomaterials for anti-aging utility. J Colloid Interface Sci 15(496):66–77

Yosef R, Pilpel N, Tokarsky-Amiel R, Biran A, Ovadya Y, Cohen S et al (2016) Directed elimination of senescent cells by inhibition of BCL-W and BCL-XL. Nat Commun 7:11190
Youle RJ, Strasser A (2008) The BCL-2 protein family: opposing activities that mediate cell death. Nat Rev Mol Cell Biol 9(1):47–59
Yousefzadeh MJ, Zhu Y, McGowan SJ, Angelini L, Fuhrmann-Stroissnigg H, Xu M et al (2018) Fisetin is a senotherapeutic that extends health and lifespan. EBioMedicine 36:18–28
Zhang X, Zhang S, Liu X, Wang Y, Chang J, Zhang X et al (2018) Oxidation resistance 1 is a novel senolytic target. Aging Cell 17(4):e12780
Zhang P, Kishimoto Y, Grammatikakis I, Gottimukkala K, Cutler RG, Zhang S et al (2019) Senolytic therapy alleviates Aβ-associated oligodendrocyte progenitor cell senescence and cognitive deficits in an Alzheimer's disease model. Nat Neurosci 22(5):719–728
Zhu Y, Tchkonia T, Pirtskhalava T, Gower AC, Ding H, Giorgadze N et al (2015) The Achilles' heel of senescent cells: from transcriptome to senolytic drugs. Aging Cell 14(4):644–658
Zhu Y, Tchkonia T, Fuhrmann-Stroissnigg H, Dai HM, Ling YY, Stout MB et al (2016) Identification of a novel senolytic agent, navitoclax, targeting the Bcl-2 family of anti-apoptotic factors. Aging Cell 15(3):428–435
Zhu Y, Doornebal EJ, Pirtskhalava T, Giorgadze N, Wentworth M, Fuhrmann-Stroissnigg H, et al (2017) New agents that target senescent cells: the flavone, fisetin, and the BCL-XL inhibitors, A1331852 and A1155463. Aging (Albany NY) 9(3):955–963

Chapter 2
Discovery of Senolytics and the Pathway to Early Phase Clinical Trials

Erin O. Wissler Gerdes, Yi Zhu, Tamar Tchkonia, and James L. Kirkland

Abstract Cellular senescence is one of the fundamental mechanisms of aging. Senescent cells accumulate at etiological sites of age-related diseases and can secrete factors that cause dysfunction at local and systemic levels. The discovery of senolytics, drugs that specifically target senescent cells, has opened an innovative pathway for treating age-related diseases. Successes in pre-clinical models have led to first-in-human trials. If effective, senolytics could have a profound impact on alleviating age-related disorders and diseases.

Keywords Senolytics · Dasatinib · Quercetin · Fisetin · SASP · Cellular senescence · SCAPs

2.1 Introduction

Aging is the leading risk factor for most serious chronic diseases and disabilities including dementias, cancers, cardiac disease, vascular diseases, atherosclerosis, osteoporosis, arthritis, diabetes, metabolic syndrome, renal failure, blindness, and frailty (Miller 2002; Kirkland 2013). Although age-related chronic conditions are the major drivers of morbidity, mortality, and health costs, most have been difficult to control. The number of chronic disease conditions *per individual* increases with aging, leading to multi-morbidity (St Sauver et al. 2015), thus circumventing the public health impact of preventing any single age-related disease (Miller 2002; Fried et al. 2009). In addition to chronic diseases, aging predisposes to geriatric syndromes and reduced resilience. Geriatric syndromes include frailty, sarcopenia, immobility, falling, depression, mild cognitive impairment, incontinence, and weight loss, among

E. O. Wissler Gerdes · Y. Zhu · T. Tchkonia · J. L. Kirkland (✉)
Cellular Senescence and Translation and Pharmacology Programs, Robert and Arlene Kogod Center on Aging, Mayo Clinic, Rochester, USA
e-mail: kirkland.james@mayo.edu

J. L. Kirkland
Division of Geriatric Medicine and Gerontology, Department of Medicine, Mayo Clinic, Rochester, USA

D. Muñoz-Espin and M. Demaria (eds.), *Senolytics in Disease, Ageing and Longevity*, Healthy Ageing and Longevity 11,
https://doi.org/10.1007/978-3-030-44903-2_2

other problems (Inouye et al. 2007). Decreased physiological resilience entails failure to respond to or recover from stresses, such as pneumonia, stroke, heart attacks, dehydration, chemotherapy, surgery, fractures, or vaccinations (Kanapuru and Ershler 2009; Bandeen-Roche et al. 2009; Qu et al. 2009; Walston et al. 2009; Leng et al. 2007; Walston et al. 2006; Fried et al. 2001; Walston et al. 2002; Bandeen-Roche et al. 2006; Rockwood et al. 2006; Kirkland et al. 2016; Hadley et al. 2017).

An ideal strategy for addressing age-related chronic diseases, geriatric syndromes, and decreased resilience may be a "root cause" approach: targeting the fundamental aging mechanisms that represent shared upstream contributors or drivers of all of these conditions. Such a strategy could delay, prevent, or alleviate these conditions as a group, instead of adhering to the traditional one-disease-at-a-time approach. By one estimate, a 2% delay in the progression of aging processes would lead to an increase of 10 million healthy, as opposed to disabled, elderly people in the US by 2060 compared to doing nothing, delaying cancer, or delaying heart disease, with a savings in US health costs of $7.1 trillion over 50 years (Goldman et al. 2013).

Aging has long been recognized as the leading risk factor for most chronic diseases, geriatric syndromes, and decreased resilience, yet the fundamental aging processes that predispose to these conditions have only recently become viewed as potentially modifiable (Kirkland 2013; Tchkonia et al. 2013; Kirkland 2016). Supporting the view that interventions targeting basic aging processes could be feasible are the findings that: (1) Maximum lifespan is extended and age-related diseases are delayed across species by a number of single gene mutations (Bartke 2011; Pilling et al. 2017), suggesting pathways affected by these mutations could be therapeutic targets; (2) Humans who live beyond age 100, a partly heritable trait, frequently have delayed onset of age-related diseases and disabilities (Lipton et al. 2010), leading to compression of morbidity and enhanced healthspan; (3) Caloric restriction, which increases maximum lifespan, is associated with delayed onset of multiple chronic diseases in animal models (Anderson and Weindruch 2012); (4) Rapamycin increases lifespan and appears to delay cancers, age-related cognitive decline, and frailty in mouse models (Harrison et al. 2009; Bitto et al. 2016); (5) Factors produced by stem cells or in blood from young individuals may alleviate dysfunction in older individuals (Conboy et al. 2005; Lavasani et al. 2012; Villeda et al. 2014); and (6) Senescent cell accumulation is associated with chronic inflammation, fibrosis, and stem and progenitor cell dysfunction that in turn promote many age-related chronic diseases and geriatric syndromes (Tchkonia et al. 2013; Zhu et al. 2014). Senolytic drugs, which selectively eliminate senescent cells, delay age-related disorders, and enhance both health and life span in mice (Tchkonia 2017; Kirkland et al. 2017; Zhu et al. 2015; Roos et al. 2016; Schafer et al. 2017; Ogrodnik et al. 2017; Farr et al. 2017), as do a number of other drugs and lifestyle interventions. While senolytics and other agents that target fundamental aging processes were discovered only recently and in some cases are not yet published, the aging field is at the point of beginning to translate these interventions from lower mammals to humans

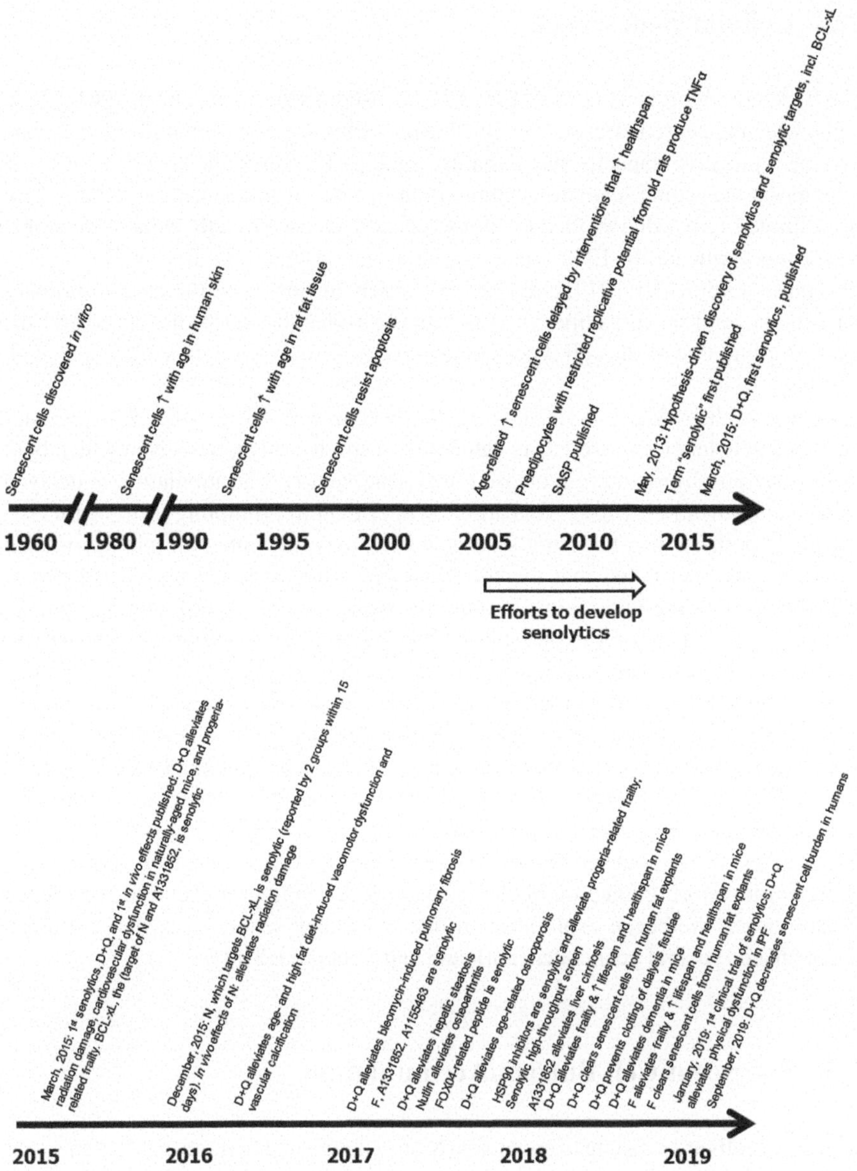

Fig. 2.1 Timeline depicting the discovery of senolytics. The discovery of senolytics began with the discovery of senescent cells in 1961 by Hayflick and Moorehead (1961). Developing senolytics began before and independently from making or studying INK-ATTAC mice

2.2 Cellular Senescence

Hayflick and Moorehead (1961) (Fig. 2.1) discovered senescent cells in 1961. These cells, which appeared after serial subculturing of human embryonic fibroblasts, had loss of replicative capacity but remained viable. This prompted work to test the hypothesis that aging leads to accumulation in vivo of pre-senescent cells, which have limited remaining replicative potential, and senescent cells, which are viable but cannot replicate. In 1979, Schneider et al. found this to be true in human skin fibroblasts (Schneider 1979). In 1990, we observed this to be the case in primary fat cell progenitors (preadipocytes) in adipose tissue cloned from rats across the age spectrum, one of the cell types employed to discover senolytics (Kirkland et al. 1990).

A key article by Sharpless et al. in 2004 showed in Ames dwarf mice with pituitary hormone deficiencies as well as in calorically restricted mice, models in which both healthspan and lifespan are increased, that senescent cell accumulation is delayed (Krishnamurthy et al. 2004). This article was critical in prompting us to begin testing the hypothesis that targeting senescent cells may alleviate multiple age-related disorders. We had noted that the preadipocytes with restricted replicative potential that accumulate with aging in adipose tissue and fail to differentiate into fat cells also have increased expression of the inflammatory cytokine, TNFα (Kirkland et al. 2002). Before their key publication in 2008 (Coppé et al. 2008), Campisi et al. presented their finding that senescent cells can secrete a range of inflammatory and pro-apoptotic factors, the senescence-associated secretory phenotype (SASP) at meetings. This suggested that senescent cells have the potential to damage tissues in vivo, further prompting us to test if the increased burden of senescent cells with aging could be a cause of local and systemic dysfunction. Thus, in 2004/5 we began to test the hypothesis that selectively targeting senescent cells is a promising strategy for restoring function in old age and for delaying or preventing age-related disease onset. We began exploratory efforts to develop senolytic agents, a strategy we published several years later (Kirkland and Tchkonia 2014).

2.3 Senescent Cell Burden/Accumulation

Senescent cells are resistant to apoptosis (Munoz-Espin and Serrano 2014). Senescent cells appear with aging in a number of tissues and develop at sites of pathogenesis of several chronic diseases. Accumulation of senescent cells can cause extensive local and systemic dysfunction due to their pro-inflammatory SASP (Coppé et al. 2008, 2010; Kuilman and Peeper 2009). In pre-clinical experiments, we showed transplanting small numbers of senescent cells around the knee joints of young mice induces an osteoarthritis-like phenotype, while transplanting non-senescent cells did not (Xu et al. 2018). Additionally, transplanting 1 million radiation- or chemotherapy-induced senescent autologous ear fibroblasts or syngeneic preadipocytes intraperitoneally into lean, adult mice, so that only 1/10,000 of all cells in the transplanted mice were senescent cells, induced impaired physical function and shortened lifespan due to early onset of all of the same age-related diseases that cause death in older naturally-aged mice (Xu et al. 2018). Transplanting non-senescent cells into control mice did not do this, nor did transplanting 500 thousand as opposed to 1 million senescent cells. However, transplanting 500 thousand senescent cells into diet-induced obese (DIO) adult mice, which have more senescent cells than lean mice, or into old mice, was sufficient to cause the accelerated aging-like state. This suggests that there is a threshold of senescent cells, pre-existing plus transplanted, above which the accelerated aging-like state occurs. These observations further suggest the potential value of interventions that target senescent cells.

2.4 Development of Senolytics

The first senolytics were ultimately identified using a mechanism-based approach informed by their modes of action and targets, rather than a random approach such as screening libraries of compounds. As stated above, our efforts to find senolytics began in 2004/5 (Fig. 2.1), with initial attempts to create fusion proteins comprising a senescent cell surface binding domain coupled to a toxin, high throughput compound library screens for agents that eliminate senescent but not non-senescent cells, and other approaches. These traditional approaches did not achieve our goal. We therefore turned to a hypothesis-driven drug discovery approach. Our hypotheses were: (1) senescent cells resist apoptotic stimuli, implying the existence of pro-survival—anti-apoptotic defenses against their own SASP and harsh metabolic internal state and (2) in some respects, senescent cells are like cancer cells that do not divide (Zhu et al. 2015). Our hypothesis driven, mechanism-based approach led to discovery of the first senolytics, the combination of Dasatinib plus Quercetin, within a month after starting this work in May, 2013 [published in March, 2015 (Zhu et al. 2015)] and more senolytics subsequently (Zhu et al. 2016, 2017; Fuhrmann-Stroissnigg 2017).

We asked how senescent cells expressing a SASP can survive, despite their own highly pro-apoptotic and metabolically-distinct, potentially damaging *milieu*. Building upon bioinformatics data derived from proteomic and transcriptomic profiles of senescent versus non-senescent cells, we searched for senescent cell anti-apoptotic pathways (SCAPs). We identified several such potential SCAPs (ephrins/dependence receptors; PI3Kδ/Akt/metabolic; Bcl-2, Bcl-xl, Bcl-w; p53/FOXO4a/p21/serpine [PAI-1&2]; HIF-1α) (Zhu et al. 2015; Baar et al. 2017) (Fig. 2.2) and then another, the HSP-90 pathway (Fuhrmann-Stroissnigg 2017). We tested whether these SCAPs are essential for senescent cell survival by targeting key proteins within these SCAP pathways using RNA interference in senescent versus non-senescent human primary preadipocytes and human umbilical vein endothelial cells (HUVECs). Of the 39 small interfering RNA's (siRNA's) targeting possible SCAPs, 17 caused death of senescent but not non-senescent cells. We noted the patterns of SCAP pathways that prevent self-induced death of senescent human preadipocytes differed considerably from

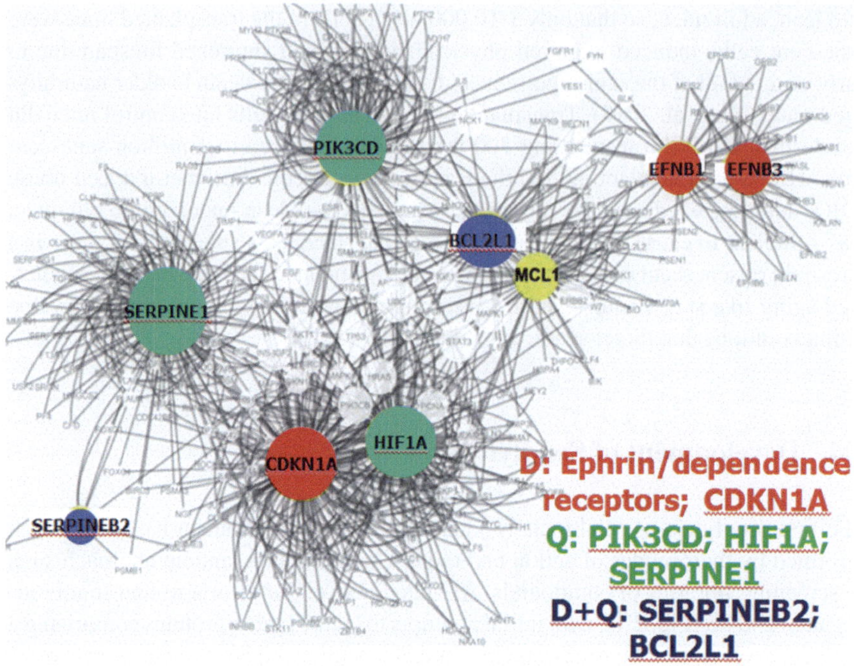

Fig. 2.2 Network analysis to test links among EFNB-1, EFNB-3, PI3KCD, p21 (CDKN1A), PAI-1 (SERPINE1), PAI-2 (SERPINB2), BCL-xL, and MCL. *Source* Author

those required for survival of senescent endothelial cells. Senescent preadipocytes rely on pathways related to tyrosine kinases involved in the apoptosis that can be caused by dependence receptors (receptors that, if present on a cell but that are unoccupied by a ligand, induce cell death by apoptosis, such as the ephrin receptors) as well as p53- and p21-related pro-survival mechanisms and metabolically-related apoptosis involving PI3-kinase, AKT, and, again, p53. Senescent preadipocytes did not depend on BCL-2 family pro-survival proteins. Conversely, senescent endothelial cells depended for survival on BCL-2 family members, particularly BCL-xL, compared to non-senescent endothelial cells, as well as components of the PI3 kinase and HIF-1α pathways.

To discover senolytic agents, senescent and non-senescent cells were treated with 46 drugs and natural products previously reported to target the key SCAP proteins that we had identified through our RNA interference studies. Based on our starting hypotheses, agents selected for testing included those reputed to have an effect against cancers. By using this mechanism-based approach, Dasatinib (D) and Quercetin (Q) were among the first senolytics we selected for more intensive investigation based on their ability to target multiple nodes in the pro-survival SCAP networks we had discovered. We also selected these compounds because D has been approved by the FDA for use in humans since 2006 and Q is a natural product with a favorable safety profile, potentially enabling progression to clinical trials. Furthermore, these agents were selected because both have short elimination half-lives, <11 h, in humans, facilitating a "hit-and-run" intermittent treatment approach. As predicted, we found D and Q are senolytic. From the RNA interference studies, D was predicted to be senolytic for senescent human cultured preadipocytes but not human endothelial cells, which turned out to be true. Also as predicted, Q was senolytic for endothelial cells but not preadipocytes. The combination of D+Q, which together targets at least 8 SCAP pathway nodes (Fig. 2.2), was senolytic for both senescent preadipocytes and endothelial cells. Thus, different types of senescent cells employ distinct SCAPs to defend themselves against their own SASP and pro-apoptotic *milieu*. In some types of senescent cells, more than one SCAP pathway is engaged and these pathways can be redundant. Therefore, for each type of senescent cell, targeting only one SCAP pathway may not be sufficient to induce apoptosis. Combinations of senolytics that target different SCAPs or individual drugs that are active against multiple targets are required to overcome this redundancy. This novel concept goes against the traditional drug development paradigm of "one drug-one target-one disease" (Fig. 2.3).

Of note, work to discover senolytics began before development of transgenic mice from which highly p16^{Ink4a}-expressing cells can be removed using a drug that acts on a cell-killing construct, ATTAC that was originally devised by P. Scherer et al. (Trujillo et al. 2005; Baker et al. 2011). Similarly, in p16-3MR mice, a drug-inducible killing construct is expressed in highly p16^{Ink4a}-expressing cells (Demaria et al. 2014). In these engineered mice, the drug-inducible killing construct is expressed in an inactive form by placing the transgene under the control of the p16^{Ink4a} promoter, either as an inserted construct or, in the case of p16-3MR mice, within a minigene. Some, but not

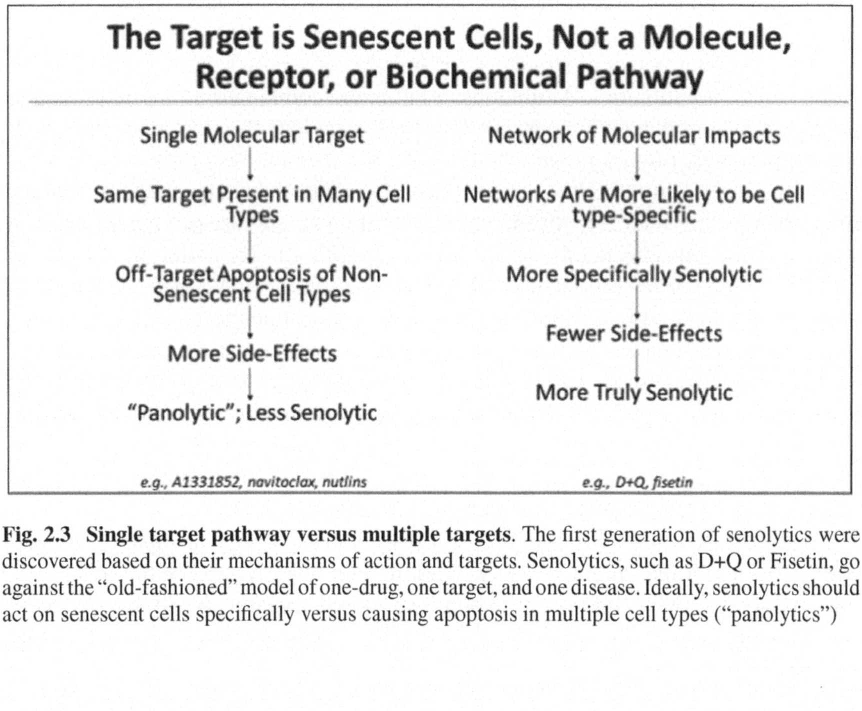

Fig. 2.3 Single target pathway versus multiple targets. The first generation of senolytics were discovered based on their mechanisms of action and targets. Senolytics, such as D+Q or Fisetin, go against the "old-fashioned" model of one-drug, one target, and one disease. Ideally, senolytics should act on senescent cells specifically versus causing apoptosis in multiple cell types ("panolytics")

all senescent cells express $p16^{Ink4a}$ and not every cell with high $p16^{Ink4a}$ expression is senescent, for example non-senescent activated macrophages (Hall et al. 2016; 2017). Unlike in the transgenic mice, senolytics do not act through targeting cells that have high expression of $p16^{Ink4a}$. Rather, senolytics selectively eliminate those senescent cells that cause tissue damage by releasing the pro-apoptotic, proinflammatory factors and proteases that are part of the SASP. Senolytics can also kill cancer cells, in particular those types of cancer cells that release pro-apoptotic factors, such as certain lymphoid malignancies. Indeed, one of the hypotheses used to develop the strategy for discovering senescent cells is that senolytics should do so (Zhu et al. 2015). Thus, senolytics act in a manner distinct from the removal of highly $p16^{Ink4a}$-expressing cells from the engineered animal models. It is therefore not a "given" that eliminating highly $p16^{Ink4a}$-expressing cells from the transgenic animal models will faithfully mimic the effects of removing senescent cells using senolytic agents. Unlike the transgenic animal approaches, senolytics are effective in wild-type mice without an inserted transgene, the discovery of senolytics did not depend on or involve use of the transgenic mice, and efforts to discover senolytics began before and independently from development of the transgenic mice.

2.5 Target of Senolytics: Senescent Cells

Completely new drug development paradigms are needed to take senolytics into clinical application, as may be the case for other types of interventions that target "root cause" fundamental aging processes (Kirkland 2013, 2016; Newman et al. 2016; Huffman et al. 2016; Justice et al. 2016; Burd et al. 2016). Perhaps the closest analogy regarding potential strategies for translating senolytics into clinical application is that of antibiotics. Antibiotics or antibiotic combinations are developed to target bacteria or other pathogens, not necessarily single molecular targets. In developing antibiotics, often a range of infections, for example respiratory and urinary tract infections, skin infections, and septicemia, are tested using candidate agents, rather than testing their effectiveness against only a single disease. Additionally, a range of pathogens is generally tested for susceptibility to the antibiotic. Effects of antibiotics alone or in combination are tested. The same may be the case for effectively developing and translating senolytics into clinical application. The key drug targets are senescent cells and the networks that sustain them, not a single molecule, not a single biochemical pathway, nor a single receptor. Multiple senescent cell types, senescence-associated diseases, and combinations of senolytics may need to be considered to successfully translate senolytics into clinical application, unlike the traditional one-drug/one-target/one-disease approach used for developing drugs that target a receptor, an enzyme, or a biochemical pathway.

Combination of senolytic agents with SASP inhibitors, such as rapamycin, metformin, or ruxolitinib, might possibly interfere with effectiveness of the senolytics, since senolytics act by allowing pro-apoptotic, proinflammatory SASP factors to kill the senescent cells from which these factors are released by transiently disabling the SCAPs that protect the senescent cells. This theoretical problem may turn out to be avoided by holding administration of SASP inhibitors for periods of time before and after senolytics are administered in the course of intermittent senolytic drug treatment regimens.

Since our report of D+Q, over a dozen more senolytics have been published using essentially the same hypothesis-driven approach [reviewed in (Tchkonia 2017; Kirkland et al. 2017)]. One such senolytic, Fisetin, is a naturally occurring flavonoid that selectively induces apoptosis in senescent but not proliferating human umbilical vein endothelial cells (HUVECs) (Zhu et al. 2017; Yousefzadeh et al. 2018). Similarly to quercetin, Fisetin selectively reduced viability and numbers of senescent HUVECs. This flavonoid is present in many fruits and vegetables such as apples, persimmon, grapes, onions, and cucumbers, with high concentrations found in strawberries (160 μg/g) (Khan et al. 2013). Its hydrophobic properties allow Fisetin to penetrate cell membranes and accumulate within cells to exert antioxidant effects (Ishige et al. 2001). It is widely available as a nutritional supplement and has a low side effect profile, which makes it an attractive option for clinical trials.

We first reported that targeting the BCL-2 pathway is a senolytic strategy in March, 2015 (Zhu et al. 2015). Knocking down BCL-xL mRNA by RNA interference killed senescent human endothelial cells, but not senescent human preadipocytes or non-senescent cells. Ten months later, we reported that Navitoclax, which targets BCL-2 family members including BCL-xL and BCL-w, induces apoptosis in senescent human endothelial cells, but not senescent human preadipocytes or non-senescent cells (Zhu et al. 2015). Within 15 days of that report, another group had found Navitoclax is senolytic and enhances bone marrow recovery following radiation in mice (Chang et al. 2016). However, Navitoclax only eliminates a subset of senescent cell types (*e.g.*, endothelial cells) and not others (*e.g.*, senescent preadipocytes) and can cause off-target side-effects through eliminating non-senescent cells, such as neutrophils and megakaryocytes or platelets. Also, unlike Dasatinib, which has been approved by the US Food and Drug Administration (FDA) for clinical use since 2006, Navitoclax is still not approved by the FDA for general clinical use. We subsequently found the more specific BCL-xL inhibitors, A1331852 and A1155463, are senolytic, at least for endothelial cells (Zhu et al. 2017), but again they have not been approved for general clinical use by the FDA.

2.6 Mouse Models of Aging and Disease

After the initial finding that the combination of D+Q is senolytic for both senescent preadipocytes and endothelial cells, we tested this first generation senolytic drug combination in naturally-aged animals. In early 2015, we published our findings that a single course of senolytics enhanced cardiac ejection fraction and improved vascular reactivity in 24 month old mice, the equivalent to 75–80 years of age in humans (Zhu et al. 2015). This was the first demonstration that targeting senescent cells enhances healthspan parameters in naturally-aged mice. Later that year, we confirmed and extended that finding by treating naturally-aged mice with a SASP inhibitor, Ruxolitinib. Metabolic healthspan parameters, including preservation of subcutaneous fat and insulin sensitivity, were enhanced by targeting senescent cells in these naturally-aged mice (Xu et al. 2015). Furthermore, total daily activity, rearing activity, ambulation, ability to remain suspended by hanging on to a wire, grip strength, and coordination were improved by targeting senescent cells in naturally-aged mice (Xu et al. 2015). Subsequently, we and others found that senolytics alleviate an impressive range of not only age-related phenotypes, but also chronic diseases in pre-clinical animal models, including high fat diet-induced vascular hyporeactivity, vascular calcification, damaged cardiac muscle, age-related cardiac hypertrophy and fibrosis, metabolic dysfunction and diabetes, chronic kidney disease, liver steatosis and fibrosis, pulmonary fibrosis, hyperoxia-induced airway disease, age-related bone loss, and obesity-related neuropsychiatric dysfunction, among others (Roos et al. 2016; Schafer et al. 2017; Ogrodnik et al. 2017; Farr et al. 2017; Xu et al. 2018; Moncsek et al. 2017; Ogrodnik et al. 2019; Parikh et al. 2018; Lewis-McDougall et al. 2019; Palmer et al. 2019; Lewis-McDougall et al. 2019; Anderson et al. 2019; Kim et al. 2019; Musi et al. 2018; Zhang et al. 2019). Senolytics can reduce chronic low-grade inflammation, protein aggregation, calcification, and fibrosis, the pathological processes active at sites of etiology in many of the major chronic diseases (Tchkonia 2017; Kirkland et al. 2017; Tchkonia and Kirkland 2018).

In a key study first demonstrating that targeting senescence is a potentially disease-modifying treatment for Alzheimer's disease and other protein aggregation-related neurodegenerative diseases, D+Q was shown to decrease brain senescence markers, SASP factors, neurofibrillary tangles, and neuro-inflammation and to partially reverse a measure of brain hypo-perfusion, decrease brain atrophy, and enhance cognition in several different Tau^+ Alzheimer's disease mouse models (Musi et al. 2018; Zhang et al. 2019). D+Q was next shown to be effective in β-amyloid-expressing mice, another Alzheimer's disease mouse model (Zhang et al. 2019). D+Q was effective even if administered to 23 month old Tau^+ mice with clinically-evident dementia (Musi et al. 2018). Thus, it may be feasible to alleviate features of Tau^+ dementia even after the dementia has become clinically manifest, a scenario that is much more translatable into initial clinical application than would be trials of preventing development of dementia, since the latter would involve discerning which subjects are most likely develop dementia before it becomes clinically evident, a difficult task and one that would entail treating many subjects unnecessarily (since many subjects with risk factors do not develop dementia or do so after a prolonged lag). Therefore, clinical trials of senolytics for clinically manifest Alzheimer's and related dementias are about to start.

Based on our findings of pro-survival SCAP networks, we and others identified the flavonoid Fisetin as senolytic in both HUVEC and IMR 90 cells. The senolytic efficacy of Fisetin was tested in both murine models (progeroid and chronological aged mice) and human tissues (Yousefzadeh et al. 2018). Intermittent treatment with Fisetin reduced abundance of some types of senescent cells in multiple tissues. Also, late intervention with Fisetin restored tissue homeostasis, alleviated age-related pathology, and extended median and maximum lifespan. These characteristics suggested the feasibility of translating Fisetin into human clinical studies, which are now underway.

The first article about senolytics demonstrated that targeting senescent cells improves function in naturally-aged animals (Zhu et al. 2015). Beneficial effects on function in naturally-aged mice were subsequently confirmed in studies using the original and later senolytics as well as SASP inhibitors (Xu et al. 2018; Yousefzadeh et al. 2018; Xu et al. 2015).

2.7 Testing If a Drug Acts as a Senolytic Using a Modified Set of Koch's Postulates

Although there is evidence that senolytics may alleviate multiple conditions as considered above, proving that drugs actually alleviate a given age-related phenotype, disorder, or disease because of senolytic as opposed to off-target effects is not trivial and has not been established beyond doubt for many such conditions so far. To prove conclusively that a candidate agent alleviates a condition because of senolytic effects, we propose a set of 8 criteria based on Koch's postulates of the type used to prove causation in the case of infectious agents. These are considered below.

To establish causality using this modified set of Koch' postulates, it would be first necessary to show that senescent cells occur in tandem with the condition in question: **(1)** Are senescent cells present in animals or humans with the condition? **(2)** Do individuals without senescent cells have the condition? Next, a way to test if cellular senescence is <u>sufficient</u> to cause a condition is to **(3)** show that the condition can be reproduced by inducing local accumulation of senescent cells. This can be achieved by transplanting senescent cells, focal irradiation, or tissue-specific genetic approaches to cause local senescent cell accumulation. For example, it was first demonstrated that senescent cells are sufficient to cause osteoarthritis by transplanting small numbers of syngeneic senescent mouse cells around the knee joints of younger mice (Xu et al. 2016). After a couple of months, this resulted in decreased mobility, knee joint pain, and radiographic changes characteristic of age-related osteoarthritis, while transplanting equal numbers of non-senescent cells did not cause this. In another example (also considered above), transplanting small numbers of senescent cells into middle-aged mice caused development of a frailty-like state after a few weeks, with decreased physical function and endurance, as well as accelerated onset of age-related diseases as a group (Xu et al. 2018). Causality can be tested further by **(4)** determining if removing these transplanted or induced senescent cells prevents or alleviates the condition. In the case of frailty or accelerated onset of age-related diseases caused by transplanting senescent cells, D+Q reduced senescent cell burden and indeed delayed or prevented these conditions. **(5)** It is then important to test if targeting naturally-occurring senescent cells alleviates the condition in question. This was achieved in the cases of frailty and age-related disease onset by treating naturally-aged, as opposed to younger transplanted mice with D+Q. (**6**) Administering the potentially senolytic drugs being investigated should have few or no effects related to the condition being tested in individuals without senescent cells, for example in young mice. (**7**) Senolytics should alleviate the condition even if given intermittently, at intervals longer than the drugs' half lives, since senescent cells can take 2–6 weeks to re-accumulate, at least in cell culture. In the case of D+Q, the elimination half life is 11 h, but the drugs are as effective if administered monthly as continuously, at least in the case of age-related osteoporosis (Farr et al. 2017). (**8**) Finally, if an agent is truly senolytic, it should alleviate multiple age-related conditions.

2.8 Clinical Trials

Senolytics were first developed using cells cultured from human subjects in order to facilitate the path to translation into human application. Senolytics were next shown to be effective in clearing senescent cells from mice and also for alleviating a range of age- and senescence-related disorders in mice, including cardiac and vascular dysfunction, insulin resistance, cognitive dysfunction, age-related osteoporosis, and frailty, among many others (see below). Furthermore, treatment with senolytics alleviated the accelerated aging-like state induced by transplanting senescent versus non-senescent cells into young mice (Xu et al. 2018). Recently, senolytics were shown to decrease senescent cell abundance in adipose tissue of humans with diabetes and obesity (Hickson et al. 2019) and to alleviate physical dysfunction in patients with idiopathic pulmonary fibrosis, a cellular senescence-driven disease [see below (Justice et al. 2019)].

Obesity with diabetes leads to accumulation of senescent cells in adipose tissue of humans (Minamino et al. 2009). To test if senolytics can clear senescent cells from human tissues, adipose tissue biopsied from obese, diabetic human subjects undergoing surgery was treated with D+Q or vehicle for 48 h (Xu et al. 2018). Within a few hours, the senescent cells in the freshly-isolated tissue began to undergo cell death through apoptosis. This was associated with reduced release of SASP factors, and increased expression of transcription factors that promote adipose tissue insulin sensitivity and of metabolically beneficial factors, including adiponectin and adipsin. Fisetin was also shown to reduce senescent cells from freshly-isolated human fat (Yousefzadeh et al. 2018).

2.8.1 *First-in-Human Trial of Senolytics: D+Q for Idiopathic Pulmonary Fibrosis*

Idiopathic pulmonary fibrosis (IPF) is a senescence-associated, progressive, fatal disease of the elderly for which treatments, other than lung transplantation, are at best modestly useful. Senescent cells accumulate in subjects with IPF and are a source of inflammatory and fibrotic mediators in this disease. In the Bleomycin inhalation mouse model of IPF, D+Q was more effective than another senolytic agent, Navitoclax, in clearing senescent cells and D+Q alleviated pulmonary dysfunction, attenuated weight loss, and improved exercise endurance in these mice (Schafer et al. 2017). A pilot, open-label clinical trial of 9 doses of oral D+Q over 3 weeks led to improved 6-min. walk distance, walking speed, ability to get up from a chair, and short physical performance battery by 5 days after the final dose in 14 subjects with IPF (Justice et al. 2019). These results led to initiation of a Phase IIb randomized, placebo-controlled, double-blind trial that is currently underway.

Approaches were devised to gain approvals of new tests for senescent cell burden, a rapidly developing field, and of other tests of fundamental aging mechanisms to be incorporated into the study without stopping the primary study each time an amendment was submitted. Subject recruitment strategies had to be developed, including adjusting and simplifying the study burden for the elderly subjects in the trial.

2.8.2 *Systemic Sclerosis*

A reanalysis of archived skin biopsies from a trial of Dasatinib administered continuously for 6 weeks to 3 subjects with systemic sclerosis with evidence of skin senescent cells before treatment was conducted to test if senescent cells were removed (Martyanov et al. 2019). Of 65 SASP factors assayed, 55 had decreased after the course of Dasatinib, as were gene signature profiles linked to senescence.

2.8.3 *Chronic Kidney Disease and Diabetes*

A Phase 1, open-label, clinical trial of D+Q for subjects with diabetic kidney disease (DKD) is underway at Mayo (ClinicalTrials.gov Identifier: NCT02848131). Interim results were encouraging, showing that a 3 day oral course of D+Q in 9 subjects with DKD caused reduced adipose tissue senescent cell burden by 11 days after the last dose. Furthermore, a composite score of 10 circulating SASP factors was significantly decreased 11 days after completing the 3 day D+Q intervention. This trial is continuing (goal = 30) to test effects of senolytics on adipose tissue and skin senescent cell abundance, blood and urine SASP factors, metabolic and renal function, inflammation, quality of life, and safety (drug toxicity) and tolerability. No serious drug side effects have emerged so far and evidence continues to show clearance of senescent cells. Each subject will be followed for 4 months after the single course of D+Q. The goal is to provide data for a larger Phase IIb randomized, placebo-controlled, double-blind trial of senolytics for DKD.

2.8.4 Alleviation of Frailty, Inflammation, and Related Measures in Older Women

This Phase IIb double-blind, placebo-controlled clinical trial of a different senolytic drug, Fisetin, to reduce senescent cell burden and alleviate frailty and inflammation in older women (AFFIRM) is underway at Mayo (ClinicalTrials.gov Identifier: NCT03430037). Our groups recently reported an association between frailty in elderly women and senescent cell burden in adipose tissue biopsies (Justice et al. 2017). In preclinical studies, we found Fisetin causes apoptosis of senescent human endothelial cells (Zhu et al. 2017), among other senescent cell types. Fisetin alleviated frailty in progeroid and naturally-aged mice and extended median and maximum lifespan (Yousefzadeh et al. 2018). The latter finding has prompted testing of continuous versus intermittent Fisetin administration in the NIA-Interventions Testing Program (ITP). We arranged for isolation of Fisetin from plants in which it is present, conducted FDA-required testing for contaminants, completed stability and toxicity testing, prepared placebo capsules, conducted dose-escalation studies in *Rhesus* monkeys, and gained an FDA-IND for testing effectiveness of Fisetin in alleviating physical dysfunction in elderly women with the Fried frailty phenotype and slow gait speed, as well as ancillary testing of senescent cell burden, metabolic function, inflammation, and bone turnover. Along with the clinical trials mentioned here, there are other trials launching in the near future, including 3 studying senolytics for Alzheimer's disease.

2.9 Translational Geroscience Network

In conducting the pre-clinical laboratory studies, completing the FDA and institutional approval processes, arranging for clinical trials to be conducted at multiple sites, enrolling subjects, coordinating specimen and data processing and storage, and harmonizing standard operating procedures (SOPs), many issues became apparent that a coordinated research network could resolve.

The Translational Geroscience Network (TGN), made up of 8 academic medical centers (Mayo, Harvard, Connecticut, Hopkins, Michigan, Wake Forest, UTHSCSA, and Minnesota), was established to develop, implement, and test SOPs for translational early phase trials of agents that target fundamental aging processes. The overall goal is to facilitate and speed translation by optimizing resource utilization, while avoiding duplication and counter-productive competition. This initial network will engage in preclinical translational research, completing the appropriate regulatory steps (*e.g.*, acquiring INDs from the FDA), and assisting in developing several "use case" proof-of-concept Phase I and II trials. By pursuing steps to translate drugs that target fundamental aging mechanisms from bench to bedside concurrently and by doing so in a coordinated way, it is hoped the TGN will accelerate development of effective treatments. With luck, such interventions might be introduced clinically within the next 5–15 years. Additionally, the creation of this network will allow researchers from different sites to select, optimize, and validate ancillary measures of fundamental aging processes to be assayed across all trials. Ultimately, one of the long term goals of the TGN is to develop a biobanking and repository network where researchers can share samples and data, allowing comparisons among interventions targeting fundamental aging processes, including different senolytic drug regimens.

2.10 Enhancing Healthspan

A method for increasing healthspan may be to remove damaging and potentially cancerous senescent cells and next to treat with trophic or anabolic factors. Removing senescent cells before administering treatments that augment stem and progenitor cell function or anabolic agents may: (1) increase effectiveness of the trophic or anabolic agents by removing the "brake" on stem and progenitor cell function and growth exerted by senescent cells and (2) reduce the risk of cancer that could complicate administering trophic or anabolic agents to elderly individuals harboring precancerous or cancerous cells. Many of the recently discovered senolytic agents have been in use for inducing apoptosis of cancer cells before their senolytic effects were discovered (Tchkonia 2017; Kirkland et al. 2017). For example, Dasatinib is used for treating certain hematological and other cancers (Keating 2017) and Quercetin can delay or alleviate cancers (Chikara et al. 2018). Thus, treating with senolytic drugs first might remove cancerous cells and senescent cells harboring potentially cancerous mutations before trophic and anabolic factors are administered. This approach for enhancing healthspan needs to be validated in pre-clinical experimental animal studies.

Perhaps better senolytics reported after the discovery that D+Q is senolytic will result in even more improvement of physical function, delay and alleviation of age-related diseases, and lifespan extension than D+Q, although it is likely that these improvements will not persist indefinitely. It seems much more likely that age-related dysfunction and death will occur at some point because of other fundamental aging processes, even if repeated administration of senolytics can completely and durably eliminate senescent cells. To achieve a more substantial increase in healthspan than is possible through senolytics alone, combining senolytic treatments with interventions that target other aging processes may be more effective.

2.11 Conclusions

Aging is a complex process that involves multiple factors that are comprised in a network driving this process. Cellular senescence is a key aging mechanism that is part of this network of fundamental aging mechanisms. Many fundamental aging processes, including genomic instability, telomere dysfunction/shortening, metabolic dysregulation, protein aggregation and misfolding, and nuclear membrane and mitochondrial dysfunction, can initiate or drive cellular senescence, and senescent cells coupled with their SASP factors can causally prime multiple aging processes, including stem cell exhaustion, disrupted intercellular communications, protein aggregation, NAD^+ depletion, and chronic inflammation, which eventually lead to aging phenotypes and chronic diseases. If senescent cells are targeted and eliminated by senolytics, many or perhaps all other aging processes might be alleviated, the Unitary Theory of Targeting Fundamental Aging Mechanisms.

Since interventions that increase health- and life-span in mammals now exist, we hypothesize that by targeting fundamental mechanisms of aging, clinical interventions can be envisaged that could delay, prevent, or alleviate age-related conditions as a group, instead of one-disease-at-a-time. If effective in humans, interventions targeting basic aging processes could have a substantially larger impact on healthspan and costs than curing any one chronic disease, a potentially fundamental transformation in health care (Miller 2002; Kirkland 2013, 2016).

Until safety and effectiveness of senolytic drugs have been established in clinical trials, we strongly emphasize that these drugs are at an early stage of development. There could be profound, serious, as yet unforeseen adverse effects of senolytics in humans. At this point, senolytic agents must not be used outside of clinical trials involving intensive monitoring for potential adverse effects. Senolytics are not ready for general prescribing or clinical use.

References

Anderson RM, Weindruch R (2012) The caloric restriction paradigm: implications for healthy human aging. Am J Hum Biol 24(2):101–6

Anderson R, Lagnado A, Maggiorani D, Walaszczyk A, Dookun E, Chapman J et al (2019) Length-independent telomere damage drives post-mitotic cardiomyocyte senescence. EMBO J 38(5)

Baar MP, Brandt RM, Putavet DA, Klein JD, Derks KW, Bourgeois BR et al (2017) Targeted apoptosis of senescent cells restores tissue homeostasis in response to chemotoxicity and aging. Cell 169(1):132–47 e16

Baker DJ, Wijshake T, Tchkonia T, LeBrasseur NK, Childs BG, van de Sluis B et al (2011) Clearance of p16Ink4a-positive senescent cells delays ageing-associated disorders. Nature 479(7372):232–6

Bandeen-Roche K, Xue QL, Ferrucci L, Walston J, Guralnik JM, Chaves P et al (2006) Phenotype of frailty: characterization in the women's health and aging studies. J Gerontol A Biol Sci Med Sci 61(3):262–6

Bandeen-Roche K, Walston JD, Huang Y, Semba RD, Ferrucci L (2009) Measuring systemic inflammatory regulation in older adults: evidence and utility. Rejuvenation Res 12(6):403–10

Bartke A (2011) Single-gene mutations and healthy ageing in mammals. Philos Trans R Soc Lond B Biol Sci 366(1561):28–34
Bitto A, Ito TK, Pineda VV, LeTexier NJ, Huang HZ, Sutlief E et al (2016) Transient rapamycin treatment can increase lifespan and healthspan in middle-aged mice. elife 5
Burd CE, Gill MS, Niedernhofer LJ, Robbins PD, Austad SN, Barzilai N et al (2016) Barriers to the preclinical development of therapeutics that target aging mechanisms. J Gerontol A Biol Sci Med Sci 71(11):1388–94
Chang J, Wang Y, Shao L, Laberge RM, Demaria M, Campisi J et al (2016) Clearance of senescent cells by ABT263 rejuvenates aged hematopoietic stem cells in mice. Nat Med 22(1):78–83
Chikara S, Nagaprashantha LD, Singhal J, Horne D, Awasthi S, Singhal SS (2018) Oxidative stress and dietary phytochemicals: role in cancer chemoprevention and treatment. Cancer Lett 413:122–34
Conboy IM, Conboy MJ, Wagers AJ, Girma ER, Weissman IL, Rando TA (2005) Rejuvenation of aged progenitor cells by exposure to a young systemic environment. Nature 433:760–4
Coppé JP, Desprez PY, Krtolica A, Campisi J (2010) The senescence-associated secretory phenotype: the dark side of tumor suppression. Annu Rev Pathol 5:99–118
Coppé JP, Patil C, Rodier F, Sun Y, Muñoz DP, Goldstein J et al (2008) Senescence-associated secretory phenotypes reveal cell-nonautonomous functions of oncogenic RAS and the p53 tumor suppressor. PLoS Biol 6:2853–68
Demaria M, Ohtani N, Youssef SA, Rodier F, Toussaint W, Mitchell JR et al (2014) An essential role for senescent cells in optimal wound healing through secretion of PDGF-AA. Dev Cell 31(6):722–33
Farr JN, Xu M, Weivoda MM, Monroe DG, Fraser DG, Onken JL et al (2017) Targeting cellular senescence prevents age-related bone loss in mice. Nat Med
Fried LP, Tangen CM, Walston J, Newman AB, Hirsch C, Gottdiener J et al (2001) Frailty in older adults: evidence for a phenotype. J Gerontol A Biol Sci Med Sci 56(3):M146–56
Fried LP, Xue QL, Cappola AR, Ferrucci L, Chaves P, Varadhan R et al (2009) Nonlinear multisystem physiological dysregulation associated with frailty in older women: implications for etiology and treatment. J Gerontol A Biol Sci Med Sci 64(10):1049–57
Fuhrmann-Stroissnigg H, Ling YY, Zhao J, McGowan SJ, Stripay JL, Gregg S et al (2017) Identification of HSP90 inhibitors as senolytics for extending healthspan. Nat Commun (in press)
Goldman DP, Cutler D, Rowe JW, Michaud PC, Sullivan J, Peneva D et al (2013) Substantial health and economic returns from delayed aging may warrant a new focus for medical research. Health Aff 32(10):1698–705
Hadley EC, Kuchel GA, Newman AB, Allore HG, Bartley JM, Bergeman CS et al (2017) Report: NIA workshop on measures of physiologic resiliencies in human aging. J Gerontol A Biol Sci Med Sci 72(7):980–90
Hall BM, Balan V, Gleiberman AS, Strom E, Krasnov P, Virtuoso LP et al (2016) Aging of mice is associated with p16(Ink4a)- and beta-galactosidase-positive macrophage accumulation that can be induced in young mice by senescent cells. Aging 8(7):1294–315
Hall BM, Balan V, Gleiberman AS, Strom E, Krasnov P, Virtuoso LP et al (2017) p16(Ink4a) and senescence-associated beta-galactosidase can be induced in macrophages as part of a reversible response to physiological stimuli. Aging
Harrison DE, Strong R, Sharp ZD, Nelson JF, Astle CM, Flurkey K et al (2009) Rapamycin fed late in life extends lifespan in genetically heterogeneous mice. Nature 460(7253):392–5
Hayflick L, Moorehead P (1961) The serial cultivation of human diploid strains. Exp Cell Res 25:585–621
Hickson LJ, Langhi Prata LGP, Bobart SA, Evans TK, Giorgadze N, Hashmi SK et al (2019) Senolytics decrease senescent cells in humans: preliminary report from a clinical trial of dasatinib plus quercetin in individuals with diabetic kidney disease. EBioMedicine 47:446–56
Huffman DM, Justice JN, Stout MB, Kirkland JL, Barzilai N, Austad SN (2016) Evaluating health span in preclinical models of aging and disease: guidelines, challenges, and opportunities for geroscience. J Gerontol A Biol Sci Med Sci 71(11):1395–406

Inouye SK, Studenski S, Tinetti ME, Kuchel GA (2007) Geriatric syndromes: clinical, research, and policy implications of a core geriatric concept. J Am Geriatr Soc 55(5):780–91
Ishige K, Schubert D, Sagara Y (2001) Flavonoids protect neuronal cells from oxidative stress by three distinct mechanisms. Free Radic Biol Med 30(4):433–46
Justice J, Miller JD, Newman JC, Hashmi SK, Halter J, Austad SN et al (2016) Frameworks for proof-of-concept clinical trials of interventions that target fundamental aging processes. J Gerontl A Biol Sci Med Sci 71(11):1415–23
Justice JN, Gregory H, Tchkonia T, LeBrasseur NK, Kirkland JL, Kritchevsky SB et al (2017) Cellular senescence biomarker p16INK4a+ cell burden in thigh adipose is associated with poor physical function in older women. J Gerontol A Biol Sci Med Sci
Justice JN, Nambiar AM, Tchkonia T, LeBrasseur NK, Pascual R, Hashmi SK et al (2019) Senolytics in idiopathic pulmonary fibrosis: results from a first-in-human, open-label, pilot study. EBioMedicine
Kanapuru B, Ershler WB (2009) Inflammation, coagulation, and the pathway to frailty. Am J Med 122(7):605–13
Keating GM (2017) Dasatinib: a review in chronic myeloid leukaemia and Ph+ acute lymphoblastic leukaemia. Drugs 77(1):85–96
Khan N, Syed DN, Ahmad N, Mukhtar H (2013) Fisetin: a dietary antioxidant for health promotion. Antioxid Redox Signal 19(2):151–62
Kim SR, Jiang K, Ogrodnik M, Chen X, Zhu XY, Lohmeier H et al (2019) Increased renal cellular senescence in murine high-fat diet: effect of the senolytic drug quercetin. Transl Res J Lab Clin Med
Kirkland JL (2013) Translating advances from the basic biology of aging into clinical application. Exp Gerontol 48(1):1–5
Kirkland JL (2016) Translating the science of aging into therapeutici interventions. Cold Spring Harb Perspect Med 6(3):a025908
Kirkland JL, Tchkonia T (2014) Clinical strategies and animal models for developing senolytic agents. Exp Gerontol
Kirkland JL, Tchkonia T (2017) Cellular senescence: a translational perspective. EBioMedicine
Kirkland JL, Hollenberg CH, Gillon WS (1990) Age, anatomic site, and the replication and differentiation of adipocyte precursors. Am J Physiol 258:C206–C10
Kirkland JL, Tchkonia T, Pirtskhalava T, Han J, Karagiannides I (2002) Adipogenesis and aging: does aging make fat go MAD? Exp Gerontol 37:757–67
Kirkland JL, Stout MB, Sierra F (2016) Resilience in aging mice. J Gerontol A Biol Sci Med Sci 71(11):1407–14
Kirkland JL, Tchkonia T, Zhu Y, Niedernhofer LJ, Robbins PD (2017) The clinical potential of senolytic drugs. J Am Geriatr Soc 65(10):2297–301
Krishnamurthy J, Torrice C, Ramsey MR, Kovalev GI, Al-Regaiey K, Su L et al (2004) Ink4a/Arf expression is a biomarker of aging. J Clin Invest 114(9):1299–307
Kuilman T, Peeper DS (2009) Senescence-messaging secretome: SMS-ing cellular stress. Nat Rev Cancer 9:81–94
Lavasani M, Robinson AR, Lu A, Song M, Feduska JM, Ahani B et al (2012) Muscle-derived stem/progenitor cell dysfunction limits healthspan and lifespan in a murine progeria model. Nat Commun 3:608
Leng SX, Xue QL, Tian J, Walston JD, Fried LP (2007) Inflammation and frailty in older women. J Am Geriatr Soc 55(6):864–71
Lewis-McDougall FC, Ruchaya PJ, Domenjo-Vila E, Shin Teoh T, Prata L, Cottle BJ et al (2019) Aged-senescent cells contribute to impaired heart regeneration. Aging Cell. e12931
Lipton RB, Hirsch J, Katz MJ, Wang C, Sanders AE, Verghese J et al (2010) Exceptional parental longevity associated with lower risk of Alzheimer's disease and memory decline. J Am Geriatr Soc 58(6):1043–9

Martyanov V, Whitfield ML, Varga J (2019) Senescence signature in skin biopsies from systemic sclerosis patients treated with senolytic therapy: potential predictor of clinical response? Arthritis Rheumatol

Miller RA (2002) Extending life: scientific prospects and political obstacles. Milbank Q 80(1):155–74

Minamino T, Orimo M, Shimizu I, Kunieda T, Yokoyama M, Ito T et al (2009) A crucial role for adipose tissue p53 in the regulation of insulin resistance. Nat Med 15(9):1082–7

Moncsek A, Al-Suraih MS, Trussoni CE, O'Hara SP, Splinter PL, Zuber C et al (2017) Targeting senescent cholangiocytes and activated fibroblasts with Bcl-xL inhibitors ameliorates fibrosis in Mdr2-/- mice. Hepatology

Munoz-Espin D, Serrano M (2014) Cellular senescence: from physiology to pathology. Nat Rev Mol Cell Biol 15(7):482–96

Musi N, Valentine JM, Sickora KR, Baeuerle E, Thompson CS, Shen Q et al (2018) Tau protein aggregation is associated with cellular senescence in the brain. Aging Cell. e12840

Newman JC, Milman S, Hashmi SK, Austad SN, Kirkland JL, Halter JB et al (2016) Strategies and challenges in clinical trials targeting human aging. J Gerontol A Biol Sci Med Sci 71(11):1424–34

Ogrodnik M, Miwa S, Tchkonia T, Tiniakos D, Wilson CL, Lahat A et al (2017) Cellular senescence drives age-dependent hepatic steatosis. Nat Commun 8:15691

Ogrodnik M, Zhu Y, Langhi LGP, Tchkonia T, Kruger P, Fielder E et al (2019) Obesity-induced cellular senescence drives anxiety and impairs neurogenesis. Cell Metab

Palmer AK, Gustafson B, Kirkland JL, Smith U (2019a) Cellular senescence: at the nexus between ageing and diabetes. Diabetologia 62(10):1835–41

Palmer AK, Xu M, Zhu Y, Pirtskhalava T, Weivoda MM, Hachfeld CM et al (2019) Targeting senescent cells alleviates obesity-induced metabolic dysfunction. Aging Cell. e12950

Parikh P, Britt RD, Jr., Manlove LJ, Wicher SA, Roesler A, Ravix J et al (2018) Hyperoxia-induced cellular senescence in fetal airway smooth muscle cells. Amer J Respir Cell Mol Biol

Pilling LC, Kuo CL, Sicinski K, Tamosauskaite J, Kuchel GA, Harries LW et al (2017) Human longevity: 25 genetic loci associated in 389,166 UK biobank participants. Aging 9(12):2504–20

Qu T, Walston JD, Yang H, Fedarko NS, Xue QL, Beamer BA et al (2009) Upregulated ex vivo expression of stress-responsive inflammatory pathway genes by LPS-challenged CD14(+) monocytes in frail older adults. Mech Ageing Dev 130(3):161–6

Rockwood K, Mitnitski A, Song X, Steen B, Skoog I (2006) Long-term risks of death and institutionalization of elderly people in relation to deficit accumulation at age 70. J Am Geriatr Soc 54(6):975–9

Roos CM, Zhang B, Palmer AK, Ogrodnik MB, Pirtskhalava T, Thalji NM et al (2016) Chronic senolytic treatment alleviates established vasomotor dysfunction in aged or atherosclerotic mice. Aging Cell. https://doi.org/10.1111/acel.12458. [Epub ahead of print]

Schafer MJ, White TA, Iijima K, Haak AJ, Ligresti G, Atkinson EJ et al (2017) Cellular senescence mediates fibrotic pulmonary disease. Nat Commun 8:14532

Schneider EL (1979) Aging and cultured human skin fibroblasts. J Invest Dermatol 73(1):15–8

St Sauver JL, Boyd CM, Grossardt BR, Bobo WV, Finney Rutten LJ, Roger VL et al (2015) Risk of developing multimorbidity across all ages in an historical cohort study: differences by sex and ethnicity. BMJ Open 5(2):e006413

Tchkonia T, Kirkland JL (2018) Aging, cell senescence, and chronic disease: emerging therapeutic strategies. J Am Med Assoc 320(13):1319–20

Tchkonia T, Zhu Y, van Deursen J, Campisi J, Kirkland JL (2013) Cellular senescence and the senescent secretory phenotype: therapeutic opportunities. J Clin Invest 123(3):966–72

Trujillo ME, Pajvani UB, Scherer PE (2005) Apoptosis through targeted activation of caspase 8 ("ATTAC-mice"): novel mouse models of inducible and reversible tissue ablation. Cell Cycle 4(9):1141–5

Villeda SA, Plambeck KE, Middeldorp J, Castellano JM, Mosher KI, Luo J et al (2014) Young blood reverses age-related impairments in cognitive function and synaptic plasticity in mice. Nat Med 20(6):659–63

Walston J, McBurnie MA, Newman A, Tracy RP, Kop WJ, Hirsch CH et al (2002) Frailty and activation of the inflammation and coagulation systems with and without clinical comorbidities: results from the cardiovascular health study. Arch Intern Med 162(20):2333–41

Walston J, Hadley E, Ferrucci L, Guralnick JM, Newman AB, Studenski SA et al (2006) Research agenda for frailty in older adults: toward a better understanding of physiology and etiology: summary from the American Geriatrics Society/National Institute on aging research conference on frailty in older adults. J Am Geriatr Soc 54:991–1001

Walston JD, Matteini AM, Nievergelt C, Lange LA, Fallin DM, Barzilai N et al (2009) Inflammation and stress-related candidate genes, plasma interleukin-6 levels, and longevity in older adults. Exp Gerontol 44(5):350–5

Xu M, Palmer AK, Ding H, Weivoda MM, Pirtskhalava T, White TA et al (2015) Targeting senescent cells enhances adipogenesis and metabolic function in old age. eLife. https://doi.org/10.7554/elife.12997

Xu M, Tchkonia T, Ding H, Ogrodnik M, Lubbers ER, Pirtskhalava T et al (2015b) JAK inhibition alleviates the cellular senescence-associated secretory phenotype and frailty in old age. Proc Natl Acad Sci (USA) 112(46):E6301–E10

Xu M, Bradley EW, Weivoda MM, Hwang SM, Pirtskhalava T, Decklever T et al (2016) Transplanted senescent cells induce an osteoarthritis-like condition in mice. J Gerontol A Biol Sci Med Sci

Xu M, Pirtskhalava T, Farr JN, Weigand BM, Palmer AK, Weivoda MM et al (2018) Senolytics improve physical function and increase lifespan in old age. Nat Med

Yousefzadeh MJ, Zhu Y, McGowan SJ, Angelini L, Fuhrmann-Stroissnigg H, Xu M et al (2018) Fisetin is a senotherapeutic that extends health and lifespan. EBioMedicine

Zhang P, Kishimoto Y, Grammatikakis I, Gottimukkala K, Cutler RG, Zhang S et al (2019) Senolytic therapy alleviates abeta-associated oligodendrocyte progenitor cell senescence and cognitive deficits in an Alzheimer's disease model. Nat Neurosci

Zhu Y, Armstrong JL, Tchkonia T, Kirkland JL (2014) Cellular senescence and the senescent secretory phenotype in age-related chronic diseases. Curr Opin Clin Nutr Metab Care 17(4):324–8

Zhu Y, Tchkonia T, Pirtskhalava T, Gower A, Ding H, Giorgadze N et al (2015a) The Achilles' heel of senescent cells: from transcriptome to senolytic drugs. Aging Cell 14:644–58

Zhu Y, Tchkonia T, Fuhrmann-Stroissnigg H, Dai HM, Ling YY, Stout MB et al (2015) Identification of a novel senolytic agent, navitoclax, targeting the Bcl-2 family of anti-apoptotic factors. Aging Cell. https://doi.org/10.1111/acel.12445. [Epub ahead of print]

Zhu Y, Tchkonia T, Fuhrmann-Stroissnigg H, Dai HM, Ling YY, Stout MB et al (2016) Identification of a novel senolytic agent, navitoclax, targeting the Bcl-2 family of anti-apoptotic factors. Aging Cell 15(3):428–35

Zhu Y, Doornebal EJ, Pirtskhalava T, Giorgadze N, Wentworth M, Fuhrmann-Stroissnigg H et al (2017) New agents that target senescent cells: the flavone, fisetin, and the BCL-XL inhibitors, A1331852 and A1155463. Aging

Part II
The Impact of Senotherapies: From Regeneration to Aging

Chapter 3
Interconnection Between Cellular Senescence, Regeneration and Ageing in Salamanders

Qinghao Yu and Maximina H. Yun

Abstract Urodele amphibians have long served as key models for regenerative, developmental and evolutionary biology research. Recent studies have uncovered the induction of cellular senescence during limb regeneration. The dynamics of senescence in this context reflects that observed in acute senescence, suggesting that senescent cells may play positive roles in regeneration. Further, salamanders possess a highly robust and efficient mechanism for senescent cell surveillance and clearance. Given the causal role of chronic senescence in ageing and age-related pathologies, it is of therapeutic interest to understand the mechanisms and regulation underlying this clearance mechanism. Here, we discuss what is known about cellular senescence in salamanders, what these organisms can offer towards understanding the roles of cellular senescence in regeneration, and how they can serve as informative models for senescence-based therapeutic approaches.

Keywords Cellular senescence · Axolotl · Newt · Regeneration · Development · Senolytics

3.1 Introduction

3.1.1 Cellular Senescence

Senescence is a stress response to severe genotoxic or cellular insults, in which cells enter a state of essentially irreversible growth arrest and acquire a set of characteristic phenotypic alterations (Hayflick and Moorhead 1961; van Deursen 2014; Campisi 2013). A variety of cell intrinsic and extrinsic stresses can trigger the senescence

Q. Yu · M. H. Yun (✉)
Technische Universität Dresden, CRTD-Center for Regenerative Therapies TU, Dresden, Germany
e-mail: maximina.yun@tu-dresden.de

M. H. Yun
Max Planck Institute of Molecular Cell Biology and Genetics, Dresden, Germany

D. Muñoz-Espin and M. Demaria (eds.), *Senolytics in Disease, Ageing and Longevity*, Healthy Ageing and Longevity 11,
https://doi.org/10.1007/978-3-030-44903-2_3

response, such as telomere attrition (Bodnar et al. 1998), DNA damage (Sedelnikova et al. 2004; Di Micco et al. 2006), oxidative damage (von Zglinicki 2002), and chronic mitogenic signalling (Moiseeva et al. 2006). These senescence-inducing stimuli engage various cellular signalling networks, but ultimately implement the characteristic growth arrest by inhibition of cyclin-Cdk complexes through the activation of the tumour suppressors p53, p16, or both (Ben-Porath and Weinberg 2005). Both pathways converge at the level of Rb hypo-phosphorylation, resulting in the continued repression of E2F-target genes required for cell-cycle progression (Stein et al. 1990; Narita et al. 2003). In addition to the senescence-associated growth arrest, senescent cells exhibit a range of distinctive phenotypic alterations, including changes in cell morphology, chromatin remodelling (Shah et al. 2013; Narita et al. 2006; Zhang et al. 2007), metabolic reprogramming (Kaplon et al. 2013; Dorr et al. 2013), and an expansion of the lysosomal and mitochondrial networks. One of the most prominent features exhibited by the majority of senescent cells is the upregulation of genes that encode secreted proteins—an array of proinflammatory cytokines and chemokines, together with various growth factors and proteases—collectively referred to as the senescence-associated secretory phenotype [SASP] (Acosta et al. 2013; Coppe et al. 2010). This phenotype may vary depending on cell type, stressor and context. The SASP is a key distinguishing feature of senescent cells from other non-proliferating states, such as quiescence and terminal differentiation, and underlies many of the physiological and pathological functions of cellular senescence by facilitating communication with the surrounding tissue microenvironment.

The irreversible cell-cycle arrest implemented during replicative senescence was recognised early on as a powerful cell-autonomous mechanism to restrict the expansion of damaged cells (Hayflick and Moorhead 1961; Bodnar et al. 1998). The subsequent discovery of oncogene-induced senescence [OIS] reinforced the notion that cellular senescence, like apoptosis, constitutes a safeguard against tumorigenesis (Serrano et al. 1997; Michaloglou et al. 2005; Chen et al. 2005; Collado et al. 2005; Braig et al. 2005). In contrast to apoptosis, however, senescent cells remain viable and metabolically active and, as such, are able to influence tissue structure and function. Methods to identify and perturb senescent cells in vivo have extended its known roles beyond tumour suppression to a wide range of biological processes. Research in the past decades has placed cellular senescence as an integral component of embryonic development (Czarkwiani and Yun 2018; Munoz-Espin et al. 2013; Davaapil et al. 2017; Storer et al. 2013), wound healing (Jun and Lau 2010; Demaria et al. 2014), and tissue repair (Krizhanovsky et al. 2008; Kim et al. 2013; Meyer et al. 2016), highlighting positive roles for cellular senescence. The process can, however, exert deleterious effects, as illustrated by its causal role in ageing, age-related loss of regenerative capacity, and neoplastic progression (van Deursen 2014; Campisi 2013; Coppe et al. 2010; Krtolica et al. 2001; Laberge et al. 2012; Munoz-Espin and Serrano 2014). The disparate physiological effects of senescence are likely reflected by diversity on the cellular level, in terms of triggering stress, the kinetics and mechanism of induction, tissue context, and a corresponding heterogeneity in SASP composition. Thus, a more nuanced understanding of the physiological

Fig. 3.1 Urodele amphibians. a The commonly used d/d axolotl strain [left] and wild-type [right] adult axolotls [*A. mexicanum*]. **b** Post-metamorphic Iberian ribbed newts [*P. waltl*]

roles of cellular senescence has emerged: a current hypothesis holds that, in addition to acting as a cell-autonomous mechanism of tumour suppression, senescent cells function primarily to restore tissue homeostasis in response to acute damage and stress (Yun 2018). Subsets of cells within the damaged tissue enter the senescent state, and, likely through various components of the SASP, coordinate responses in the surrounding microenvironment [which may range from modulation of cellular plasticity to pro-regenerative ECM remodelling and vascularisation] to restore tissue and organ function. This response, referred to as 'acute senescence', culminates with the recruitment of components of the immune system that mediate the elimination of senescent cells. Among the critical findings contributing to this hypothesis are those obtained using unconventional yet rapidly developing model organisms, the salamanders [Urodele amphibians; Fig. 3.1] (Yun et al. 2015).

3.1.2 Salamanders as model organisms for senescence studies

Urodeles such as the Mexican axolotl [*Ambystoma mexicanum*] and Iberian ribbed newt [*Pleurodeles waltl*] possess a remarkable capacity for regeneration. A salamander can restore its limbs and tail, upper and lower jaws, ocular tissues such as the lens and retina, the intestine and portions of the heart and brain (Brockes and Kumar 2005; Brockes 1997). Following limb amputation, the amputation plane is sealed by the wound epidermis, a transient epithelium formed by rapid migration of epithelial cells from the wound circumference (Brockes and Kumar 2005; Brockes 1997). Regeneration then proceeds through the formation of a blastema, a transient mass of progenitor cells that proliferate and differentiate to restore the missing structure with remarkable fidelity and morphogenic autonomy (Brockes and Kumar 2005;

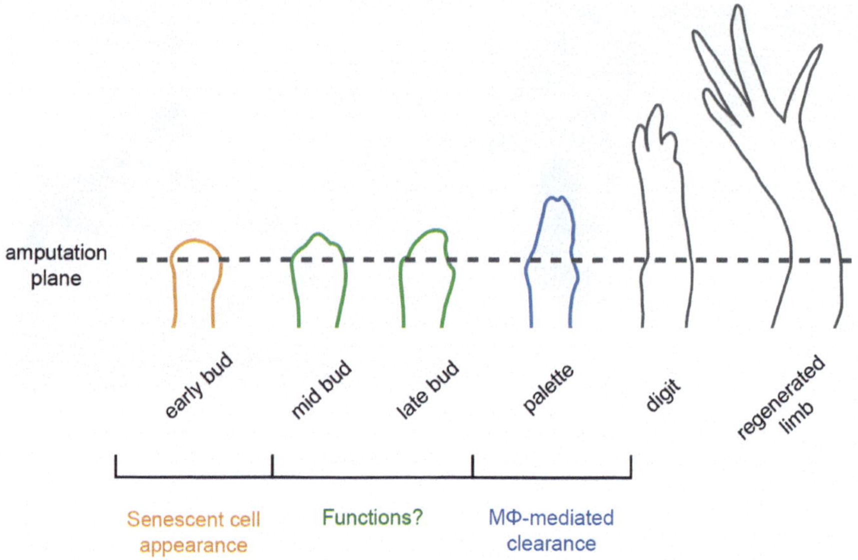

Fig. 3.2 The events and kinetics of senescent cells during salamander limb regeneration. Senescent cells appear during the early to intermediate stages of regeneration and accumulate in the blastema. They produce an array of secreted factors that can affect neighbouring cells, which could directly or indirectly contribute to different aspects of the regenerative programme

Tanaka 2016; Nacu and Tanaka 2011). It was recently reported that limb regeneration is accompanied by induction of a senescent population within the blastema, a process which is specific to regeneration and absent in limb development, and occurs recurrently with multiple rounds of regeneration (Yun et al. 2015). The events and kinetics of cellular senescence in this context are highly reminiscent of those observed during acute senescence, and proceeds with a timely induction and concludes with macrophage-dependent elimination (Fig. 3.2). These observations raise the intriguing possibility that senescent cells may be functionally important for regeneration of complex structures, such as—but not limited to—the salamander limb. Indeed, various lines of research have placed cellular senescence as an integral component in the response to injury, and highlighted its capacity for promoting tissue remodelling and maintenance (Yun 2018). The mechanisms by which senescent cells exert these effects may also be pertinent to urodele regeneration, and will be discussed in depth below.

Beyond the role of senescence in regeneration, salamanders also offer a model of relevance to senescence in ageing contexts. In contrast to the temporally-controlled induction of senescence during processes such as wound healing and embryonic development, the accumulation of senescent cells during organismal ageing appears to be stochastic and unscheduled, and likely occurs due to a combination of different stresses [chronic senescence] and a failure in immune-mediated clearance (van Deursen 2014; Wang et al. 2011). The increased prevalence of senescent cells with age represents a driving factor in tissue deterioration and age-related pathologies (van Deursen 2014). Notably, salamanders do not appear to accumulate senescent cells with age, possess long lifespans without any obvious signs of tissue deterioration, and retain the capacity for repetitive rounds of regeneration throughout the entirety of their lifespans (Yun 2015). Thus, these organisms offer an interesting tool for comparative approaches towards understanding mechanisms that limit age-related senescent cell accumulation. In this review, we examine the interplay between cellular senescence, regeneration, and organismal ageing in salamanders. We discuss how insights from cellular senescence in other processes can inform us about the contribution of senescence towards the remarkable features of salamanders, and highlight what these organisms can offer as models to investigate the physiological roles of senescence.

3.2 Physiological Roles of Cellular Senescence

3.2.1 Physiological Roles of Cellular Senescence

3.2.1.1 Senescence in Wound Healing and Tissue Repair

Research into the functional roles of cellular senescence in different physiological contexts have illustrated that senescent cells hold tremendous potential in influencing tissue organisation and architecture. Through paracrine signalling mediated by the SASP, senescent cells can impact various biological processes, including cell proliferation, differentiation, angiogenesis, cellular plasticity, inflammation, and immune-modulation (Rajagopalan and Long 2012; Lujambio et al. 2013). The potential of senescent cells to coordinate tissue remodelling in vivo has been appreciated in the context of embryonic development, wound healing and tissue repair. In mammals, the response to tissue and organ damage often proceeds through fibrosis instead of regeneration. Excessive fibrosis can lead to scar formation and tissue dysfunction, and the failure to regenerate imposes a major clinical burden (Gurtner et al. 2008). Research over the past decade has revealed that cellular senescence is a central component of the response to tissue injury and damage, and plays an important role in limiting excessive fibrosis at injury sites through the production of SASP factors that promote matrix degradation (Yun 2018). Senescence-mediated restriction of fibrosis is observed in various systems, including the liver (Kim et al. 2013; Borkham-Kamphorst et al. 2014), skin (Jun and Lau 2010; Demaria et al. 2014), and heart (Meyer et al. 2016) and constitutes a conserved response during tissue repair and wound healing.

The extracellular matrix protein CCN1 is upregulated at injury sites and coordinates multiple aspects of wound healing (Kim et al. 2018). Through genetic and biochemical analyses, CCN1 has been found to promote tissue repair in numerous contexts, including skin injury and liver damage, by stimulating senescence in cells at sites of injury (Jun and Lau 2010; Kim et al. 2013). CCN1 acts through integrin a6ß1 and HSPS-mediated activation of the RAC1-dependent NADPH1 oxidase, resulting in sustained reactive oxygen species [ROS] accumulation and consequent p53 and p16 activation (Jun and Lau 2010). Recombinant CCN1 is able to induce senescence in fibroblasts in vitro, and stimulate the expression of matrix degrading enzymes [MMP1, MMP3], pro-inflammatory cytokines, and significantly downregulate type I collagen expression. Indeed, when the integrin interaction of wild-type CCN1 is impaired in vivo, wound healing following skin injury proceeds with exacerbated fibrosis due to a failure of myofibroblasts to undergo senescence. Treatment of wounds with recombinant CCN1 in this background was able to limit excessive fibrosis and reduce collagen deposition. In the case of liver injury, CCN1 is produced by damaged hepatocytes, and is required for senescence induction in activated HSCs. Accordingly, mice with hepatocyte-specific deletions of CCN1 exhibit exacerbated liver fibrosis in response to damage (Kim et al. 2013; Borkham-Kamphorst et al. 2014).

A more recent study from Campisi and co-workers further examined the role of the secretome in the pro-regenerative effects of senescence during cutaneous wounding. Using a mouse model that enables inducible elimination of senescent cells, the authors showed that senescent cell depletion resulted in poor formation of granulation tissue at the wound site and reduced angiogenesis (Demaria et al. 2014). Characterisation of the senescent population revealed that these cells largely comprise fibroblasts and endothelial cells, and show elevated expression of the SASP components PDGF-A and VEGF. Senescent cell-depletion significantly delayed the kinetics of wound closure, a process which depends on the induction of contractile myofibroblasts. Histological analysis revealed a reduction in the number of myofibroblasts in the mid-region of wound sites, suggesting a potential decrease in myofibroblast differentiation. This idea was supported by the capacity of PDGF-AA to stimulate the differentiation of fibroblasts to myofibroblasts in vitro. Indeed, topical treatment of senescence-free wounds with recombinant PDGF-AA restored normal kinetics of wound closure, with concomitant restoration of myofibroblast numbers (Demaria et al. 2014). It should be noted that wound healing under these conditions still concluded with excessive fibrosis, consistent with additional components of the SASP being required for full wound resolution (Demaria et al. 2014).

The aforementioned studies support the idea that transient induction of senescent cells can promote certain types of wound healing. However, further research suggest that the timely elimination of those cells is equally important for the outcome of the process. Particularly, studies of the senescence response during liver damage have emphasised the importance of senescent cell clearance in tissue repair contexts (Krizhanovsky et al. 2008). Gene expression analysis of senescent HSCs revealed up-regulation of pathways involved in immune surveillance, including stimulating receptors for natural killer cell function such as MICA, ULBP2 and PVR2. Upon abrogation of NK cell-mediated clearance, senescent cells accumulated and treated livers displayed significantly more fibrosis as compared with controls (Krizhanovsky et al. 2008). Clearly, cellular senescence has an important role in fibrosis restriction in response to injury, and may also represent an important factor in the response to amputation through regeneration rather than fibrotic scarring in salamanders (Yun 2018).

3.2.1.2 Developmental Senescence

Surveys for senescent markers in axolotl, *Xenopus*, mice, chick and quail embryos has revealed that senescence induction occurs consistently in discrete time windows in numerous structures during development (Munoz-Espin et al. 2013; Davaapil et al. 2017; Storer et al. 2013; Villiard et al. 2017). The major function of cellular senescence during development is to promote the regression of transient embryonic structures, through the recruitment of immune components. For example, senescence is induced in the tubules of the amphibian pronephros and the mammalian mesonephros [precursor embryonic kidney forms], spreads throughout the structure, culminating in monocyte/macrophage recruitment and subsequent senescent cell clearance and associated structural degeneration (Munoz-Espin et al. 2013; Davaapil et al. 2017). In addition, senescent cells are thought to modulate tissue patterning and morphogenesis, as exemplified by patterning defects in the cement gland and the oral and nasal cavities when TGF-ß signalling is perturbed in *Xenopus* embryos. In the murine apical epidermal ridge [AER], a key signalling centre during limb development, SASP-derived FGF8 and FGF4 serve as key proliferation-inducing signals to the adjacent mesenchyme, and loss of cellular senescence in p21 null embryos results in reduced proliferation in the underlying stroma and disrupts the normal expression of key patterning genes (Storer et al. 2013). Further, cellular senescence has been proposed to modulate the differential arrest and expansion of different cell populations. In the endolymphatic sac epithelium of the inner ear, senescence induction occurs in a subset of epithelial cells, and coincides with a robust expansion of a minor, pendrin-positive population beginning at E14.5 (Kim and Wangemann 2011). During murine development, the loss of p21-dependent senescence results in abnormal expansion of pendrin-negative cells, and the robust in pendrin-positive population is notably reduced, resulting in aberrant infoldings of the epithelium into the lumen (Munoz-Espin et al. 2013). In many cases analysed, developmental abnormalities are corrected at later stages through compensatory mechanisms such as apoptosis and late macrophage infiltration. In contrast to damage-induced senescence, growth arrest during development appears to be implemented mainly through p21, and occurs in the absence of DNA damage, although p15 is detected in the mouse mesonephros and endolymphatic sac, and p53 in the axolotl pronephros. The signalling networks underlying developmental senescence induction, however, appear to be less conserved. For example, TGFB plays a key role in senescence induction in the mouse mesonephros, salamander pronephros and the *Xenopus* cement gland (Davaapil et al. 2017). However, whilst ERK signalling is required for senescence induction in the mouse AER, ERK inhibition has no discernible effects on amphibian senescence (Davaapil et al. 2017).

These observations have opened several important questions. First, they demonstrate that cellular senescence is an intrinsic component of vertebrate development, and can occur in a programmed manner in response to developmental cues outside of pathology. They also raise the question as to whether senescence initially evolved to orchestrate tissue remodeling and morphogenesis during embryonic development, perhaps predating other forms of senescence, and was co-opted for its roles in adult life later in evolution. Further, the species-to-species variation between developmental senescence [such as its presence and absence during mouse and amphibian limb development, respectively] and its involvement in clade-specific structures [such as the mammalian Wolffian duct and the amphibian cement gland] suggest that developmental senescence arose multiple times independently during vertebrate evolution (Czarkwiani and Yun 2018). The contributions of cellular senescence to tissue patterning and remodeling in these developmental contexts could also be of relevance to regeneration in salamanders, a process which requires careful control of patterning and coordination of multiple cell types in order to reconstitute a functional structure (Yun 2018).

3.2.1.3 The Interplay Between Cellular Senescence and Plasticity

Recently, an interesting interplay between senescence and the control of cellular plasticity has come to light. The ectopic expression of the transcription factors OCT4, SOX2, KLF4, and cMYC [OSKM factors] in vivo leads to reprogramming of adult cells into induced pluripotent stem cells [iPSCs] and the formation of teratomas [tumours derived from iPSCs] (Abad et al. 2013). Using a murine model for in vivo reprogramming, Serrano and colleagues uncovered a strong correlation between pluripotency induction and cellular senescence (Abad et al. 2013; Mosteiro et al. 2016). OSKM expression led to reprogramming in a subset of cells, and senescence in many others in close proximity. Using genetic and pharmaceutical analysis, it was found that abrogation of senescence through deletion of p16 or the senolytic agent navitoclax [ABT-263] severely compromised the efficiency of reprogramming in vivo. In contrast, under conditions whereby senescence is elevated, such as upon damage, in tissues of progeroid mice, or even in naturally-aged mice, reprogramming and teratoma formation increased. This effect is dependent on the SASP and, in particular, IL-6 was identified as a critical factor in mediating the interplay between cellular senescence and reprogramming (Mosteiro et al. 2016). In support of this, using a transplant model, Keyes and colleagues showed that exposure of primary keratinocytes to OIS-derived SASP enhanced the expression of stem-cell markers and increased their proliferative capacity after grafting (Ritschka et al. 2017). However, prolonged exposure to OIS-derived medium led to the acquisition of senescent traits and a loss of proliferative capacity, reemphasising the importance of the transient nature of senescence to its beneficial effects (Ritschka et al. 2017).

Although these studies have illustrated the potential for cellular senescence to impact cellular plasticity, whether the same mechanism applies to normal in vivo contexts is not yet clear. However, salamander models may shed light into this important question. Post-metamorphic salamanders such as *Notophthalmus viridescens* rely on the tightly-controlled dedifferentiation of adult tissues to form regenerative progenitors (Brockes and Kumar 2005). Two informative systems have been used to analyse reversal of the differentiated state in urodeles: lens and limb regeneration in adult newts. Early work from Eguchi and co-workers showed that following lens removal, pigmented epithelial cells [PECs] of the dorsal iris re-enter the cell cycle, lose their pigmentation, and transdifferentiate into lens cells (Eguchi and Shingai 1971; Eguchi et al. 1974; Del and Tsonis 2003). In the case of muscle, regeneration proceeds through dedifferentiation of muscle fibres to generate proliferative mononucleate progenitors (Echeverri et al. 2001; Kumar et al. 2000; Kumar et al. 2004; Lo et al. 1993). The implantation of purified myofibres from culture or of iris tissue fragments into a blastema leads to the generation of mononucleate cells or lens, respectively (Lo et al. 1993; Reyer et al. 1973). The blastema thus provides an environment capable of destabilising the differentiated state without fully erasing cell identity and of promoting return to the cell cycle (Brockes 1998); whether this property is a function of cellular senescence is currently unknown. However, the advent of pharmaceutical and genetic approaches to perturb the blastemal senescent population (Box 3.1) in vivo positions salamander regeneration as a model to assess the interplay between naturally-occurring cellular senescence and plasticity. Indeed, the induction of senescent cells in the limb blastema coincides with the period of progenitor generation and expansion (Yun et al. 2015) and is consistent with a potential role in modulating cellular plasticity.

The enhancement of cellular plasticity through dedifferentiation and increased 'stemness' in the blastema raises the issue of tumorigenesis in urodeles. The early stages of regeneration share many parallels with tumorigenesis; yet, experimental evidence has shown that salamanders possess not an increased susceptibility, but rather a remarkable resistance to cancer formation, particularly in regenerative tissues (Brockes 1998). The local administration of chemical carcinogens during regeneration in these animals results in markedly low tumour incidence (Tsonis 1983; Tsonis and Eguchi 1981), and in particular, tumours of mesenchymal origin never arise (Tsonis and Eguchi 1981; Tsonis and Eguchi 1982; Zilakos et al. 1992). In cases when abnormal regeneration does occur, carcinogen treatment manifests instead in supernumerary regenerates (Eguchi and Watanabe 1973), such as supernumerary appendages or lens formation from the ventral iris, a process never observed during normal lens regeneration. On the basis of these observations, it has been proposed that cells harbouring tumorigenic mutations are tightly constrained within the framework imposed by epimorphic regeneration (Brockes 1998). In addition, cellular senescence could conceivably serve to spatiotemporally restrict enhanced cellular plasticity to within the blastema, and prevent its occurrence outside of regenerative contexts. As such, their elimination by the immune system would represent an important mechanism to constrain tumourigenesis within the highly plastic environment required for regeneration.

3.2.1.4 What About Senescence in Salamander Regeneration?

The discovery of cellular senescence in the blastema represents the first evidence of senescence in *bona fide* regeneration (Yun et al. 2015). Although currently a phenomenological description, the transient induction of cellular senescence in this context shares many features with acute senescence, and as such, raises the possibility that it may be functionally important for regeneration. Senescent cells are able to exert a strong influence on the surrounding microenvironment of the blastema through paracrine signalling via the SASP. Indeed, a strong SASP signature is detected within the blastema coinciding with the peak of senescent cell induction (Yun et al. 2015). Drawing from the known understanding of how cellular senescence is able to impact numerous biological processes, it is possible to formulate a number of hypotheses for how cellular senescence could contribute towards epimorphic regeneration (Fig. 3.3) (Yun 2018). Firstly, senescent cells could directly modulate the behaviour of progenitors, such as mobilising stem cell reserves or enhancing their generation through increased dedifferentiation and proliferation. The fact that senescence induction coincides with the period of generation and expansion of regenerative progenitors is consistent with such a hypothesis, and it will be interesting to assess whether the interplay between cellular senescence and plasticity extends to the context of natural reprogramming. Alternatively, senescent cells could contribute indirectly by establishing

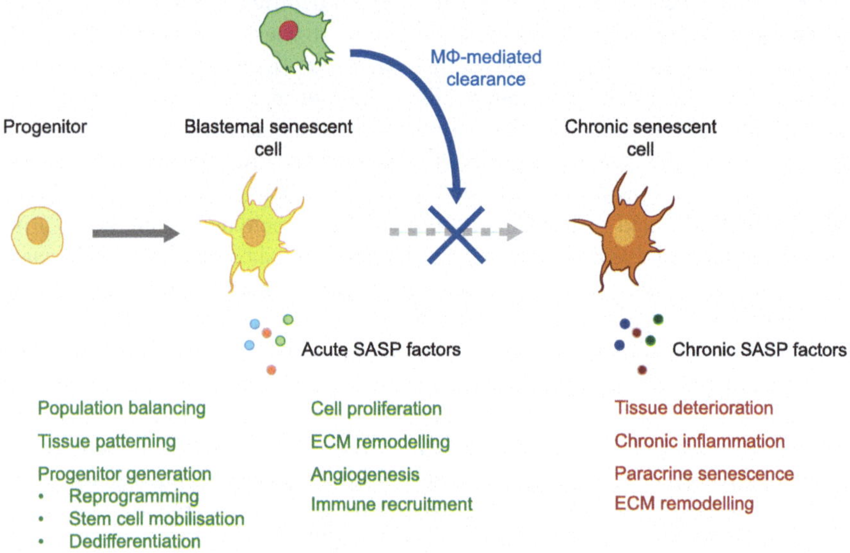

Fig. 3.3 Regenerative effects of senescent cells. Senescent cells are able to impact a variety of biological processes. In the context of epimorphic regeneration, senescence could modulate tissue patterning and morphogenesis, population balancing, contribute to generation of regenerative progenitors, increase in cell proliferation, and angiogenesis. Acute and chronic SASP factors comprise molecules of different nature, or similar factors whose effect may differ due to the strength or kinetics of their expression, or the biological context

a microenvironment conducive for regeneration, such as through matrix remodelling, by stimulating angiogenesis, or through recruiting components of the immune system that in turn exert pro-regenerative functions. Specifically, macrophages are an essential cell type required for salamander regeneration (Godwin et al. 2013) and are found in close proximity to senescent cells in the blastema (Yun et al. 2015). In line with this, it is possible that senescent cells elicit changes in macrophage polarisation and functionality during regeneration, as observed in the context of hepatocellular carcinogenesis (Lujambio et al. 2013). Lastly, senescence could serve as a population balancing mechanism, either in a cell-autonomous manner by restricting the expansion of particular populations, or by acting on neighbouring cells, as observed in the context of developmental senescence in the murine endolymphatic sac (Munoz-Espin et al. 2013). Such balancing would be important for proper patterning, in which cells of multiple lineages must proliferate and differentiate co-ordinately to couple growth and patterning to reconstitute a functional limb.

An expansion in the available experimental tools for both salamander and senescence research (Box 3.1) means it is now possible to directly address these hypotheses experimentally, and thereby delineate the relationship between cellular senescence and regeneration on the molecular and cellular levels. In order to do so, it is necessary to deplete the blastemal senescent population and assess the impact on the outcome of regeneration. The development of pharmacological (Childs et al. 2017) and genetic (Demaria et al. 2014; Baker et al. 2011; Baker et al. 2016) approaches to specifically target senescent cells in vivo enables ablation of the senescent population, making it possible to assess how regeneration occurs in the absence of senescent cells through conventional methods such as histology and live imaging. The same methods can be employed for isolation and detailed characterisation of the blastemal senescent cell population. Such studies will provide important insights into their nature, and address key questions such as their ontogeny, the signalling networks underlying induction of the senescent state, and how they evolve throughout different stages of regeneration. Transcriptional profiling will further provide a platform for functional studies, and identify candidate pro-regenerative SASP factors for functional analysis and senescent-derived signals that mediate immune recruitment and their associated clearance. Moreover, the recent sequencing of the axolotl and newt genomes (Nowoshilow et al. 2018; Elewa et al. 2017) will enable in-depth comparative analysis of senescent-related genes, as well as allowing their genetic manipulation to generate knock-out or knock-in organisms. A better understanding of these processes on the molecular and cellular level is of interest not only to elucidating the mechanism of epimorphic regeneration, but also for therapeutic approaches for regenerative medicine and senescence-based interventions.

3.3 Senescence in Ageing

3.3.1 A Causal Link Between Senescence and Organismal Ageing

An important aspect of the beneficial effects of acute senescence is its transient nature. However, cellular senescence can also have deleterious effects, especially if allowed to persist. It has long been postulated that cellular senescence drives ageing phenotypes, an idea supported by observations of senescent cell accumulation in rodent, primate and human tissues with age (Dimri et al. 1995; Lawless et al. 2010; Wang et al. 2009; Jeyapalan et al. 2007; Herbig et al. 2006). A causal role for senescence in ageing and age-related decline was provided by seminal studies by van Deursen and colleagues. Through the use of a transgene termed INK-ATTAC, which specifically induces apoptosis in p16-expressing cells upon administration of the synthetic compound AP20187, it was shown that selective elimination of senescent cells in BubR1 progeroid mice delayed the onset of several age-related diseases (Baker et al. 2011). Subsequently, the beneficial effects of senescent cell clearance were extended to natural ageing: elimination of p16-positive cells in naturally-aged mice extended median lifespan and attenuated the functional decline of heart, kidney and fat (Baker et al. 2016). Furthermore, treatment delayed cancer progression and resulted in higher spontaneous activity and exploratory behaviour (Baker et al. 2016). Other studies have further reinforced the contribution of cellular senescence to a wide range of age-related diseases, including astherosclerosis, sarcopenia, neurodegeneration and osteoarthritis among others (Jeon et al. 2017; Childs et al. 2016; Sousa-Victor et al. 2014; Bussian et al. 2018).

Mechanistically, cellular senescence is thought to drive ageing through two mechanisms. Firstly, replicative arrest of stem-cell and progenitor pools has been proposed to prevent their participation in tissue regeneration (van Deursen 2014), an idea supported by the observation that progenitor cells of skeletal muscle and fat tissue of BbuR1 progeroid mice are more prone to undergo senescence (Baker et al. 2013) and the demonstration that geriatric muscle stem cells lose their reversible quiescent state during ageing through the induction of senescence (Sousa-Victor et al. 2014; Garcia-Prat et al. 2016). The induction of senescence in muscle satellite cells is coupled with a decline in autophagy during ageing, and consequent loss of proteostasis, mitochondrial dysfunction and the generation of ROS. Specific silencing of p16 in geriatric satellite cells and re-establishment of autophagy enabled cell-cycle re-entry and restored their regenerative functions (Garcia-Prat et al. 2016). Secondly, cellular senescence is able to drive organismal ageing through adverse effects of the SASP. Chronic secretion of cytokines and chemokines can drive sterile inflammation, a hallmark of ageing (Lopez-Otin et al. 2013). SASP-derived proteases are able to cleave membrane-bound receptors, ligands, extracellular matrix proteins or other components in the tissue. In addition, senescence induction can spread across tissues through a mechanism known as paracrine senescence, which is dependent on cytokines such as IL-1ß, TGFß and chemokines (Acosta et al. 2013; Nelson et al.

2012). Together, these processes could disrupt local stem cell niches and overall tissue architecture with age, driving ageing-associated deterioration. Another important consideration is that senescent cells can remain viable in culture for months and continually evolve following the initial cell-cycle arrest, and enter a state termed 'deep' or 'late' senescence (De Cecco et al. 2013). Senescence progression is accompanied by extrusion of chromatin into the cytoplasm to form cytoplasmic chromatin fragments [CCFs] (Ivanov et al. 2013). Lysosome-mediated proteolysis drives histone loss, and is thought to contribute to epigenomic remodelling and SASP diversification in chronic senescent cells. In addition, chronic senescent cells are characterised by a dramatic increase in the transcription of retrotransposable elements [RTEs] (De Cecco et al. 2013). Recent evidence has functionally linked RTE activation with adverse effects of the late SASP (De Cecco et al. 2019). Transcriptional de-repression of L1 RTEs and reduced exonuclease activity results in the accumulation of L1 cytosolic cDNA, which activates the type-I IFN response through the cGAS-STING pathway (De Cecco et al. 2019). Treatment with the nucleoside reverse transcriptase inhibitor lamivudine resulted in dampening of the late SASP response [e.g. expression of CCL2, IL-6 and MMP3] and alleviate several ageing phenotypes in vivo, without impacting cell-cycle arrest or the early SASP response (De Cecco et al. 2019).

3.3.2 Senescence and Ageing in Salamanders

In contrast to mammals, a survey of adult salamander tissues showed a remarkable absence of senescent cells (Yun et al. 2015). Notably, salamanders possess very long lifespans, lack obvious signs of ageing, and are able to sustain indefinite rounds of regeneration throughout their lifetime (Eguchi et al. 2011). Key questions are how salamanders maintain such low systemic levels of senescent cells during ageing, and whether this underlies their lack of age-related deterioration and sustained regenerative capacity. The paucity of senescent cells in salamanders can be conceivably attributed to (Hayflick and Moorhead 1961) active mechanisms to restrict the induction of cellular senescence and/or (van Deursen 2014) efficient clearance mechanisms relative to mammals that persist throughout their lifespan (Fig. 3.3). Clearance refers to the mechanisms by which components of the immune system detect senescent cells and mediate their elimination. Experimental data suggest that both mechanisms contribute towards the low basal levels of senescent cells observed in salamanders (Yun et al. 2015; Ferretti and Brockes 1988). For example, salamander blastemal cells do not undergo crisis or replicative senescence, and can be maintained for more than 200 generations in culture (Ferretti and Brockes 1988), indicating the presence of active mechanisms to circumvent senescence. This property of urodele cells is thought to underlie their capacity to sustain an indefinite number of regeneration cycles.

Studies of cellular senescence in salamanders uncovered a highly robust mechanism for senescent cell surveillance and clearance (Yun et al. 2015). It has been proposed the increase in senescent cell accumulation in mammals with age is driven by an increased decline in immune function (van Deursen 2014). This notion is supported by studies data from mice with impaired immune-mediated cytotoxicity; perforin-knockout mice display premature accumulation of senescent cells and accelerated ageing phenotypes (Ovadya et al. 2018). In salamanders, implanted senescent cells are efficiently detected and eliminated from adult tissues (Yun et al. 2015). This mechanism is dependent on the innate immune system, as evidenced by the persistence of implanted cells following macrophage depletion with chlodrosome treatment (Yun et al. 2015). Owing to their amenability to live imaging and transplantations, together with the use of transgenic reporter lines or pharmaceutic approaches to label senescent cells and components of the immune system, the salamanders constitute an experimental system to investigate the mechanisms by which the immune system efficiently detects and disposes of senescent cells in vivo (Box 3.1). Molecular characterisation of the senescent population may reveal senescent-derived signals that mediate immune recruitment. A better understanding of the pathways governing this process will lead to the identification of molecular targets for experimental perturbation. By experimentally preventing immune-surveillance of senescent cells, it will be possible to evaluate the contribution of immune-mediated clearance towards maintaining the low levels of senescent cells in the salamander during homeostasis. Furthermore, given the low base-line level of endogenous senescent cells in these organisms, it will be interesting to assess whether experimental accumulation of senescent cells would mirror the age-related deterioration caused by chronic senescence observed in mammals or restrict regenerative capacity.

3.4 Conclusions

Taken together, recent studies addressing the roles of cellular senescence under physiological contexts have underscored the capacity of senescent cells to modulate tissue structure and function. Although not yet established experimentally, we speculate that cellular senescence may be functionally important for regeneration. The dynamics of senescence induction and clearance in the blastema are consistent with that observed during development, wound healing and tissue repair; as such, the blastemal senescent population may influence regeneration through similar mechanisms. It will be of both biological and translational interest to delineate the interactions between the senescent population and other cells in the blastema. Furthermore, the causal link between chronic senescence and age-related decline has galvanized efforts to develop novel therapies against senescent cells. One potential strategy is to enhance the immune response against senescent cells. As salamanders possess a remarkably efficient mechanism for immune-mediated surveillance and disposal of senescent cells, they offer an experimental system for mechanistic investigations into the interaction between such cells and the immune system. We expect that a better understanding of the molecular and cellular mechanisms underlying this process will inform rational interventions for senescence-cell removal. Lastly, research in salamanders will provide deeper insights into the evolutionary origins of cellular senescence, and allow for comparisons across the animal kingdom.

Box 3.1 Molecular toolbox for salamander research

- **Germline transgenesis**. Tools for germline transgenesis in both axolotl (Khattak et al. 2013) and Iberian ribbed newts (Hayashi et al. 2013), based on the I-SceI meganuclease and the Tol2 transposon system, have been essential for obtaining a mechanistic understanding of limb regeneration. Owing to the ease with which salamanders can be bred in the laboratory, it is possible to obtain a substantial number of F_0 individuals harbouring transgenes for experimental investigation.
- **Genome sequence and assembly and CRISPR-mediated gene editing**. The recent sequencing and assembly of the 32-Gb axolotl genome (Nowoshilow et al. 2018) and the 20-Gb *P. waltl* genome (Elewa et al. 2017) provides a rich platform for investigations into the molecular basis of regeneration. Together with CRISPR-mediated gene editing (Elewa et al. 2017; Fei et al. 2018), it is possible to assess candidate genes for functional analysis.
- **Somatic gene delivery methods**. Several different technologies exist for gene delivery in salamander cells and tissues, including electroporation (Yun et al. 2013; Echeverri and Tanaka 2003), and different viral transfection methods (Khattak et al. 2013; Whited et al. 2013; Oliveira et al. 2018).
- **Transplantation**. Salamanders are highly receptive to transplantations without graft rejection. The use of surgical manipulation to transplant cells or tissues combined with several molecular and transgenic technologies have been informative towards understanding key aspects of regeneration, including the identity of cell types in the blastema (Kragl et al. 2009). In addition, such methods can be used to implant exogenously-induced senescent cells into salamanders, as described in Yun et al. 2015.
- **Live imaging**. Many salamander tissues are optically transparent, and highly amenable to live imaging (Currie et al. 2016). Through these approaches, it is possible to follow interactions between senescent cells [endogenously induced or implanted] and the immune system.
- **Chemical screening**. Pre-feeding salamander larvae can be reared in microtitre plates to perform moderate-throughput screening for pharmaceutical compounds (Ponomareva et al. 2015) such as senolytics.

References

Abad M, Mosteiro L, Pantoja C, Canamero M, Rayon T, Ors I et al (2013) Reprogramming in vivo produces teratomas and iPS cells with totipotency features. Nature 502(7471):340–345

Acosta JC, Banito A, Wuestefeld T, Georgilis A, Janich P, Morton JP et al (2013) A complex secretory program orchestrated by the inflammasome controls paracrine senescence. Nat Cell Biol 15(8):978–990

Baker DJ, Wijshake T, Tchkonia T, LeBrasseur NK, Childs BG, van de Sluis B et al (2011) Clearance of p16(Ink4a)-positive senescent cells delays ageing-associated disorders. Nature 479(7372):232–236

Baker DJ, Weaver RL, van Deursen JM (2013) p21 both attenuates and drives senescence and aging in BubR1 Progeroid Mice. Cell Rep 3(4):1164–1174

Baker DJ, Childs BG, Durik M, Wijers ME, Sieben CJ, Zhong J et al (2016) Naturally occurring p16(Ink4a)-positive cells shorten healthy lifespan. Nature 530(7589):184–189

Ben-Porath I, Weinberg RA (2005) The signals and pathways activating cellular senescence. Int J Biochem Cell Biol 37(5):961–976

Bodnar AG, Ouellette M, Frolkis M, Holt SE, Chiu CP, Morin GB et al (1998) Extension of life-span by introduction of telomerase into normal human cells. Science 279(5349):349–352

Borkham-Kamphorst E, Schaffrath C, Van de Leur E, Haas U, Tihaa L, Meurer SK et al (2014) The anti-fibrotic effects of CCN1/CYR61 in primary portal myofibroblasts are mediated through induction of reactive oxygen species resulting in cellular senescence, apoptosis and attenuated TGF-beta signaling. Bba-Mol Cell Res 1843(5):902–914

Braig M, Lee S, Loddenkemper C, Rudolph C, Peters AHFM, Schlegelberger B et al (2005) Oncogene-induced senescence as an initial barrier in lymphoma development. Nature 436(7051):660–665

Brockes JP (1997) Amphibian limb regeneration: Rebuilding a complex structure. Science 276(5309):81–97

Brockes JP (1998) Regeneration and cancer. Bba-Rev Cancer 1377(1):M1–M11

Brockes JP, Kumar A (2005) Appendage regeneration in adult vertebrates and implications for regenerative medicine. Science 310(5756):1919–1923

Bussian TJ, Aziz A, Meyer CF, Swenson BL, van Deursen JM, Baker DJ (2018) Clearance of senescent glial cells prevents tau-dependent pathology and cognitive decline. Nature 562(7728):578–582

Campisi J (2013) Aging, cellular senescence, and cancer. Annu Rev Physiol 75:685–705

Chen ZB, Trotman LC, Shaffer D, Lin HK, Dotan ZA, Niki M et al (2005) Crucial role of p53-dependent cellular senescence in suppression of Pten-deficient tumorigenesis. Nature 436(7051):725–730

Childs BG, Baker DJ, Wijshake T, Conover CA, Campisi J, van Deursen JM (2016) Senescent intimal foam cells are deleterious at all stages of atherosclerosis. Science 354(6311):472–477

Childs BG, Gluscevic M, Baker DJ, Laberge RM, Marquess D, Dananberg J et al (2017) Senescent cells: an emerging target for diseases of ageing. Nat Rev Drug Discov 16(10):718–735

Collado M, Gil J, Efeyan A, Guerra C, Schuhmacher AJ, Barradas M et al (2005) Tumour biology: senescence in premalignant tumours. Nature 436(7051):642

Coppe JP, Desprez PY, Krtolica A, Campisi J (2010) The senescence-associated secretory phenotype: the dark side of tumor suppression. Annu Rev Pathol 5:99–118

Currie JD, Kawaguchi A, Traspas RM, Schuez M, Chara O, Tanaka EM (2016) Live Imaging of Axolotl digit Regeneration reveals Spatiotemporal Choreography of Diverse Connective Tissue Progenitor Pools. Dev Cell 39(4):411–423

Czarkwiani A, Yun MH (2018) Out with the old, in with the new: senescence in development. Curr Opin Cell Biol 55:74–80

Davaapil H, Brockes JP, Yun MH (2017) Conserved and novel functions of programmed cellular senescence during vertebrate development. Development 144(1):106–114

De Cecco M, Criscione SW, Peckham EJ, Hillenmeyer S, Hamm EA, Manivannan J et al (2013) Genomes of replicatively senescent cells undergo global epigenetic changes leading to gene silencing and activation of transposable elements. Aging Cell 12(2):247–256

De Cecco M, Ito T, Petrashen AP, Elias AE, Skvir NJ, Criscione SW et al (2019) L1 drives IFN in senescent cells and promotes age-associated inflammation (vol 566, pg 73, 2019). Nature 572(7767):73–78
Del R-TK, Tsonis PA (2003) Eye regeneration at the molecular age. Dev Dyn 226(2):211–224
Demaria M, Ohtani N, Youssef SA, Rodier F, Toussaint W, Mitchell JR et al (2014) An essential role for senescent cells in optimal wound healing through secretion of PDGF-AA. Dev Cell 31(6):722–733
Di Micco R, Fumagalli M, Cicalese A, Piccinin S, Gasparini P, Luise C et al (2006) Oncogene-induced senescence is a DNA damage response triggered by DNA hyper-replication. Nature 444(7119):638–642
Dimri GP, Lee XH, Basile G, Acosta M, Scott C, Roskelley C et al (1995) A biomarker that identifies senescent human-cells in culture and in aging skin in vivo. Proc Natl Acad Sci USA 92(20):9363–9367
Dorr JR, Yu Y, Milanovic M, Beuster G, Zasada C, Dabritz JH et al (2013) Synthetic lethal metabolic targeting of cellular senescence in cancer therapy. Nature 501(7467):421–425
Echeverri K, Tanaka EM (2003) Electroporation as a tool to study in vivo spinal cord regeneration. Dev Dynam 226(2):418–425
Echeverri K, Clarke JDW, Tanaka EM (2001) In vivo imaging indicates muscle fiber dedifferentiation is a major contributor to the regenerating tail blastema. Dev Biol 236(1):151–164
Eguchi G, Shingai R (1971) Cellular analysis on localization of lens forming potency in newt iris epithelium. Dev Growth Differ 13(4):337–349
Eguchi G, Watanabe K (1973) Elicitation of Lens Formation from Ventral Iris Epithelium of Newt by a Carcinogen, N-Methyl-N'-Nitro-N-Nitrosoguanidine. J Embryol Exp Morph 30(1):63–71
Eguchi G, Abe SI, Watanabe K (1974) Differentiation of lens-like structures from newt iris epithelial-cells In vitro. Proc Natl Acad Sci USA 71(12):5052–5056
Eguchi G, Eguchi Y, Nakamura K, Yadav MC, Millan JL, Tsonis PA (2011) Regenerative capacity in newts is not altered by repeated regeneration and ageing. Nat Commun 2:384
Elewa A, Wang H, Talavera-Lopez C, Joven A, Brito G, Kumar A et al (2017) Reading and editing the Pleurodeles waltl genome reveals novel features of tetrapod regeneration. Nat Commun 8(1):2286
Fei JF, Lou WPK, Knapp D, Murawala P, Gerber T, Taniguchi Y et al (2018) Application and optimization of CRISPR-Cas9-mediated genome engineering in axolotl (Ambystoma mexicanum). Nat Protoc 13(12):2908–2943
Ferretti P, Brockes JP (1988) Culture of Newt Cells from different tissues and their expression of a regeneration-associated antigen. J Exp Zool 247(1):77–91
Garcia-Prat L, Martinez-Vicente M, Perdiguero E, Ortet L, Rodriguez-Ubreva J, Rebollo E et al (2016) Autophagy maintains stemness by preventing senescence. Nature 529(7584):37–42
Godwin JW, Pinto AR, Rosenthal NA (2013) Macrophages are required for adult salamander limb regeneration. Proc Natl Acad Sci USA 110(23):9415–9420
Gurtner GC, Werner S, Barrandon Y, Longaker MT (2008) Wound repair and regeneration. Nature 453(7193):314–321
Hayashi T, Yokotani N, Tane S, Matsumoto A, Myouga A, Okamoto M et al (2013) Molecular genetic system for regenerative studies using newts. Dev Growth Differ 55(2):229–236
Hayflick L, Moorhead PS (1961) The serial cultivation of human diploid cell strains. Exp Cell Res 25(3):585–621
Herbig U, Ferreira M, Condel L, Carey D, Sedivy JM (2006) Cellular senescence in aging primates. Science 311(5765):1257
Ivanov A, Pawlikowski J, Manoharan I, van Tuyn J, Nelson DM, Rai TS et al (2013) Lysosome-mediated processing of chromatin in senescence. J Cell Biol 202(1):129–143
Jeon OH, Kim C, Laberge RM, Demaria M, Rathod S, Vasserot AP et al (2017) Local clearance of senescent cells attenuates the development of post-traumatic osteoarthritis and creates a pro-regenerative environment. Nat Med 23(6):775–781
Jeyapalan JC, Ferreira M, Sedivy JA, Herbig U (2007) Accumulation of senescent cells in mitotic tissue of aging primates. Mech Ageing Dev 128(1):36–44

Jun JI, Lau LF (2010) The matricellular protein CCN1 induces fibroblast senescence and restricts fibrosis in cutaneous wound healing. Nat Cell Biol 12(7):676–685
Kaplon J, Zheng L, Meissl K, Chaneton B, Selivanov VA, Mackay G et al (2013) A key role for mitochondrial gatekeeper pyruvate dehydrogenase in oncogene-induced senescence. Nature 498(7452):109–112
Khattak S, Schuez M, Richter T, Knapp D, Haigo SL, Sandoval-Guzman T et al (2013a) Germline transgenic methods for tracking cells and testing gene function during regeneration in the axolotl. Stem Cell Rep 1(1):90–103
Khattak S, Sandoval-Guzman T, Stanke N, Protze S, Tanaka EM, Lindemann D (2013b) Foamy virus for efficient gene transfer in regeneration studies. BMC Dev Biol 13:17
Kim HM, Wangemann P (2011) Epithelial cell stretching and luminal acidification lead to a retarded development of stria Vascularis and deafness in mice lacking Pendrin. PLoS ONE 6(3):e17949
Kim KH, Chen CC, Monzon RI, Lau LF (2013) Matricellular Protein CCN1 promotes regression of liver fibrosis through induction of cellular senescence in hepatic Myofibroblasts. Mol Cell Biol 33(10):2078–2090
Kim KH, Won JH, Cheng N, Lau LF (2018) The matricellular protein CCN1 in tissue injury repair. J Cell Commun Signal 12(1):273–279
Kragl M, Knapp D, Nacu E, Khattak S, Maden M, Epperlein HH et al (2009) Cells keep a memory of their tissue origin during axolotl limb regeneration. Nature 460(7251):60–65
Krizhanovsky V, Yon M, Dickins RA, Hearn S, Simon J, Miething C et al (2008) Senescence of activated stellate cells limits liver fibrosis. Cell 134(4):657–667
Krtolica A, Parrinello S, Lockett S, Desprez PY, Campisi J (2001) Senescent fibroblasts promote epithelial cell growth and tumorigenesis: a link between cancer and aging. Proc Natl Acad Sci USA 98(21):12072–12077
Kumar A, Velloso CP, Imokawa Y, Brockes JP (2000) Plasticity of retrovirus-labelled myotubes in the newt limb regeneration blastema. Dev Biol 218(2):125–136
Kumar A, Velloso CP, Imokawa Y, Brockes JP (2004) The regenerative plasticity of isolated urodele myofibers and its dependence on Msx1. PLoS Biol 2(8):1168–1176
Laberge RM, Awad P, Campisi J, Desprez PY (2012) Epithelial-mesenchymal transition induced by senescent fibroblasts. Cancer Microenviron 5(1):39–44
Lawless C, Wang CF, Jurk D, Merz A, von Zglinicki T, Passos JF (2010) Quantitative assessment of markers for cell senescence. Exp Gerontol 45(10):772–778
Lo DC, Allen F, Brockes JP (1993) Reversal of muscle differentiation during urodele limb regeneration. Proc Natl Acad Sci USA 90(15):7230–7234
Lopez-Otin C, Blasco MA, Partridge L, Serrano M, Kroemer G (2013) The hallmarks of aging. Cell 153(6):1194–1217
Lujambio A, Akkari L, Simon J, Grace D, Tschaharganeh DF, Bolden JE et al (2013) Non-cell-autonomous tumor suppression by p53. Cell 153(2):449–460
Meyer K, Hodwin B, Ramanujam D, Engelhardt S, Sarikas A (2016) Essential role for premature senescence of Myofibroblasts in myocardial fibrosis. J Am Coll Cardiol 67(17):2018–2028
Michaloglou C, Vredeveld LC, Soengas MS, Denoyelle C, Kuilman T, van der Horst CM et al (2005) BRAFE600-associated senescence-like cell cycle arrest of human naevi. Nature 436(7051):720–724
Moiseeva O, Mallette FA, Mukhopadhyay UK, Moores A, Ferbeyre G (2006) DNA damage signaling and p53-dependent senescence after prolonged beta-interferon stimulation. Mol Biol Cell 17(4):1583–1592
Mosteiro L, Pantoja C, Alcazar N, Marion RM, Chondronasiou D, Rovira M, et al (2016) Tissue damage and senescence provide critical signals for cellular reprogramming in vivo. Science 354(6315):aaf4445
Munoz-Espin D, Serrano M (2014) Cellular senescence: from physiology to pathology. Nat Rev Mol Cell Bio 15(7):482–496

Munoz-Espin D, Canamero M, Maraver A, Gomez-Lopez G, Contreras J, Murillo-Cuesta S et al (2013) Programmed cell senescence during mammalian embryonic development. Cell 155(5):1104–1118

Nacu E, Tanaka EM (2011) Limb Regeneration: a new development? Annu Rev Cell Dev Bi 27:409–440

Narita M, Nunez S, Heard E, Narita M, Lin AW, Hearn SA et al (2003) Rb-mediated heterochromatin formation and silencing of E2F target genes during cellular senescence. Cell 113(6):703–716

Narita M, Narita M, Krizhanovsky V, Nunez S, Chicas A, Hearn SA et al (2006) A novel role for high-mobility group a proteins in cellular senescence and heterochromatin formation. Cell 126(3):503–514

Nelson G, Wordsworth J, Wang C, Jurk D, Lawless C, Martin-Ruiz C et al (2012) A senescent cell bystander effect: senescence-induced senescence. Aging Cell 11(2):345–349

Nowoshilow S, Schloissnig S, Fei JF, Dahl A, Pang AWC, Pippel M et al (2018) The axolotl genome and the evolution of key tissue formation regulators. Nature 554(7690):50–55

Oliveira CR, Lemaitre R, Murawala P, Tazaki A, Drechsel DN, Tanaka EM (2018) Pseudo typed baculovirus is an effective gene expression tool for studying. Dev Biol 433(2):262–275

Ovadya Y, Landsberger T, Leins H, Vadai E, Gal H, Biran A, et al (2018) Impaired immune surveillance accelerates accumulation of senescent cells and aging. Nat Commun 9(1):5435

Ponomareva LV, Athippozhy A, Thorson JS, Voss SR (2015) Using Ambystoma mexicanum (Mexican axolotl) embryos, chemical genetics, and microarray analysis to identify signaling pathways associated with tissue regeneration. Comp Biochem Phys C 178:128–135

Rajagopalan S, Long EO (2012) Cellular senescence induced by CD158d reprograms natural killer cells to promote vascular remodeling. Proc Natl Acad Sci USA 109(50):20596–20601

Reyer RW, Woolfitt RA, Withersty LT (1973) Stimulation of lens regeneration from newt dorsal iris when implanted into blastema of regenerating limb. Dev Biol 32(2):258–281

Ritschka B, Storer M, Mas A, Heinzmann F, Ortells MC, Morton JP et al (2017) The senescence-associated secretory phenotype induces cellular plasticity and tissue regeneration. Genes Dev 31(2):172–183

Sedelnikova OA, Horikawa I, Zimonjic DB, Popescu NC, Bonner WM, Barrett JC (2004) Senescing human cells and ageing mice accumulate DNA lesions with unrepairable double-strand breaks. Nat Cell Biol 6(2):168–170

Serrano M, Lin AW, McCurrach ME, Beach D, Lowe SW (1997) Oncogenic ras provokes premature cell senescence associated with accumulation of p53 and p16INK4a. Cell 88(5):593–602

Shah PP, Donahue G, Otte GL, Capell BC, Nelson DM, Cao K et al (2013) Lamin B1 depletion in senescent cells triggers large-scale changes in gene expression and the chromatin landscape. Genes Dev 27(16):1787–1799

Sousa-Victor P, Gutarra S, Garcia-Prat L, Rodriguez-Ubreva J, Ortet L, Ruiz-Bonilla V et al (2014) Geriatric muscle stem cells switch reversible quiescence into senescence. Nature 506(7488):316–421

Stein GH, Beeson M, Gordon L (1990) Failure to phosphorylate the retinoblastoma gene product in senescent human fibroblasts. Science 249(4969):666–669

Storer M, Mas A, Robert-Moreno A, Pecoraro M, Ortells MC, Di Giacomo V et al (2013) Senescence is a developmental mechanism that contributes to embryonic growth and patterning. Cell 155(5):1119–1130

Tanaka EM (2016) The molecular and cellular choreography of appendage regeneration. Cell 165(7):1598–1608

Tsonis PA (1983) Effects of carcinogens on regenerating and non-regenerating limbs in amphibia. Anticancer Res 3(3):195–202

Tsonis PA, Eguchi G (1981) Carcinogens on regeneration—effects of N-Methyl-N'-Nitro-N-Nitrosoguanidine and 4-Nitroquinoline-1-Oxide on limb regeneration in adult newts. Differentiation 20(1):52–60

Tsonis PA, Eguchi G (1982) Abnormal limb regeneration without tumor production in adult newts directed by Carcinogens, 20-Methylcholanthrene and Benzo (Alpha) Pyrene. Dev Growth Differ 24(2):183–190

van Deursen JM (2014) The role of senescent cells in ageing. Nature 509(7501):439–446

Villiard E, Denis JF, Hashemi FS, Igelmann S, Ferbeyre G, Roy S (2017) Senescence gives insights into the morphogenetic evolution of anamniotes. Biol Open 6(6):891–896

von Zglinicki T (2002) Oxidative stress shortens telomeres. Trends Biochem Sci 27(7):339–344

Wang CF, Jurk D, Maddick M, Nelson G, Martin-Ruiz C, von Zglinicki T (2009) DNA damage response and cellular senescence in tissues of aging mice. Aging Cell 8(3):311–323

Wang JW, Geiger H, Rudolph KL (2011) Immunoaging induced by hematopoietic stem cell aging. Curr Opin Immunol 23(4):532–536

Whited JL, Tsai SL, Beier KT, White JN, Piekarski N, Hanken J et al (2013) Pseudotyped retroviruses for infecting axolotl in vivo and in vitro. Development 140(5):1137–1146

Yun MH (2015) Changes in regenerative capacity through lifespan. Int J Mol Sci 16(10):25392–25432

Yun MH (2018) Cellular senescence in tissue repair: every cloud has a silver lining. Int J Dev Biol 62(6–8):591–604

Yun MH, Gates PB, Brockes JP (2013) Regulation of p53 is critical for vertebrate limb regeneration. Proc Natl Acad Sci USA 110(43):17392–17397

Yun MH, Davaapil H, Brockes JP (2015) Recurrent turnover of senescent cells during regeneration of a complex structure. Elife 4:e05505

Zhang R, Chen W, Adams PD (2007) Molecular dissection of formation of senescence-associated heterochromatin foci. Mol Cell Biol 27(6):2343–2358

Zilakos NP, Tsonis PA, Delriotsonis K, Parchment RE (1992) Newt squamous carcinoma proves phylogenetic conservation of tumors as caricatures of tissue renewal. Cancer Res 52(18):4858–4865

Chapter 4
Senolytics Target Senescent Cells and Improve Aging and Age-Related Diseases

Tony Buffard and Gerardo Ferbeyre

Abstract Accumulation of senescent cells has emerged as a major pathogenic factor in aging and multiple age-related diseases. Studies on the biology of senescent cells have identified vulnerabilities to eliminate them using a novel class of drugs called senolytics. These drugs kill senescent cells by blocking their resistance to apoptosis, by reactivating latent p53 or by increasing oxidative stress. Other compounds inhibit the senescence associated secretory phenotype or SASP. Senolytics and SASP modulators have been effective to improve natural aging and age-related diseases in mice models leading to ongoing clinical trials in humans.

Keywords Senolytics · Senescence associated secretory phenotype (SASP) · BCL2 family · Apoptosis · P53 · Metformin

4.1 Introduction

Cellular senescence is a programmed response triggered by both physiological or pathological factors that results in a phenotype characterized by an inability to respond to proliferative signals, resistance to apoptosis and the secretion of a variety of proteins and lipids with potent proinflammatory activity (Ferbeyre 2018; Lopes-Paciencia et al. 2019). In vivo, senescent cells can be divided into three distinct categories: embryonic, acute and chronic. Embryonic senescent cells help to shape developing tissues in mammals, fish and amphibia (Davaapil et al. 2017; Yun et al. 2015; Storer et al. 2013; Munoz-Espin et al. 2013; Villiard et al. 2017) whereas acute senescent cells are a protective response to abrupt stress such as a wound or an oncogenic signal. Both acute and embryonic senescent cells are beneficial and are eliminated through the immune system. Chronic senescence, on the other hand, may result from slowly accumulating damage at the macromolecular level and is associated to

T. Buffard · G. Ferbeyre (✉)
Department of Biochemistry and Molecular Medicine, Université de Montréal, Succ. Centre-Ville, C.P. 6128, Montréal, QC H3C 3J7, Canada
e-mail: g.ferbeyre@umontreal.ca

CRCHUM, 900 Saint-Denis St, Montréal, QC H2X 0A9, Canada

D. Muñoz-Espin and M. Demaria (eds.), *Senolytics in Disease, Ageing and Longevity*, Healthy Ageing and Longevity 11,
https://doi.org/10.1007/978-3-030-44903-2_4

aging, cancer and age-related diseases (Childs et al. 2015). Accumulation of DNA damage, including damage in telomeric regions, have been linked to aging (Sedelnikova et al. 2004; Liu et al. 2005; Lombard et al. 2005; Herbig et al. 2006; Sahin and Depinho 2010; Hewitt et al. 2012). DNA damage activates the DNA damage response, which is measurable using antibodies that recognize proteins phosphorylated by DNA damage-activated kinases and includes the histone variant γ-H2AX (Mallette et al. 2007; Mallette et al. 2007; Halazonetis et al. 2008; Di Micco et al. 2006). However, the precise triggers of DNA damage and senescence during aging are still unknown.

4.2 Evidence of Senescent Cell Accumulation in Vivo

A lot of research confirms that accumulation of senescent cells is a hallmark of aging (López-Otín et al. 2013; He and Sharpless 2017). This might be the consequence of an increased generation of senescent cells with aging and/or a decrease in senescent cell clearance as the immune system gets older (Childs et al. 2015). In fact, senescent cells have a longer half-life in old animals (Karin et al. 2019), suggesting that their clearance becomes less effective with aging. Accumulation of senescent cells has been demonstrated using several senescence biomarkers in old zebrafish (Kishi et al. 2008) and many mammals (Jeyapalan and Sedivy 2008).

A biomarker commonly used to detect senescent cells is the staining for the Senescence-Associated β-Galactosidase (SA-β-Gal), a lysosomal enzyme upregulated in senescent cells (Kurz et al. 2000; Bandyopadhyay et al. 2005). The other standard biomarker is the induction of the cyclin-dependent kinase inhibitor (CKI) $p16^{Ink4a}$ mRNA levels. Expression of this tumor suppressor is undetectable in young rodents, but it increases with age in older tissues (Krishnamurthy et al. 2006; Berkenkamp et al. 2014; Burd et al. 2013) including stem cells (Janzen et al. 2006; Molofsky et al. 2006). $P16^{Ink4a}$ is mechanistically connected to senescence since inhibition of its expression in stem cells reduces their aging phenotype and allows faster tissue repair (Janzen et al. 2006; Molofsky et al. 2006). Telomere shortening, another well-known cause of senescence, can be measured using in situ hybridization with telomeric probes and can be used as a biomarker in some tissues. Telomere length decreases with age in the gut and liver in mice (Hewitt et al. 2012) and primates (Jeyapalan et al. 2007). Given the lack of a universal marker for senescence, quantification of senescent cells should use methods combining several biomarkers. For example, by combining the SA-β-Gal assay, DNA damage response markers and the depletion of HMGB1 from the cell nucleus, Biran et al. found senescent cells to be 10–20 times more abundant in old than in young mice (Biran et al. 2017). Using a similar strategy based on several biomarkers, Herbig et al. showed that the percentage of senescent cells in baboons' skin increased exponentially from 2% in young individuals to more than 15% in aged ones (Herbig et al. 2006). This was confirmed by another study and it was suggested that it would be the case for any mitotic tissue (Jeyapalan et al. 2007).

In humans, the levels of p16^{Ink4a} and p27 (another CKI) are accurate biomarkers of aging in kidneys (Chkhotua et al. 2003; Melk et al. 2004), while the abundance of SA-β-Gal positive cells correlated with age in skin samples (Dimri et al. 1995). This has been confirmed using other senescence-associated markers in the skin (Wang and Dreesen 2018), bones, mesenchymal stem cells (Zhou et al. 2008; Farr and Khosla 2019) and human peripheral blood T lymphocytes (Liu et al. 2009). All of this research convincingly shows that senescent cells do accumulate with age in mammals. Importantly, senescent cells are also more readily detected in many ailing tissues and age-related conditions (Table 4.1) (Jaul and Barron 2017; Franceschi et al. 2018). This suggests that drugs acting on senescent cells will have a major impact in gerontology and healthy aging.

4.3 Elimination of Senescent Cells: Senolytics

Elimination of senescent cells using suicide genes in genetically modified mice or drugs that kill senescent cells, called senolytics, improve many age-related diseases (Table 4.2). Senolytics are thus posed to have broad medical applications. Here, we will discuss them according to their mechanism of action, illustrated in Fig. 4.1.

4.3.1 Senolytics that Inhibit the Bcl2 Family

In multicellular organisms, cells can be eliminated by a process of programmed cell death called apoptosis. Apoptosis can be triggered in two ways. The extrinsic pathway involves a death receptor situated on the cytoplasmic membrane that can be activated by several death effector cytokines. The intrinsic pathway is triggered by endogenous damage that engages mitochondria to release pro-apoptotic factors such as cytochrome c. In both cases, Bcl-2 family proteins (Bcl-2, Bcl-xl, Bcl-w, Bfl-1 or Mcl-1) antagonize this process (Azmi et al. 2011).

Senescent cells are particularly resistant to apoptosis. For example, senescent human fibroblasts express high levels of Bcl-2 family members and can last as long as four weeks in media lacking serum without signs of apoptosis (Wang 1995). Senescent cells also secrete many cytokines and lipids, collectively known as the Senescence-Associated Secretory Phenotype (SASP), that may have anti-apoptotic functions. Together these Senescent Cell Anti-apoptotic Pathways protect senescent cells from cell death (Kirkland and Tchkonia 2017) and targeting them could be a promising way to selectively kill senescent cells.

Recent publications have shown several compounds that could act as effective senolytics via inhibition of antiapoptotic pathways. ABT-263, also known as Navitoclax, preferentially induces apoptosis in senescent fibroblasts and vein epithelial cells by inhibiting Bcl-2, Bcl-xl and Bcl-w (Zhu et al. 2016). These results were also observed in vivo in different mice models (Chang et al. 2016, 2016; Pan et al. 2017).

Table 4.1 Senescent cells in human diseases. ABT-263: inhibitor of the Bcl2 family, D + Q: Dasatinib + Quercetin, Ink-ATTAC: mouse model for clearance of p16Ink4a positive cells, HSV-TK: human herpes simplex virus thymidine kinase, JAKi: JAK kinase inhibitor, Rapamycin: mTOR inhibitor to suppress IL6 and the senescence-associate secretory phenotype

Disease	System affected	References
Presbycusis	Hearing	Watson et al. (2017)
Sarcopenia	Muscle	Snijders and Parise (2017), Sousa-Victor et al. (2014), Sousa-Victor et al. (2014)
Immunosenescence and AIDS	Immune	Fülöp et al. (2017), Palacio et al. (2019), Lanna et al. (2014), Bestilny et al. (2000)
Heart diseases	Cardiovascular	Rotter Sopasakis et al. (2019), Shimizu et al. (2019), Balint et al. (2019)
Hypertension	Cardiovascular	Guzik and Touyz (2017)
Atherosclerosis	Cardiovascular	Minamino et al. (2002), Sasaki et al. (2019), Thorin and Thorin-Trescases (2009), Voghel et al. (2010), Childs et al. (2016), Garrido and Bennett (2016), Roos et al. (2016)
Benign Neoplasia	Multiple	Choi et al. (2000), Castro et al. (2003), Nakamura and Nishioka (2003), Castro et al. (2004), Maldonado et al. (2004), Michaloglou et al. (2005), Vernier et al. (2011), Deschenes-Simard et al. (2013, 2019), Burd et al. (2013)
Osteoarthritis	Skeletal	Franceschi et al. (2018), Jeon et al. (2017), Xu et al. (2016)
Osteoporosis	Skeletal	Saeed et al. (2011), Farr et al. (2017)
Rheumatoid arthritis	Skeletal	Chalan et al. (2015), Fessler et al. (2018), Petersen et al. (2019)
Intervertebral disc degeneration	Skeletal	Patil et al. (2019)
Alzheimer's disease	Nervous	Franceschi et al. (2018), Bhat et al. (2012), Garwood et al. (2014), Zhang et al. (2019)
Parkinson's	Nervous	Franceschi et al. (2018), Chinta et al. (2013)
Diabetes	Metabolic	Palmer et al. (2015), Aguayo-Mazzucato et al. (2019)

(continued)

Table 4.1 (continued)

Disease	System affected	References
Obesity	Adipose and Metabolic	Palmer et al. (2019), Tchkonia et al. (2010), Yoshimoto et al. (2013), Loo et al. (2017)
Chronic lung disease	Respiratory	Franceschi et al. (2018), Barnes et al. (2019), Noureddine et al. (2011)
Hepatitis and fatty liver	Liver	Papatheodoridi et al. (2019), Paradis et al. (2001)
Cirrhosis	Liver	Wiemann et al. (2002), Gutierrez-Reyes et al. (2010)

TW-37, an inhibitor of Bcl-2, Bcl-xl and Mcl-1, was less senolytic than ABT-263 (Zhu et al. 2016), suggesting that Bcl-w plays an important role in protecting senescent cells from apoptosis. ABT-263 was ineffective against human senescent primary preadipocytes (Zhu et al. 2016), demonstrating that senolytics act in a tissue specific manner, a factor that should be taken into account for their use. ABT-737, an inhibitor of Bcl-xl and Bcl-w, preferentially kills senescent cells induced by DNA damage in the lung and senescent cells induced by $p14^{ARF}$ expression in the epidermis (Yosef et al. 2016). Interestingly, the elimination of senescent cells in the epidermis led to an increase in hair follicle stem cell proliferation (Yosef et al. 2016). Of note, the anti-Bcl2 family of drugs cause neutropenia and thrombocytopenia (Roberts et al. 2012), side effects that could limit their application in healthy old individuals.

Other inhibitors of the Bcl2 family with senolytic activity include fisetin, a flavone molecule that induces apoptosis in senescent fibroblasts and endothelial cells but not in senescent preadipocytes (Zhu et al. 2017). In progeroid $Ercc1^{-/\Delta}$ mice, fisetin killed senescent cells and reduced senescence biomarkers. In old naturally aged C57BL/6 mice, a 5-day diet of fisetin was able to significantly reduce the proportion of senescent cells in different tissues and extend median and maximal lifespan even when the treatment was initiated in old animals (Yousefzadeh et al. 2018). In cancer cells, fisetin can cause apoptosis by activating both the intrinsic and the extrinsic pathways and had beneficial effects to treat inflammation and metastasis (Kashyap et al. 2018). Fisetin is present in many fruits and vegetables, suggesting that it can be safely used as a senolytic and anti-aging agent in humans (Kashyap et al. 2018). Epigallocatechin gallate, a phytochemical found in green tea, inhibits both the anti-apoptotic Bcl-2 family and mTOR. The latter controls the SASP by regulating the translation of mRNAs coding for inflammatory cytokines (Herranz et al. 2015; Laberge et al. 2015). Epigallocatechin can thus act both as a SASP modulator by inhibiting mTOR and as a senolytic (Kumar et al. 2019).

Panobinostat is a deacetylase inhibitor used to treat multiple myeloma. Panobinostat is particularly potent against all deacetylases of class I, II and IV (Laubach

Table 4.2 Diseases improved by senolytics

Disease	Models		Senolytic used	References
	In vitro	In vivo		
Diabetes	Mice β cells Human β cells	Ink-ATTAC Mice	Ink-ATTAC and ABT-263	Aguayo-Mazzucato et al. (2019)
Age-related bone loss		Mice	Ink-ATTAC and D + Q	Farr and Khosla (2019)
Cancer		Mice	siRNA against HSP47, Ink-ATTAC, Drug delivery system	Muñoz-Espín et al. (2018), Yoshimoto et al. (2013), Baker et al. (2016)
	Human cervical cancer		Rapamycin (to suppress IL-6 from SASP)	Laberge et al. (2015)
Pulmonary fibrosis	Human and murine lung cells	Ink-ATTAC mice	ABT-263, D + Q	Pan et al. (2017, Muñoz-Espín et al. (2018), Schafer et al. (2017)
		Human	D + Q	Justice et al. (2019)
Hepatic steatosis		Mice	Ink-ATTAC and D + Q	Cellular senescence drives age-dependent hepatic steatosis. Nat Commun.(2017)
Atherosclerosis		Mice	Ink-ATTAC and ABT-263	Childs et al. (2016)
Osteoarthritis		Mice	Ganciclovir with HSV-TK	Jeon et al. (2017)
		Human	Fenofibrate	Nogueira-Recalde et al.(2019)
Alzheimer's disease		Mice	D + Q	Zhang et al. (2019)
Tau-mediated neurodegeneration		Mice	Ink-ATTAC, ABT-263, D + Q	Mendelsohn and Larrick (2018, Bussian et al. (2018)
Osteoporosis		Ink-ATTAC Mice	Ink-ATTAC, JAKi	Farr et al. (2017)
Dysglycemia		Mice	JAKi	Xu et al. (2015)
Cardiovascular diseases		Mice	Ink-ATTAC, D + Q, ABT-263	Roos et al. (2016, Walaszczyk et al. 2019)

(continued)

Table 4.2 (continued)

Disease	Models		Senolytic used	References
	In vitro	In vivo		
Glomerulosclerosis		Ink-ATTAC Mice	Ink-ATTAC	Baker et al. (2016)
Fatty liver disease		Mice	JAKi, D + Q	Papatheodoridi et al. (2019)
Renal diseases		Mice	Ink-ATTAC	Valentijn et al. (2018)

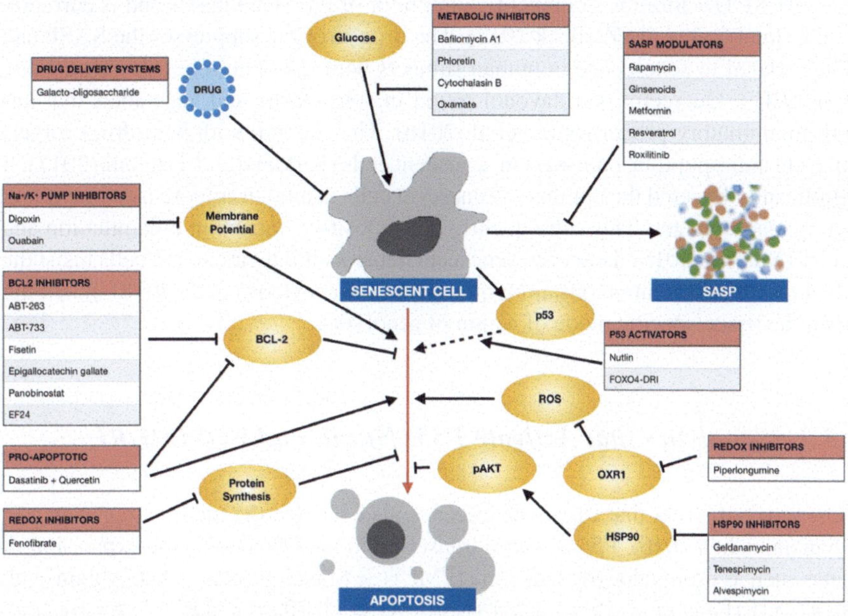

Fig. 4.1 Schematic overview of the different strategies pursued to eliminate senescent cells or alleviate the detrimental effects of the SASP

et al. 2015). In non-small cell lung cancer and head and neck squamous cell carcinoma, senescent cells have altered H3 acetylation and Bcl-xl expression. Panobinostat inhibits Bcl-xl and kill senescent cells induced by chemotherapy (Samaraweera et al. 2017). However, this drug may cause a few adverse effects, including diarrhea, asthenia and a lower count of immune blood cells (Van Veggel et al. 2018; Hennika et al. 2017).

Finally, EF24, a natural compound found in turmeric (*Curcuma longa*) that is similar to curcumin, can kill senescent cells by downregulating Bcl-xl (Li et al. 2019). EF24 could be used in synergy with ABT-263 to kill senescent cells more effectively, and at the same time, prevent ABT-263's cytotoxic effects (Li et al. 2019).

4.3.2 Proapoptotic Cocktail Dasatinib + Quercetin

Based on the fact that senescent cells are resistant to apoptosis induced by serum deprivation and other stresses, Zhu et al. hypothesized that it could be possible to kill them by inhibiting their antiapoptotic pathways. A screening using molecules that block these pathways showed that the drugs dasatinib and quercetin were particularly efficient to kill senescent cells and improves health span in progeroid mice (Zhu et al. 2015). The synergistic combination proved to be efficient to counteract the effect of irradiation in vivo, and to extend lifespan in naturally-aged mice (Xu et al. 2018). Dasatinib is a competitive inhibitor of tyrosine kinases and is currently used to treat leukemia (Zarbock 2012). The drug strongly suppresses the SASP and genes related to senescence in human subjects with systemic sclerosis (Martyanov et al. 2019). Quercetin is a flavonol found in some fruits and vegetables that has anti-inflammatory properties (Li et al. 2016). The cocktail with both drugs targets different anti-apoptotic pathways in senescent cells (Kirkland and Tchkonia 2017). It significantly lowered the number of senescent cells in human adipose tissues without killing macrophages. This subsequently had a positive impact on inflammation and frailty (Xu et al. 2018). However, senescent hepatocellular carcinoma cells resist the effect of the combination dasatinib/quercetin (Kovacovicova et al. 2018) indicating again the tissue specific mode of action of senolytics.

4.3.3 Senolytics that Activate P53: Nutlin and FOXO4DRI

Most types of stress that trigger senescence also activate the tumor suppressor p53 (Qian and Chen 2010). P53 acts as a transcription factor inducing the expression of genes such as promyelocytic leukemia (PML) (de Stanchina et al. 2004; Ablain et al. 2014), p21 (Fang et al. 1999), PAI1 (Kortlever et al. 2006) and E2F7 (Aksoy et al. 2012) that mediate the growth arrest phenotype of senescent cells. P53 can also trigger cell death but the key molecular switches that control cell fate downstream of p53 activation are not well understood (Macip et al. 2003). Reactive oxygen species (ROS) can convert a p53-dependent senescence response into apoptosis but the mechanism has not been totally elucidated (Macip et al. 2003). It has been proposed that p53 must overcome a concentration threshold to trigger apoptosis (Kracikova et al. 2013). Another pertinent characteristic of p53 in this regard is its ability to translocate to mitochondria (Mihara et al. 2003) and inhibit Bcl-2 and Bcl-xl, preventing their anti-apoptotic activities (Hagn et al. 2010). P53 translocation to mitochondria is inhibited by the nucleolar protein nucleophosmin (Dhar and St Clair 2009) and the E3 ubiquitin ligase TRAF6 (Zhang et al. 2016).

Nutlin (Nutley inhibitor) was designed by Vassilev and colleagues to prevent the binding of Mdm2 to p53. The nutlin binding site is situated in a deep hydrophobic pocket in the Mdm2 protein (Vassilev et al. 2004). Mdm2, an E3 ubiquitin ligase, inhibits p53 through three mechanisms: (1) degradation of p53 by poly-ubiquitylation and targeting it to the proteasome; (2) export of p53 out of the nucleus by monoubiquitylation and (3) direct binding to p53 preventing its activity as a transcription factor (Wu and Prives 2018). Since p53 is needed to express Mdm2, levels of p53 are autoregulated in a feedback loop (Lessel et al. 2017).

Nutlin is not genotoxic and does not promote p53 modifications associated to DNA damage, it only stabilizes p53 by protecting it from Mdm2. Although senescent cells activate p53, this activation is limited (Huang and Vassilev 2009) suggesting that p53 cannot attain the levels needed to trigger apoptosis. It is then quite possible that nutlin could stabilize latent p53 in senescent cells and trigger apoptosis. This has been shown in cultures of senescent chondrocytes obtained from patients with osteoarthrosis or in vivo in a mouse model of osteoarthrosis triggered by transection of the anterior cruciate ligament (Jeon et al. 2017). However, p53 reactivation could lead to toxic effects in normal cells, including death by apoptosis (Burgess et al. 2016), or could generate a selective pressure for mutations of p53 (Aziz et al. 2011). Also, nutlin could potentially bind to other protein pockets with similar shapes and physicochemical properties, leading to potential toxic effects (Nguyen et al. 2019).

Proteolysis Targeting Chimera (PROTAC) consist of two protein binding fragments, one capable of binding to a target protein and another that binds to an E3 ubiquitin ligase (Bondeson et al. 2018). PROTACs having an Mdm2-binding fragment such as nutlin could have a double effect. First, they can activate p53 by preventing its inhibition by Mdm2. Second, they could target another protein for degradation by the proteasome by promoting its interaction with Mdm2. Such a PROTAC targeting BRD4 was shown to be very effective against cancer cells with wild type p53 (Hines et al. 2019) but it could also be modified by coupling nutlin to anti-Bcl2 family compounds such as ABT-273 to better kill senescent cells.

FOXO4 is a member of the Forkhead box O (FOXO) family of transcription factors which are negatively regulated by insulin or IGF-1 via AKT-dependent phosphorylation and cytoplasmic retention (Martins et al. 2016). In the nucleus, FOXO factors can localize to PML nuclear bodies (Trotman et al. 2006) which are particularly induced in senescent cells (Bourdeau et al. 2009). FOXO factors are involved in resistance to stress, metabolism, cell cycle arrest and apoptosis (Martins et al. 2016).

In response to acute damage, FOXO4 was shown to favor senescence instead of apoptosis. During senescence, p53 is phosphorylated by ATM (Ataxia-Telangiectasia Mutated), which would prevent its inhibition by Mdm2 (Mallette et al. 2007). In this situation, p53 localizes to chromatin having persistent DNA damage (DNA-SCARS), next to PML bodies containing FOXO4. FOXO4 would then limit p53's ability to promote apoptosis by sequestering p53 in PML bodies (Baar et al. 2017). Based on this, Baar and colleagues synthesized a FOXO4-derived peptide that would prevent the binding of FOXO4 to p53 (Baar et al. 2017). This peptide was designed as a D-retro inverse isoform (DRI) so that it would have a better potency than its natural L-isoform counterpart. The FOXO4-DRI peptide was effective in relocating p53 to mitochondria, promoting apoptosis selectively in senescent cells both in vitro and in vivo. This peptide had potent anti-aging effects both in progeroid and wild-type mice illustrating once again the causal relationship between senescent cells and aging (Baar et al. 2017). FOXO4-DRI showed a tenfold selectivity to senescent cells compared to normal cells. Although the peptide was safe in rodents upon repeated administration, long living humans may require even more injections. It would be preferable to optimize this peptide to achieve a higher degree of selectivity to avoid potential toxicities in humans (Baar et al. 2018).

4.3.4 Metabolic Inhibitors: Targeting Glycolysis and REDOX Metabolism

Senescent cells have a dramatic upregulation of glucose utilization in association to their mitochondrial dysfunction (Moiseeva et al. 2009). For instance, therapy-induced senescence (TIS) in lymphoma cells is accompanied by an increase in glucose utilization and autophagy that together support the ATP and metabolic demands of senescent cells. A combination of deoxyglucose with the autophagy inhibitor bafilomycin A1 selectively killed these senescent cells (Dörr et al. 2013). The glucose transport inhibitors phloretin and cytochalasin B or the lactate dehydrogenase inhibitor oxamate were also selectively toxic for TIS cells (Dörr et al. 2013). This was actually the first demonstration of a pharmacological approach to kill senescent cells.

The screening of a library of small molecules supposed to target important pathways for senescent cells showed that piperlongumine was a promising candidate as a senolytic (Wang et al. 2016). The drug induced apoptosis selectively in senescent fibroblasts compared to control cells. A significant synergy between piperlongumine and ABT-263 could allow for a lower dose of ABT-263. The latter causes thrombocytopenia and neutropenia because of inhibition of Bcl-xl in platelets (Wang et al. 2016). Although piperlongumine's mode of action is not fully understood, the drug was shown to bind the protein Oxidation Resistance 1 (OXR1) leading to its degradation by the ubiquitin-proteasome system (Liu et al. 2018). Targeting OXR1 kills senescent cells by promoting oxidative stress (Yang et al. 2014; Zhang et al. 2018).

Senescent chondrocytes have been linked to osteoarthritis, a disease for which there is no cure. Nogueira-Recalde and colleagues found fenofibrate, an agonist of PPARα (peroxisome proliferator-activated receptor alpha) as a senolytic after interrogating the Prestwick chemical library for molecules that kill senescent cells. They showed that fenofibrate effectively and selectively killed senescent chondrocytes by promoting apoptosis in vitro. Since this drug was found by screening chemical compounds it is not yet clear how it selectively kills senescent cells. The authors correlated the effects of fenofibrate with inhibition of the mTOR effector S6 kinase (Nogueira-Recalde et al. 2019). Since mTOR is required for protein synthesis, this strategy may work by inhibiting the expression of anti-apoptotic proteins. In a retrospective study, human osteoarthritis patients taking fenofibrate reported a significant decrease in disability and pain leading to fewer joint surgeries (Nogueira-Recalde et al. 2019). The use of fenofibrate as a senolytic could be however limited by its hepatotoxicity (Hedrington and Davis 2018).

4.3.5 HSP90 Inhibitors

Robbins and colleagues screened a library of autophagic regulators for compounds that killed senescent *ercc1* null murine embryonic fibroblasts. They identified the HSP90 inhibitors geldanamycin, 17-AAG (tenespimycin) and 17-DMAG (alvespimycin) as potent senolytics that triggered apoptosis in senescent cells (Fuhrmann-Stroissnigg et al. 2017). The mechanism of senolytic activity of HSP90 inhibitors included the destabilization of phospho-AKT (Fuhrmann-Stroissnigg et al. 2017). Of note, treating progeroid $ercc1^{-/\Delta}$ mice with 17-DMAG reduced the expression of senescence biomarkers in the kidneys and delayed the onset of age-related phenotypes (Fuhrmann-Stroissnigg et al. 2017).

4.3.6 Sodium/Potassium ATPase Inhibitors

Produced naturally by many plants, cardiac glycosides have been recently found to be senolytic agents. They prevent cytoplasmic transmembrane Na^+/K^+ pumps to maintain the resting potential across the membrane by binding to their alpha 1 subunit (Langford and Boor 1996). Using a high throughput screening method on compounds found in the Prestwick library, Triana-Martinez et al. identified 9 cardiac glycosides as senolytics. Among them, digoxin had the highest senolytic index (Triana-Martinez et al. 2019). Cardiac glycosides selectively trigger apoptosis of senescent cells by increasing intracellular concentration of Na^+ ions. This in turn would inhibit Na^+/Ca^{2+} and Na^+/H^+ exchangers, leading to increasing concentrations of Ca^{2+} and H^+. Since senescent cells already have a lower cytosolic pH than normal cells, digoxin could activate both the intrinsic and the extrinsic apoptosis pathways in these cells only (Majdi et al. 2016). In a similar fashion, ouabain, another cardiac glycoside, was also found to have senolytic properties (Guerrero et al. 2019). Interestingly, in this study ouabain was shown to induce the expression of the proapoptotic protein Noxa, suggesting that changes in gene expression underpin the senolytic activity of cardiac glycosides.

4.3.7 Immunotherapy

Human senescent fibroblasts express higher levels of the cell surface marker dipeptidyl peptidase 4 (DPP4) (Kim et al. 2017). An antibody-dependent cell-mediated cytotoxicity (ADCC) assay showed that NK cells can recognize and selectively kill these DPP4 positive senescent fibroblasts (Kim et al. 2017). As senescence can also occur in immune cells, the immune system can become impaired, leading to accumulation of senescent cells. Thus, targeting specifically immune senescent cells could have an indirect senolytic effect on senescent cells from other tissues. In mice, removal of senescent hematopoietic stem cells had a rejuvenating effect on aged tissues (Chang et al. 2016). Also, clearance of senescent cells in irradiated mouse spleen restored the functions of T cells and macrophages (Palacio et al. 2019). It seems also possible to engineer immune cells against senescent cells expressing IL-6, and then promote their death via cell fusion (Qudrat et al. 2017).

4.3.8 Drug Delivery System Targeting Senescent Cells

Another interesting way of killing senescent cells is to encapsulate a cytotoxic compound in a shell that would preferentially target these cells. Since they have a high level of lysosomal β-galactosidase compared to normal cells, Muñoz-Espín et al. used nanoparticles covered with galacto-oligosaccharides on a silica scaffold. These beads are integrated by most cells via endocytosis and then quickly released via exocytosis. However, in the case of senescent cells, β-galactosidase will digest the polysaccharide allowing the release of drugs inside the nanoparticles before exocytosis. Fluorophores were used to show that these nanoparticles identified senescent cells in vivo and when loaded with doxorubicin they selectively killed senescent cells (Muñoz-Espín et al. 2018).

4.3.9 SASP Modulation

Another way to fight the detrimental effects of senescent cells is to attenuate their inflammatory secretions. The hundreds of cytokines, chemokines, growth factors and metalloproteases come mainly from two distinct pathways: NF-κB and C/EBPβ (Paez-Ribes et al. 2019). Targeting these pathways or their upstream regulators could help reduce inflammation linked to aging. Rapamycin decreases secretion of interleukin-6 (IL-6) and other inflammatory cytokines by inhibiting their translation (Herranz et al. 2015; Laberge et al. 2015). MAPK pathway inhibitors, such as ginsenosides, were able to suppress the SASP in senescent astrocytes (Hou et al. 2018) or hematopoietic stem cells (Tang et al. 2015). The NF-κB pathway could also be inhibited by the antidiabetic drug metformin, preventing the expression of several SASP cytokines (Moiseeva et al. 2013; Oubaha et al. 2016). Resveratrol, a polyphenol, was able to inhibit the SASP through SIRT1/NF-κB pathway on melanoma cells (Menicacci et al. 2017) or in the gut of the fish *N. guentheri* (Liu et al. 2018). Administration of a JAK inhibitor, ruxolitinib, reduced inflammation in aged mice by down-regulating the C/EBPβ pathway (Xu et al. 2015). Glucocorticoids such as corticosterone or cortisol were shown to suppress IL-6 as well as several other SASP components. They inhibit IL-1α signaling upstream of NF-κB, which in turn, stimulates the expression of IL-1α, in a positive feedback loop (Laberge et al. 2012).

4.4 Clinical Trials and Future Directions

Since aging is still not recognized as a disease, very few of the aforementioned senolytics have made their ways to clinical studies. Unity biotechnology is currently testing the effect of a compound called UBX0101 on osteoarthritis. After a successful phase I (NCT03513016) they are recruiting patients for a phase II trial (NCT04129944). UBX0101 is supposed to eliminate senescent cells that accumulate in joints, which should decrease local inflammation and alleviate the pain. The Mayo clinic is testing the cocktail D + Q in 2 different clinical trials. One is aimed at chronic kidney diseases (NCT02848131) and the other one is targeted against Alzheimer's disease (NCT04063124). Mayo clinic is also investigating the senolytic effect of fisetin on frail elderly syndrome (NCT03430037). They are currently recruiting for phase II on these 3 studies. A preliminary report of the phase I trial with D + Q in 9 patients with diabetic kidney disease claims a reduction in adipose tissue senescent cells burden 11 days after completion of a 3-days senolytic treatment (Hickson et al. 2019). If confirmed in a large number of patients, this study suggests that all the beneficial effects observed in mice treated with senolytics will be also attained in humans.

Acknowledgements We thank members of the Ferbeyre laboratory for their critical insights reading this manuscript. G.F. is supported by the CIBC chair for breast cancer research at the CRCHUM.

References

Ablain J, Rice K, Soilihi H, de Reynies A, Minucci S, de The H (2014) Activation of a promyelocytic leukemia-tumor protein 53 axis underlies acute promyelocytic leukemia cure. Nat Med 20(2):167–174

Aguayo-Mazzucato C, Andle J, Lee TB, Jr., Midha A, Talemal L, Chipashvili V, et al (2019) Acceleration of beta cell aging determines diabetes and senolysis improves disease outcomes. Cell Metabolism 30(1):129–42 e4

Aguayo-Mazzucato C, Andle J, Lee TB, Midha A, Talemal L, Chipashvili V, et al (2019) Acceleration of β cell aging determines diabetes and senolysis improves disease outcomes. Cell Metabolism 30(1):129–42 e4

Aksoy O, Chicas A, Zeng T, Zhao Z, McCurrach M, Wang X et al (2012) The atypical E2F family member E2F7 couples the p53 and RB pathways during cellular senescence. Genes Dev 26(14):1546–1557

Aziz MH, Shen H, Maki CG (2011) Acquisition of p53 mutations in response to the non-genotoxic p53 activator Nutlin-3. Oncogene 30(46):4678–4686

Azmi AS, Wang Z, Philip PA, Mohammad RM, Sarkar FH (2011) Emerging Bcl-2 inhibitors for the treatment of cancer. Expert Opin Emerg Drugs 16(1):59–70

Baar MP, Brandt RMC, Putavet DA, Klein JDD, Derks KWJ, Bourgeois BRM et al (2017) Targeted apoptosis of senescent cells restores tissue homeostasis in response to Chemotoxicity and aging. Cell 169(1):132–47.e16

Baar MP, Perdiguero E, Munoz-Canoves P, de Keizer PL (2018) Musculoskeletal senescence: a moving target ready to be eliminated. Curr Opin Pharmacol 40:147–155

Baker DJ, Childs BG, Durik M, Wijers ME, Sieben CJ, Zhong J et al (2016) Naturally occurring p16Ink4a-positive cells shorten healthy lifespan. Nature 530(7589):184–189

Balint B, Yin H, Nong Z, Arpino J-M, O'Neil C, Rogers SR et al (2019) Seno-destructive smooth muscle cells in the ascending aorta of patients with bicuspid aortic valve disease. EBioMedicine 43:54–66

Bandyopadhyay D, Gatza C, Donehower LA, Medrano EE (2005) Analysis of cellular senescence in culture in vivo: the senescence-associated beta-galactosidase assay. Current protocols in cell biology/editorial board, Juan S Bonifacino [et al. Chapter 18: Unit 18 9

Barnes PJ, Baker J, Donnelly LE (2019) Cellular senescence as a mechanism and target in chronic lung diseases. Am J Respir Crit Care Med

Berkenkamp B, Susnik N, Baisantry A, Kuznetsova I, Jacobi C, Sörensen-Zender I et al (2014) In vivo and in vitro analysis of age-associated changes and somatic cellular senescence in renal epithelial cells. PLoS ONE 9(2):e88071

Bestilny LJ, Gill MJ, Mody CH, Riabowol KT (2000) Accelerated replicative senescence of the peripheral immune system induced by HIV infection. AIDS. 14(7):771–780

Bhat R, Crowe EP, Bitto A, Moh M, Katsetos CD, Garcia FU et al (2012) Astrocyte senescence as a component of Alzheimer's disease. PLoS ONE 7(9):e45069

Biran A, Zada L, Karam PA, Vadai E, Roitman L, Ovadya Y et al (2017) Quantitative identification of senescent cells in aging and disease. Aging Cell 16(4):661–671

Bondeson DP, Smith BE, Burslem GM, Buhimschi AD, Hines J, Jaime-Figueroa S et al (2018) Lessons in PROTAC design from selective degradation with a promiscuous warhead. Cell Chem Biol 25(1):78–87.e5

Bourdeau V, Baudry D, Ferbeyre G (2009) PML links aberrant cytokine signaling and oncogenic stress to cellular senescence. Front Biosci 14:475–485

Burd CE, Sorrentino JA, Clark KS, Darr DB, Krishnamurthy J, Deal AM et al (2013a) Monitoring tumorigenesis and senescence in vivo with a p16INK4a-luciferase model. Cell 152:340–351

Burd CE, Sorrentino JA, Clark KS, Darr DB, Krishnamurthy J, Deal AM et al (2013b) Monitoring tumorigenesis and senescence in vivo with a p16(INK4a)-luciferase model. Cell 152(1–2):340–351

Burgess A, Chia KM, Haupt S, Thomas D, Haupt Y, Lim E (2016) Clinical overview of MDM2/X-Targeted therapies. Front Oncol 6:7

Bussian TJ, Aziz A, Meyer CF, Swenson BL, van Deursen JM, Baker DJ (2018) Clearance of senescent glial cells prevents tau-dependent pathology and cognitive decline. Nature 562(7728):578–582

Castro P, Giri D, Lamb D, Ittmann M (2003) Cellular senescence in the pathogenesis of benign prostatic hyperplasia. Prostate 55(1):30–38

Castro P, Xia C, Gomez L, Lamb DJ, Ittmann M (2004) Interleukin-8 expression is increased in senescent prostatic epithelial cells and promotes the development of benign prostatic hyperplasia. Prostate 60(2):153–159

Chalan P, van den Berg A, Kroesen B-J, Brouwer L, Boots A (2015) Rheumatoid arthritis, immunosenescence and the hallmarks of aging. Curr Aging Sci 8(2):131–146

Chang J, Wang Y, Shao L, Laberge R-M, Demaria M, Campisi J et al (2016) Clearance of senescent cells by ABT263 rejuvenates aged hematopoietic stem cells in mice. Nat Med 22(1):78–83

Childs BG, Durik M, Baker DJ, van Deursen JM (2015) Cellular senescence in aging and age-related disease: from mechanisms to therapy. Nat Med 21(12):1424–1435

Childs BG, Baker DJ, Wijshake T, Conover CA, Campisi J, van Deursen JM (2016) Senescent intimal foam cells are deleterious at all stages of atherosclerosis. Science (New York, NY) 354(6311):472–477

Chinta SJ, Lieu CA, Demaria M, Laberge RM, Campisi J, Andersen JK (2013) Environmental stress, ageing and glial cell senescence: a novel mechanistic link to Parkinson's disease? J Intern Med 273(5):429–436

Chkhotua AB, Gabusi E, Altimari A, D'Errico A, Yakubovich M, Vienken J et al (2003) Increased expression of p16(INK4a) and p27(Kip1) cyclin-dependent kinase inhibitor genes in aging human kidney and chronic allograft nephropathy. Am J Kidney Dis 41(6):1303–1313

Choi J, Shendrik I, Peacocke M, Peehl D, Buttyan R, Ikeguchi EF et al (2000) Expression of senescence-associated beta-galactosidase in enlarged prostates from men with benign prostatic hyperplasia [In Process Citation]. Urology 56(1):160–166

Davaapil H, Brockes JP, Yun MH (2017) Conserved and novel functions of programmed cellular senescence during vertebrate development. Development (Cambridge, England). 144(1):106–114

de Stanchina E, Querido E, Narita M, Davuluri RV, Pandolfi PP, Ferbeyre G et al (2004) PML is a direct p53 target that modulates p53 effector functions. Mol Cell 13(4):523–535

Deschenes-Simard X, Gaumont-Leclerc MF, Bourdeau V, Lessard F, Moiseeva O, Forest V et al (2013) Tumor suppressor activity of the ERK/MAPK pathway by promoting selective protein degradation. Genes Dev 27(8):900–915

Deschenes-Simard X, Parisotto M, Rowell MC, Le Calve B, Igelmann S, Moineau-Vallee K et al (2019) Circumventing senescence is associated with stem cell properties and metformin sensitivity. Aging Cell 18(2):e12889

Dhar SK, St Clair DK (2009) Nucleophosmin blocks mitochondrial localization of p53 and apoptosis. J Biol Chem 284(24):16409–16418

Di Micco R, Fumagalli M, Cicalese A, Piccinin S, Gasparini P, Luise C et al (2006) Oncogene-induced senescence is a DNA damage response triggered by DNA hyper-replication. Nature 444(7119):638–642

Dimri GP, Lee X, Basile G, Acosta M, Scott G, Roskelley C et al (1995) A biomarker that identifies senescent human cells in culture and in aging skin in vivo. Proc Natl Acad Sci USA 92(20):9363–9367

Dörr JR, Yu Y, Milanovic M, Beuster G, Zasada C, Däbritz JHM et al (2013) Synthetic lethal metabolic targeting of cellular senescence in cancer therapy. Nature 501:421

Fang L, Igarashi M, Leung J, Sugrue MM, Lee SW, Aaronson SA (1999) p21Waf1/Cip1/Sdi1 induces permanent growth arrest with markers of replicative senescence in human tumor cells lacking functional p53. Oncogene 18(18):2789–2797

Farr JN, Khosla S (2019) Cellular senescence in bone. Bone 121:121–133

Farr JN, Xu M, Weivoda MM, Monroe DG, Fraser DG, Onken JL et al (2017) Targeting cellular senescence prevents age-related bone loss in mice. Nat Med 23(9):1072–1079

Ferbeyre G (2018) Aberrant signaling and senescence associated protein degradation. Exp Gerontol 107:50–54

Fessler J, Husic R, Schwetz V, Lerchbaum E, Aberer F, Fasching P et al (2018) Senescent T-cells promote bone loss in rheumatoid arthritis. Front Immunol 9:95

Franceschi C, Garagnani P, Morsiani C, Conte M, Santoro A, Grignolio A et al (2018) The continuum of aging and age-related diseases: common mechanisms but different rates. Front Med (Lausanne) 5:61

Fuhrmann-Stroissnigg H, Ling YY, Zhao J, McGowan SJ, Zhu Y, Brooks RW et al (2017) Identification of HSP90 inhibitors as a novel class of senolytics. Nat Commun 8(1):422

Fülöp T, Herbein G, Cossarizza A, Witkowski JM, Frost E, Dupuis G et al (2017) Cellular senescence, Immunosenescence and HIV. Interdiscip Top Gerontol Geriatr 42:28–46

Garrido AM, Bennett M (2016) Assessment and consequences of cell senescence in atherosclerosis. Curr Opin Lipidol 27(5):431–438

Garwood CJ, Simpson JE, Al Mashhadi S, Axe C, Wilson S, Heath PR et al (2014) DNA damage response and senescence in endothelial cells of human cerebral cortex and relation to Alzheimer's neuropathology progression: a population-based study in the Medical Research Council Cognitive Function and Ageing Study (MRC-CFAS) cohort. Neuropathol Appl Neurobiol 40(7):802–814

Guerrero A, Herranz N, Sun B, Wagner V, Gallage S, Guiho R et al (2019) Cardiac glycosides are broad-spectrum senolytics. Nature Metabolism 1(11):1074–1088

Gutierrez-Reyes G, del Carmen Garcia de Leon M, Varela-Fascinetto G, Valencia P, Perez Tamayo R, Rosado CG, et al (2010) Cellular senescence in livers from children with end stage liver disease. PloS One 5(4):e10231

Guzik TJ, Touyz RM (2017) Oxidative Stress, inflammation, and vascular aging in hypertension. Hypertension 70(4):660–667

Hagn F, Klein C, Demmer O, Marchenko N, Vaseva A, Moll UM et al (2010) BclxL changes conformation upon binding to wild-type but not mutant p53 DNA binding domain. J Biol Chem 285(5):3439–3450

Halazonetis TD, Gorgoulis VG, Bartek J (2008) An oncogene-induced DNA damage model for cancer development. Science 319(5868):1352–1355

He S, Sharpless NE (2017) Senescence in health and disease. Cell 169(6):1000–1011

Hedrington MS, Davis SN (2018) Peroxisome proliferator-activated receptor alpha-mediated drug toxicity in the liver. Expert Opin Drug Metab Toxicol. 14(7):671–677

Hennika T, Hu G, Olaciregui NG, Barton KL, Ehteda A, Chitranjan A et al (2017) Pre-clinical study of Panobinostat in xenograft and genetically engineered murine diffuse intrinsic pontine glioma models. PLoS ONE 12(1):e0169485

Herbig U, Ferreira M, Condel L, Carey D, Sedivy JM (2006a) Cellular senescence in aging primates. Science 311(5765):1257

Herbig U, Ferreira M, Condel L, Carey D, Sedivy JM (2006b) Cellular senescence in aging primates. Science (New York, NY). 311(5765):1257

Herranz N, Gallage S, Mellone M, Wuestefeld T, Klotz S, Hanley CJ et al (2015) mTOR regulates MAPKAPK2 translation to control the senescence-associated secretory phenotype. Nat Cell Biol 17(9):1205–1217

Hewitt G, Jurk D, Marques FDM, Correia-Melo C, Hardy T, Gackowska A et al (2012) Telomeres are favoured targets of a persistent DNA damage response in ageing and stress-induced senescence. Nature Commun 3:708

Hickson LJ, Langhi Prata LGP, Bobart SA, Evans TK, Giorgadze N, Hashmi SK et al (2019) Senolytics decrease senescent cells in humans: preliminary report from a clinical trial of Dasatinib plus Quercetin in individuals with diabetic kidney disease. EBioMedicine 47:446–456

Hines J, Lartigue S, Dong H, Qian Y, Crews CM (2019) MDM2-Recruiting PROTAC offers superior, synergistic Antiproliferative activity via simultaneous degradation of BRD4 and stabilization of p53. Can Res 79(1):251–262

Hou J, Cui C, Kim S, Sung C, Choi C (2018) Ginsenoside F1 suppresses astrocytic senescence-associated secretory phenotype. Chem Biol Interact 283:75–83

Huang B, Vassilev LT (2009) Reduced transcriptional activity in the p53 pathway of senescent cells revealed by the MDM2 antagonist nutlin-3. Aging (Albany NY) 1(10):845–854

Janzen V, Forkert R, Fleming HE, Saito Y, Waring MT, Dombkowski DM et al (2006) Stem-cell ageing modified by the cyclin-dependent kinase inhibitor p16INK4a. Nature 443(7110):421–426

Jaul E, Barron J (2017) Age-related diseases and clinical and public health implications for the 85 years old and over population. Front Public Health 5:335

Jeon OH, Kim C, Laberge R-M, Demaria M, Rathod S, Vasserot AP et al (2017) Local clearance of senescent cells attenuates the development of post-traumatic osteoarthritis and creates a pro-regenerative environment. Nat Med 23(6):775

Jeyapalan JC, Sedivy JM (2008) Cellular senescence and organismal aging. Mech Ageing Dev 129(7–8):467–474

Jeyapalan JC, Ferreira M, Sedivy JM, Herbig U (2007) Accumulation of senescent cells in mitotic tissue of aging primates. Mech Ageing Dev 128(1):36–44

Justice JN, Nambiar AM, Tchkonia T, LeBrasseur NK, Pascual R, Hashmi SK et al (2019) Senolytics in idiopathic pulmonary fibrosis: results from a first-in-human, open-label, pilot study. EBioMedicine 40:554–563

Karin O, Agrawal A, Porat Z, Krizhanovsky V, Alon U (2019) Senescent cells and the dynamics of aging. bioRxiv, 470500

Kashyap D, Sharma A, Sak K, Tuli HS, Buttar HS, Bishayee A (2018) Fisetin: a bioactive phytochemical with potential for cancer prevention and pharmacotherapy. Life Sci 194:75–87

Kim KM, Noh JH, Bodogai M, Martindale JL, Yang X, Indig FE et al (2017) Identification of senescent cell surface targetable protein DPP4. Genes Dev 31(15):1529–1534

Kirkland JL, Tchkonia T (2017) Cellular senescence: a translational perspective. EBioMedic 21:21–28

Kishi S, Bayliss PE, Uchiyama J, Koshimizu E, Qi J, Nanjappa P et al (2008) The identification of zebrafish mutants showing alterations in senescence-associated biomarkers. PLoS Genet 4(8):e1000152

Kortlever RM, Higgins PJ, Bernards R (2006) Plasminogen activator inhibitor-1 is a critical downstream target of p53 in the induction of replicative senescence. Nat Cell Biol 8(8):877–884

Kovacovicova K, Skolnaja M, Heinmaa M, Mistrik M, Pata P, Pata I, et al (2018) Senolytic Cocktail Dasatinib + Quercetin (D + Q) does not enhance the efficacy of senescence-inducing chemotherapy in liver cancer. Front Oncol 8:459

Kracikova M, Akiri G, George A, Sachidanandam R, Aaronson SA (2013) A threshold mechanism mediates p53 cell fate decision between growth arrest and apoptosis. Cell Death Differ 20(4):576–588

Krishnamurthy J, Ramsey MR, Ligon KL, Torrice C, Koh A, Bonner-Weir S et al (2006) p16INK4a induces an age-dependent decline in islet regenerative potential. Nature 443(7110):453–457

Kumar R, Sharma A, Kumari A, Gulati A, Padwad Y, Sharma R (2019) Epigallocatechin gallate suppresses premature senescence of preadipocytes by inhibition of PI3K/Akt/mTOR pathway and induces senescent cell death by regulation of Bax/Bcl-2 pathway. Biogerontology 20(2):171–189

Kurz DJ, Decary S, Hong Y, Erusalimsky JD (2000) Senescence-associated (beta)-galactosidase reflects an increase in lysosomal mass during replicative ageing of human endothelial cells. J Cell Sci 113(Pt 20):3613–3622

Laberge RM, Zhou L, Sarantos MR, Rodier F, Freund A, de Keizer PL et al (2012) Glucocorticoids suppress selected components of the senescence-associated secretory phenotype. Aging Cell 11(4):569–578

Laberge RM, Sun Y, Orjalo AV, Patil CK, Freund A, Zhou L et al (2015) MTOR regulates the pro-tumorigenic senescence-associated secretory phenotype by promoting IL1A translation. Nat Cell Biol 17(8):1049–1061

Langford SD, Boor PJ (1996) Oleander toxicity: an examination of human and animal toxic exposures. Toxicology 109(1):1–13

Lanna A, Henson SM, Escors D, Akbar AN (2014) The kinase p38 activated by the metabolic regulator AMPK and scaffold TAB1 drives the senescence of human T cells. Nat Immunol 15(10):965–972

Laubach JP, Moreau P, San-Miguel JF, Richardson PG (2015) Panobinostat for the treatment of multiple Myeloma. Clin Cancer Res 21(21):4767–4773

Lessel D, Wu D, Trujillo C, Ramezani T, Lessel I, Alwasiyah MK et al (2017) Dysfunction of the MDM2/p53 axis is linked to premature aging. J Clin Investig 127(10):3598–3608

Li Y, Yao J, Han C, Yang J, Chaudhry MT, Wang S, et al (2016) Quercetin, Inflammation and Immunity. Nutrients 8(3):167

Li W, He Y, Zhang R, Zheng G, Zhou D (2019) The curcumin analog EF24 is a novel senolytic agent. Aging (Albany NY) 11(2):771–782

Liu B, Wang J, Chan KM, Tjia WM, Deng W, Guan X et al (2005) Genomic instability in laminopathy-based premature aging. Nat Med 11(7):780–785

Liu Y, Sanoff HK, Cho H, Burd CE, Torrice C, Ibrahim JG et al (2009) Expression of p16INK4a in peripheral blood T-cells is a biomarker of human aging. Aging Cell 8(4):439–448

Liu X, Wang Y, Zhang X, Gao Z, Zhang S, Shi P et al (2018a) Senolytic activity of piperlongumine analogues: Synthesis and biological evaluation. Bioorg Med Chem 26(14):3925–3938

Liu S, Zheng Z, Ji S, Liu T, Hou Y, Li S et al (2018b) Resveratrol reduces senescence-associated secretory phenotype by SIRT1/NF-kappaB pathway in gut of the annual fish Nothobranchius guentheri. Fish Shellfish Immunol 80:473–479

Lombard DB, Chua KF, Mostoslavsky R, Franco S, Gostissa M, Alt FW (2005) DNA repair, genome stability, and aging. Cell 120(4):497–512

Loo TM, Kamachi F, Watanabe Y, Yoshimoto S, Kanda H, Arai Y et al (2017) Gut microbiota promotes obesity-associated liver cancer through PGE2-mediated suppression of antitumor immunity. Cancer Discov 7(5):522–538

Lopes-Paciencia S, Saint-Germain E, Rowell MC, Ruiz AF, Kalegari P, Ferbeyre G (2019) The senescence-associated secretory phenotype and its regulation. Cytokine 117:15–22

López-Otín C, Blasco MA, Partridge L, Serrano M, Kroemer G (2013) The hallmarks of aging. Cell 153(6):1194–1217

Macip S, Igarashi M, Berggren P, Yu J, Lee SW, Aaronson SA (2003) Influence of induced reactive oxygen species in p53-mediated cell fate decisions. Mol Cell Biol 23(23):8576–8585

Majdi A, Mahmoudi J, Sadigh-Eteghad S, Golzari SEJ, Sabermarouf B, Reyhani-Rad S (2016) Permissive role of cytosolic pH acidification in neurodegeneration: A closer look at its causes and consequences. J Neurosci Res 94(10):879–887

Maldonado JL, Timmerman L, Fridlyand J, Bastian BC (2004) Mechanisms of cell-cycle arrest in Spitz nevi with constitutive activation of the MAP-kinase pathway. Am J Pathol 164(5):1783–1787

Mallette FA, Ferbeyre G (2007) The DNA damage signaling pathway connects oncogenic stress to cellular senescence. Cell cycle (Georgetown, Tex 6(15):1831–6

Mallette FA, Gaumont-Leclerc MF, Ferbeyre G (2007) The DNA damage signaling pathway is a critical mediator of oncogene-induced senescence. Genes Dev 21(1):43–48

Martins R, Lithgow GJ, Link W (2016) Long live FOXO: unraveling the role of FOXO proteins in aging and longevity. Aging Cell 15(2):196–207

Martyanov V, Whitfield ML, Varga J (2019) Senescence signature in skin biopsies from systemic sclerosis patients treated with senolytic therapy: potential predictor of clinical response? Arthritis Rheumatol

Melk A, Schmidt BMW, Takeuchi O, Sawitzki B, Rayner DC, Halloran PF (2004) Expression of p16INK4a and other cell cycle regulator and senescence associated genes in aging human kidney. Kidney Int 65(2):510–520

Mendelsohn AR, Larrick JW (2018) Cellular senescence as the key intermediate in tau-mediated neurodegeneration. Rejuvenation Res 21(6):572–579

Menicacci B, Laurenzana A, Chilla A, Margheri F, Peppicelli S, Tanganelli E et al (2017) Chronic Resveratrol treatment inhibits MRC5 Fibroblast SASP-related protumoral effects on melanoma cells. J Gerontol A Biol Sci Med Sci 72(9):1187–1195

Michaloglou C, Vredeveld LC, Soengas MS, Denoyelle C, Kuilman T, van der Horst CM et al (2005) BRAFE600-associated senescence-like cell cycle arrest of human naevi. Nature 436(7051):720–724

Mihara M, Erster S, Zaika A, Petrenko O, Chittenden T, Pancoska P et al (2003) p53 has a direct apoptogenic role at the mitochondria. Mol Cell 11(3):577–590

Minamino T, Miyauchi H, Yoshida T, Ishida Y, Yoshida H, Komuro I (2002) Endothelial cell senescence in human atherosclerosis: role of telomere in endothelial dysfunction. Circulation 105(13):1541–1544

Moiseeva O, Bourdeau V, Roux A, Deschenes-Simard X, Ferbeyre G (2009) Mitochondrial dysfunction contributes to oncogene-induced senescence. Mol Cell Biol 29(16):4495–4507

Moiseeva O, Deschênes-Simard X, St-Germain E, Igelmann S, Huot G, Cadar AE et al (2013) Metformin inhibits the senescence-associated secretory phenotype by interfering with IKK/NF-κB activation. Aging Cell 12(3):489–498

Molofsky AV, Slutsky SG, Joseph NM, He S, Pardal R, Krishnamurthy J et al (2006) Increasing p16INK4a expression decreases forebrain progenitors and neurogenesis during ageing. Nature 443(7110):448–452

Munoz-Espin D, Canamero M, Maraver A, Gomez-Lopez G, Contreras J, Murillo-Cuesta S et al (2013) Programmed cell senescence during mammalian embryonic development. Cell 155(5):1104–1118

Muñoz-Espín D, Rovira M, Galiana I, Giménez C, Lozano-Torres B, Paez-Ribes M et al (2018) A versatile drug delivery system targeting senescent cells. EMBO Molecul Medic 10(9):e9355

Nakamura S, Nishioka K (2003) Enhanced expression of p16 in seborrhoeic keratosis; a lesion of accumulated senescent epidermal cells in G1 arrest. BC J Dermatol 149(3):560–565

Nguyen MN, Sen N, Lin M, Joseph TL, Vaz C, Tanavde V et al (2019) Discovering putative protein targets of small molecules: a study of the p53 activator nutlin. J Chem Inf Model 59(4):1529–1546

Nogueira-Recalde U, Lorenzo-Gomez I, Blanco FJ, Loza MI, Grassi D, Shirinsky V et al (2019) Fibrates as drugs with senolytic and autophagic activity for osteoarthritis therapy. EBioMedicine 45:588–605

Noureddine H, Gary-Bobo G, Alifano M, Marcos E, Saker M, Vienney N et al (2011) Pulmonary artery smooth muscle cell senescence is a pathogenic mechanism for pulmonary hypertension in chronic lung disease. Circ Res 109(5):543–553

Ogrodnik M, Miwa S, Tchkonia T, Tiniakos D, Wilson CL, Lahat A, et al (2017) Cellular senescence drives age-dependent hepatic steatosis. Nat Commun 8:15691

Oubaha M, Miloudi K, Dejda A, Guber V, Mawambo G, Germain M-A, et al (2016) Senescence-associated secretory phenotype contributes to pathological angiogenesis in retinopathy. Sci Transl Medic 8(362):362ra144

Paez-Ribes M, Gonzalez-Gualda E, Doherty GJ, Munoz-Espin D (2019) Targeting senescent cells in translational medicine. EMBO Mol Med, e10234

Palacio L, Goyer ML, Maggiorani D, Espinosa A, Villeneuve N, Bourbonnais S et al (2019) Restored immune cell functions upon clearance of senescence in the irradiated splenic environment. Aging Cell 18(4):e12971

Palmer AK, Tchkonia T, LeBrasseur NK, Chini EN, Xu M, Kirkland JL (2015) Cellular senescence in type 2 diabetes: a therapeutic opportunity. Diabetes 64(7):2289–2298

Palmer AK, Xu M, Zhu Y, Pirtskhalava T, Weivoda MM, Hachfeld CM et al (2019) Targeting senescent cells alleviates obesity-induced metabolic dysfunction. Aging Cell 18(3):e12950

Pan J, Li D, Xu Y, Zhang J, Wang Y, Chen M et al (2017) Inhibition of Bcl-2/xl With ABT-263 selectively kills senescent Type II Pneumocytes and reverses persistent pulmonary fibrosis induced by ionizing radiation in mice. Int J Radiat Oncol Biol Phys 99(2):353–361

Papatheodoridi A-M, Chrysavgis L, Koutsilieris M, Chatzigeorgiou A (2019) The role of senescence in the development of non-alcoholic fatty liver disease and progression to non-alcoholic steatohepatitis. Hepatology (Baltimore, Md)

Paradis V, Youssef N, Dargere D, Ba N, Bonvoust F, Deschatrette J et al (2001) Replicative senescence in normal liver, chronic hepatitis C, and hepatocellular carcinomas. Hum Pathol 32(3):327–332

Patil P, Dong Q, Wang D, Chang J, Wiley C, Demaria M et al (2019) Systemic clearance of p16(INK4a) -positive senescent cells mitigates age-associated intervertebral disc degeneration. Aging Cell 18(3):e12927

Petersen LE, Schuch JB, de Azeredo LA, Baptista TSA, Motta JG, do Prado AD, et al (2019) Characterization of senescence biomarkers in rheumatoid arthritis: relevance to disease progression. Clin Rheumatol

Qian Y, Chen X (2010) Tumor suppression by p53: making cells senescent. Histol Histopathol 25(4):515–526

Qudrat A, Wong J, Truong K (2017) Engineering mammalian cells to seek senescence-associated secretory phenotypes. J Cell Sci 130(18):3116–3123

Roberts AW, Seymour JF, Brown JR, Wierda WG, Kipps TJ, Khaw SL et al (2012) Substantial susceptibility of chronic lymphocytic leukemia to BCL2 inhibition: results of a phase I study of navitoclax in patients with relapsed or refractory disease. J Clin Oncol 30(5):488–496

Roos CM, Zhang B, Palmer AK, Ogrodnik MB, Pirtskhalava T, Thalji NM et al (2016) Chronic senolytic treatment alleviates established vasomotor dysfunction in aged or atherosclerotic mice. Aging Cell 15(5):973–977

Rotter Sopasakis V, Sandstedt J, Johansson M, Lundqvist A, Bergström G, Jeppsson A, et al (2019) Toll-like receptor-mediated inflammation markers are strongly induced in heart tissue in patients with cardiac disease under both ischemic and non-ischemic conditions. Int J Cardiol
Saeed H, Abdallah BM, Ditzel N, Catala-Lehnen P, Qiu W, Amling M et al (2011) Telomerase-deficient mice exhibit bone loss owing to defects in osteoblasts and increased osteoclastogenesis by inflammatory microenvironment. J Bone Miner Res 26(7):1494–1505
Sahin E, Depinho RA (2010) Linking functional decline of telomeres, mitochondria and stem cells during ageing. Nature 464(7288):520–528
Samaraweera L, Adomako A, Rodriguez-Gabin A, McDaid HM (2017) A Novel Indication for Panobinostat as a Senolytic Drug in NSCLC and HNSCC. Scie Reports 7(1):1900
Sasaki Y, Ikeda Y, Miyauchi T, Uchikado Y, Akasaki Y, Ohishi M (2019) Estrogen-SIRT1 axis plays a pivotal role in protecting arteries against menopause-induced senescence and atherosclerosis. J Atheroscler Thromb
Schafer MJ, White TA, Iijima K, Haak AJ, Ligresti G, Atkinson EJ et al (2017) Cellular senescence mediates fibrotic pulmonary disease. Nat Commun 8:14532
Sedelnikova OA, Horikawa I, Zimonjic DB, Popescu NC, Bonner WM, Barrett JC (2004) Senescing human cells and ageing mice accumulate DNA lesions with unrepairable double-strand breaks. Nat Cell Biol 6(2):168–170
Shimizu I, Minamino T (2019) Cellular senescence in cardiac diseases. J Cardiol
Snijders T, Parise G (2017) Role of muscle stem cells in sarcopenia. Curr Opin Clin Nutrition Metabolic Care 20(3):186–190
Sousa-Victor P, Gutarra S, García-Prat L, Rodriguez-Ubreva J, Ortet L, Ruiz-Bonilla V et al (2014a) Geriatric muscle stem cells switch reversible quiescence into senescence. Nature 506(7488):316–321
Sousa-Victor P, Perdiguero E, Munoz-Canoves P (2014) Geroconversion of aged muscle stem cells under regenerative pressure. Cell cycle (Georgetown, Tex. 13(20):3183–90
Storer M, Mas A, Robert-Moreno A, Pecoraro M, Ortells MC, Di Giacomo V et al (2013) Senescence is a developmental mechanism that contributes to embryonic growth and patterning. Cell 155(5):1119–1130
Tang YL, Zhou Y, Wang YP, Wang JW, Ding JC (2015) SIRT6/NF-κB signaling axis in ginsenoside Rg1-delayed hematopoietic stem/progenitor cell senescence. Int J Clin Exp Pathol 8(5):5591–5596
Tchkonia T, Morbeck DE, Von Zglinicki T, Van Deursen J, Lustgarten J, Scrable H et al (2010) Fat tissue, aging, and cellular senescence. Aging Cell 9(5):667–684
Thorin E, Thorin-Trescases N (2009) Vascular endothelial ageing, heartbeat after heartbeat. Cardiovasc Res 84(1):24–32
Triana-Martinez F, Picallos-Rabina P, Da Silva-Alvarez S, Pietrocola F, Llanos S, Rodilla V et al (2019) Identification and characterization of Cardiac Glycosides as senolytic compounds. Nat Commun 10(1):4731
Trotman LC, Alimonti A, Scaglioni PP, Koutcher JA, Cordon-Cardo C, Pandolfi PP (2006) Identification of a tumour suppressor network opposing nuclear Akt function. Nature 441(7092):523–7
Valentijn FA, Falke LL, Nguyen TQ, Goldschmeding R (2018) Cellular senescence in the aging and diseased kidney. J Cell Commun Signal 12(1):69–82
Van Veggel M, Westerman E, Hamberg P (2018) Clinical pharmacokinetics and pharmacodynamics of Panobinostat. Clin Pharmacokinet 57(1):21–29
Vassilev LT, Vu BT, Graves B, Carvajal D, Podlaski F, Filipovic Z et al (2004) In vivo activation of the p53 pathway by small-molecule antagonists of MDM2. Science (New York, NY) 303(5659):844–848
Vernier M, Bourdeau V, Gaumont-Leclerc MF, Moiseeva O, Begin V, Saad F et al (2011) Regulation of E2Fs and senescence by PML nuclear bodies. Genes Dev 25(1):41–50
Villiard E, Denis JF, Hashemi FS, Igelmann S, Ferbeyre G, Roy S (2017) Senescence gives insights into the morphogenetic evolution of anamniotes. Biol Open 6(6):891–896

Voghel G, Thorin-Trescases N, Mamarbachi AM, Villeneuve L, Mallette FA, Ferbeyre G et al (2010) Endogenous oxidative stress prevents telomerase-dependent immortalization of human endothelial cells. Mech Ageing Dev 131(5):354–363

Walaszczyk A, Dookun E, Redgrave R, Tual-Chalot S, Victorelli S, Spyridopoulos I et al (2019) Pharmacological clearance of senescent cells improves survival and recovery in aged mice following acute myocardial infarction. Aging Cell 18(3):e12945

Wang E (1995) Senescent human fibroblasts resist programmed cell death, and failure to suppress bcl2 is involved. Cancer Res 55(11):2284–2292

Wang AS, Dreesen O (2018) Biomarkers of cellular senescence and skin aging. Front Genet 9:247

Wang Y, Chang J, Liu X, Zhang X, Zhang S, Zhang X et al (2016) Discovery of piperlongumine as a potential novel lead for the development of senolytic agents. Aging (Albany NY) 8(11):2915–2926

Watson N, Ding B, Zhu X, Frisina RD (2017) Chronic inflammation—inflammaging—in the ageing cochlea: a novel target for future presbycusis therapy. Ageing Res Rev 40:142–148

Wiemann SU, Satyanarayana A, Tsahuridu M, Tillmann HL, Zender L, Klempnauer J et al (2002) Hepatocyte telomere shortening and senescence are general markers of human liver cirrhosis. FASEB J 16(9):935–942

Wu D, Prives C (2018) Relevance of the p53–MDM2 axis to aging. Cell Death Differ 25(1):169–179

Xu M, Tchkonia T, Ding H, Ogrodnik M, Lubbers ER, Pirtskhalava T et al (2015) JAK inhibition alleviates the cellular senescence-associated secretory phenotype and frailty in old age. Proc Natl Acad Sci 112(46):E6301–E6310

Xu M, Bradley EW, Weivoda MM, Hwang SM, Pirtskhalava T, Decklever T et al (2016) Transplanted senescent cells induce an osteoarthritis-like condition in mice. J Gerontol A Biol Sci Med Sci 72(6):780–785

Xu M, Pirtskhalava T, Farr JN, Weigand BM, Palmer AK, Weivoda MM et al (2018) Senolytics improve physical function and increase lifespan in old age. Nat Med 24(8):1246–1256

Yang M, Luna L, Sorbo JG, Alseth I, Johansen RF, Backe PH et al (2014) Human OXR1 maintains mitochondrial DNA integrity and counteracts hydrogen peroxide-induced oxidative stress by regulating antioxidant pathways involving p21. Free Radic Biol Med 77:41–48

Yosef R, Pilpel N, Tokarsky-Amiel R, Biran A, Ovadya Y, Cohen S et al (2016) Directed elimination of senescent cells by inhibition of BCL-W and BCL-XL. Nat Commun 7:11190

Yoshimoto S, Loo TM, Atarashi K, Kanda H, Sato S, Oyadomari S et al (2013) Obesity-induced gut microbial metabolite promotes liver cancer through senescence secretome. Nature 499(7456):97–101

Yousefzadeh MJ, Zhu Y, McGowan SJ, Angelini L, Fuhrmann-Stroissnigg H, Xu M et al (2018) Fisetin is a senotherapeutic that extends health and lifespan. EBio Medic 36:18–28

Yun MH, Davaapil H, Brockes JP (2015) Recurrent turnover of senescent cells during regeneration of a complex structure. eLife; 4

Zarbock A (2012) The shady side of dasatinib. Blood 119(21):4817–4818

Zhang X, Li CF, Zhang L, Wu CY, Han L, Jin G et al (2016) TRAF6 Restricts p53 mitochondrial translocation, apoptosis, and tumor suppression. Mol Cell 64(4):803–814

Zhang X, Zhang S, Liu X, Wang Y, Chang J, Zhang X, et al (2018) Oxidation resistance 1 is a novel senolytic target. Aging Cell, e12780

Zhang P, Kishimoto Y, Grammatikakis I, Gottimukkala K, Cutler RG, Zhang S et al (2019) Senolytic therapy alleviates Abeta-associated oligodendrocyte progenitor cell senescence and cognitive deficits in an Alzheimer's disease model. Nat Neurosci 22(5):719–728

Zhou S, Greenberger JS, Epperly MW, Goff JP, Adler C, LeBoff MS et al (2008) Age-related intrinsic changes in human bone-marrow-derived mesenchymal stem cells and their differentiation to osteoblasts. Aging Cell 7(3):335–343

Zhu Y, Tchkonia T, Pirtskhalava T, Gower AC, Ding H, Giorgadze N et al (2015) The Achilles' heel of senescent cells: from transcriptome to senolytic drugs. Aging Cell 14(4):644–658

Zhu Y, Tchkonia T, Fuhrmann-Stroissnigg H, Dai HM, Ling YY, Stout MB et al (2016) Identification of a novel senolytic agent, navitoclax, targeting the Bcl-2 family of anti-apoptotic factors. Aging Cell 15(3):428–435

Zhu Y, Doornebal EJ, Pirtskhalava T, Giorgadze N, Wentworth M, Fuhrmann-Stroissnigg H et al (2017) New agents that target senescent cells: the flavone, fisetin, and the BCL-XL inhibitors, A1331852 and A1155463. Aging (Albany NY) 9(3):955–963

Chapter 5
Senotherapy of Cancer

Pilar Picallos-Rabina, Sabela Da Silva-Álvarez, Lucía Antelo-Iglesias, Francisco Triana-Martínez, and Manuel Collado

Abstract Cellular senescence is a stress and repair response that protects us from cancer and contributes to tissue homeostasis by inducing a stable cell cycle arrest and imposing a secretory phenotype. Senescent cells are held in check to avoid their aberrant proliferation while at the same time they serve as new signaling nodes to orchestrate tissue repair and to reestablish homeostasis in damaged tissue. Chemotherapeutic drugs can induce senescence in cancer cells and, although restricting tumor growth, senescence can also have negative consequences for cancer therapy. Senescent cancer cells remaining after chemotherapy might represent a risk of tumor relapse and secrete a huge number of soluble factors known as SASP with detrimental activities that can alter the tumor microenvironment. In addition, induction of senescence in the normal surrounding tissue can produce severe side effects. In recent years, we have deciphered many aspects of the process of cellular senescence that can help us apply this response in our benefit. In particular, we have identified anticancer drugs that can induce a potent senescence response in cancer cells, we are learning how to modulate the SASP to avoid the negative effects, and we have found vulnerabilities in senescent cancer cells that allow their specific cell killing by senolytic compounds. In summary, we are in a position to start considering the possibility of developing effective senotherapies of cancer.

Keywords Cellular Senescence · Cancer · Senolytics · Senotherapy

P. Picallos-Rabina · S. Da Silva-Álvarez · L. Antelo-Iglesias · F. Triana-Martínez · M. Collado (✉)
Laboratory of Stem Cells in Cancer and Aging, Health Research Institute of Santiago de Compostela (IDIS), Xerencia de Xestión Integrada de Santiago (XXIS/SERGAS), Santiago de Compostela E15706, Spain
e-mail: manuel.collado.rodriguez@sergas.es

D. Muñoz-Espin and M. Demaria (eds.), *Senolytics in Disease, Ageing and Longevity*, Healthy Ageing and Longevity 11,
https://doi.org/10.1007/978-3-030-44903-2_5

5.1 Cellular Senescence in Cancer

Senescence is a cellular state defined by a stable cell cycle arrest and the acquisition of characteristic molecular and morphological features. Senescent cells are enlarged and flattened, contain numerous lysosomes with increased beta-galactosidase activity, lack proliferation due to the expression of high levels of cell cycle inhibitors, secrete huge amounts of soluble factors, and show heterochromatic foci and DNA damage. It was originally described after prolonged in vitro culture of human normal diploid cells and considered to represent the exhaustion of the predetermined replicative potential of cells (Hayflick and Moorhead 1961). As such, senescence was linked to the process of aging. The accumulation of senescent cells in multiple tissues is associated with pathological and normal aging (Childs et al. 2017; McHugh and Gil 2018). Already in the initial description of cellular senescence it was clear that this process took place in normal non-transformed cells, but not in tumor derived cancer cells, suggesting that cancer cells were not under the control of senescence. This differential response of normal and tumor cells suggested that senescence was a process involved in limiting the indefinite proliferation of cells to avoid the emergence of cells accumulating unrepaired damage.

5.1.1 Tumor Suppressive Function of Senescence

The observation of a sudden induction of senescence in normal primary cells after the introduction of an activated HRAS oncogene led to the notion of cellular senescence as a tumor suppressor mechanism (Serrano et al. 1997). The molecular players involved in this response, p16 and p53, were the same ones described to operate during the so-called replicative senescence on in vitro serially passaged cells, and the morphological change undergone by the RAS-expressing cells was identical to the one experienced after replicative exhaustion. Cells expressing single activated oncogenes do not show any of the typical features of transformed cells and the combination of two (in the case of mouse cells) or more (in human) oncogenic events are required to achieve fully transformed cells (Hahn and Weinberg 2002). Cultures of cells expressing oncogenes tend to show an initial phase of hyperproliferation followed by massive induction of apoptosis and/or senescence. Overcoming these cellular protective responses is a requisite of cancer cells to successfully become a tumor. Indeed, two of the initially recognized hallmarks of cancer in the classical review by Hanahan and Weinberg are the bypass of apoptosis and the acquisition of indefinite replicative potential, highlighting the crucial role played by apoptosis and senescence as barriers against cancer progression (Hanahan and Weinberg 2000). The identification of the senescence response in vivo on mouse models of oncogenic activation and in human samples from preneoplastic tissue, as well as the tumor prone phenotype of animals lacking some of the crucial regulators of senescence (e.g. p16 and p53) clearly established cellular senescence as a robust barrier restricting tumor development (Braig et al. 2005; Chen et al. 2005; Collado et al. 2005; Lazzerini Denchi et al. 2005; Michaloglou et al. 2005). Oncogene activation in vivo does not lead to the immediate growth of a tumor. After an initial phase of hyperproliferation, oncogenically-damaged cells enter senescence, restricting cancer development. Premalignant lesions are composed of a mixed population of cells undergoing senescence and/or apoptosis and bypass of these defensive responses is absolutely required for tumors to advance on their road to full malignant transformation (Collado and Serrano 2006).

5.1.2 Pro-Tumor Activity of Senescence

Despite being growth arrested, senescent cells are metabolically active and are characterized by a particular secretory phenotype known as SASP (Senescence-Associated Secretory Phenotype) (Coppé et al. 2008). Among the different molecules released by senescent cells there are a great number of pro-inflammatory cytokines and chemokines (e.g. IL6, IL8, CXCL1…), matrix remodeling enzymes (e.g. MMP1, MMP3, PAI-1…), and growth promoting factors (e.g. HGF, Epiregulin, Amphiregulin…). The putative pro-tumoral nature of these secreted factors seemed perplexing when considering the known antitumor effect of senescence. Reports from different laboratories have confirmed this pro-tumorigenic activity of the SASP (Krtolica et al. 2001; Angelini et al. 2013). Already when SASP was first identified, it was shown that premalignant epithelial cells exposed to SASP factors from genotoxic-induced senescence experienced an epithelial-mesenchyme transition and enhanced invasiveness, both hallmarks of malignant tumor cells (Coppé et al. 2008). Similarly, stromal senescent cells that accumulate during aging seem to be capable of conditioning the niche in the bone through secreted factors to promote metastasis development (Luo et al. 2016). This is in contrast with other reports showing induction of paracrine senescence in cells exposed to secretions of senescent cells (Acosta et al. 2008). We should be cautious when considering the putative role of SASP regarding tumor promotion or prevention; these relative effects could be influenced by cell type, senescence-inducing stimulus or time scale at which the different SASP factors are produced (Chan and Narita 2019). As already mentioned, the SASP is a complex and dynamic phenotype that evolves in composition with time from immunosuppressive and profibrotic to proinflammatory and fibrolytic activities, affecting surrounding cells in a beneficial or detrimental fashion according to the nature of these signals and the state of the recipient cells (Ito et al. 2017).

5.2 Chemotherapy-Induced Senescence

Even though cancer cells require bypassing the senescence barrier on their road to malignant transformation to achieve unlimited proliferative potential, the cellular senescence response can still be engaged if cancer cells receive the appropriate stimulus. This is evidenced for example using mouse models of inducible oncogenic activation in which shutting off the oncogene in an already established tumor leads to the control of tumor progression through induction of senescence (Wu et al. 2007). Similarly, tumors developed in the absence of crucial tumor suppressor genes such as p53 are restricted and even disappear after induction of senescence (Xue et al. 2007; Ventura et al. 2007). Interestingly, this senescence response is triggered only on the tumor cells and not in normal cells that are also deficient in p53. This potent tumor restrictive activity of senescence induction on tumor cells led many to consider the possibility of developing a prosenescence therapy of cancer (Serrano 2007).

5.2.1 Drugs Inducing Senescence Limit Tumor Growth

We now know that senescence can account for the beneficial response of cancer cells to chemotherapy. Establishing a senescence index by measuring a set of senescence markers on samples obtained at the time of colorectal cancer diagnosis demonstrated that the presence of senescence in the tumor lesion, previous to the treatment, correlated with a better therapeutic response (Haugstetter et al. 2010). On the other hand, the identification of senescence markers in vitro using cellular models of chemotherapy-induced senescence was further corroborated on samples derived from tumors from cancer patients that had been subjected to chemotherapy prior to surgery (te Poele et al. 2002; Roberson et al. 2005). In support of these data, the reevaluation of the effects on senescence of classic chemotherapeutic agents using mouse models of cancer revealed that this was a crucial component of the therapeutic effects of these drugs (Schmitt et al. 2002). This pro-senescence response was not intended when these anticancer drugs were originally developed because chemotherapy of cancer has traditionally been designed with the basic idea of killing tumor cells in the most effective way, and as selective as possible.

However, cytotoxic chemotherapy might be limited by the extend of cancer cell death achieved with the use of these drugs, the heterogenous state at which cancer cells are present within the tumor, and many other factors influencing chemotherapy success. The induction of cancer cell senescence could represent an alternative approach with beneficial effects for cancer therapy. For this, we would need to identify drugs inducing a potent cellular senescence response or develop new ones selected for their senescence-inducing potential (Fig. 5.1). Although there are a number of known chemotherapeutic agents that have been identified as capable of inducing cellular senescence in cancer cells, such as doxorubicin, etoposide, bleomycin, etc. (Liu et al. 2019), we still lack in depth information regarding the mechanisms involved and the clinical relevance of inducing senescence in the context of cancer treatment. Reevaluating known anticancer drugs for their potential senescence-inducing activity could also provide clues about their potential use on a prosenescence therapy of cancer. This is the case, for example, of recently developed CDK4/6i (Palbociclib and Abemaciclib) (Llovet et al. 2016; Torres-Guzmán et al. 2017). These drugs have been shown to elicit a senescence response in pre-clinical and clinical settings (Llovet et al. 2016; Gong et al. 2017). A more recent example is provided by poly (ADP-ribose) polymerase 1 inhibitors (PARPi). These drugs have been developed and are in clinical use for ovarian and breast cancer and have been recently shown to induce a strong senescence response in cancer cell lines of these tumor types (Fleury et al. 2019).

To discover new drugs inducing cancer cell senescence we will need to screen for compounds with this activity or identify potential targets that can be drugged to trigger senescence in cancer cells. There are some recent elegant examples of high-throughput genetic and compound screenings to develop cancer cell senescence-inducing drugs. Using a CRISPR library targeting enzymes involved in remodeling of chromatin and in the modulation of epigenetic marks it was possible to show that suppression of SMARCB1, a component of the SWI/SNF nucleosome remodeling complex, induces senescence of melanoma cells (Wang et al. 2017). Similarly, a CRISPR library targeting kinases identified several kinases required for proliferation of p53-mutant liver cancer cells. Among them, inhibition of CDC7 using a chemical inhibitor induced cellular senescence specifically in liver cancer cells, suggesting that this kinase could be used as a potential target to develop a senogenic cancer therapy (Wang et al. 2019). On the other hand, using compound screening, multiple aurora kinase inhibitors were shown to possess strong senescence inducing activity on RAS mutant lung cancer cell lines (Wang et al. 2017).

5.2.2 Senescent Cancer Cells Fuel Malignant Growth

Apart from the proposed non-cell autonomous negative effect of senescence over premalignant or malignant cells in their vicinity by SASP factors, tumor cells can also be fueled by the senescence program when they escape from the tight cell cycle control (Milanovic et al. 2018; Achuthan et al. 2011). Chemotherapy-induced senescence of cancer cells has been shown to induce pathways similar to those that are typical of a stemness state. When these cells escape the control of senescence, they manifest an increased malignancy with respect to their parental tumor cells. This reinforced tumorigenic potential seems to be related with the activation of the Wnt pathway during chemotherapy-induced senescence and, accordingly, inhibition of Wnt leads to a reduced tumorigenicity of these "cancer stem cell-like" escapees. A similar pro-stemness activity of senescence cells has been linked to the SASP. Senescence induction alters surrounding cells increasing their plasticity and promoting their regenerative potential and cell reprogramming to pluripotency (Rhinn et al.

2019; Mosteiro et al. 2016). This state can be triggered by chemotherapeutic drugs but can also result from oncogenic induction (Rhinn et al. 2019; Ferreirós et al. 2019).

5.2.3 Secondary Effects of Cancer Therapy Mediated by Senescence Induction

Chemotherapy and radiotherapy of cancer is unfortunately associated with devastating secondary effects due to the unspecific targeting of normal cells. Tissues subjected to the harsh action of anticancer therapies frequently accumulate damaged cells that have undergone apoptosis or senescence. There is now mounting evidence that these senescent cells contribute to the adverse effects of chemotherapy and radiotherapy. The analysis of p16INK4A expression levels, a classical marker of cellular senescence, in CD3+ lymphocytes in the blood of breast cancer survivors revealed that adjuvant chemotherapy of breast cancer promotes aging by inducing cellular senescence (Sanoff et al. 2014). Similarly, irradiation of normal tissues is known to induce cellular senescence and to cause for example pulmonary fibrosis or loss of salivary gland function during radiotherapy of lung or head and neck tumors (He et al. 2019; Marmary et al. 2016). These negative effects are considered to be mediated by the secretion of pro-inflammatory factors present in the SASP that cause a low level chronic inflammatory environment. Transgenic mice that allow visualization of senescent cells revealed the induction of senescence after treatment with the anthracycline Doxorubicin (Demaria et al. 2017). These animals suffer from systemic inflammation that results in bone marrow suppression, cardiac dysfunction, fatigue and even cancer recurrence. All these adverse effects derive from senescence induced by chemotherapy on normal cells and model the response of cancer patients to treatments (Fig. 5.1).

5.3 Senolytics as Antitumor Adjuvant Therapy

Senescent cells present a particular expression profile, part of which consists on the secreted factors that form the SASP and on the expression of cell cycle regulators responsible for the cell cycle arrest. Apart from these typical hallmarks of the process, cellular senescence is characterized by an increased resistance to apoptosis (Childs et al. 2014). This seems to rely, at least in some settings, on an upregulated expression of the BCL-2 family of antiapoptotic proteins (Wang 1995). The use of chemical inhibitors of these antiapoptotic proteins, for example using ABT-263 or ABT-737, demonstrated the feasibility of inducing selectively the killing of senescent cells (Chang et al. 2016; Zhu et al. 2016; Yosef et al. 2016). There are other compounds with this activity, now collectively known as senolytics. Senolytic compounds such as Quercetin and Dasatinib are also thought to work by disarming senescent cells (Zhu

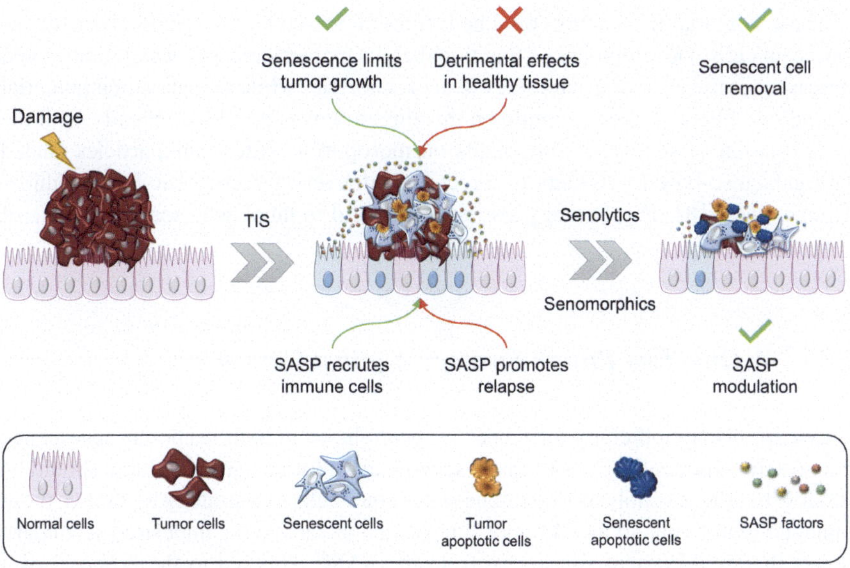

Fig. 5.1 Senotherapy of cancer. Chemotherapy and radiotherapy can induce cellular senescence as part of their action on tumor cells and this can have positive and negative impacts on cancer therapy derived from the cell cycle arrest and the SASP. Using senolytics to kill specifically the tumor and normal senescent cells, and senomorphics to modulate the SASP can contribute to the success of the therapy while reducing the secondary effects at the same time. TIS: Therapy-Induced Senescence. SASP: Senescence-Associated Secretory Phenotype

et al. 2015). In contrast, a novel family of broad spectrum senolytic compounds, the Cardiac Glycosides, seem to induce senescent cell killing by taking advantage of an acquired vulnerability of senescent cells derived from their membrane depolarization and disbalanced electrochemical gradient (Guerrero et al. 2019; Triana-Martínez et al. 2019).

Importantly, in vivo senolytic treatment has proved to provide therapeutic benefit in a number of age-related diseases and ultimately to lead to improved physical function and increased lifespan in mice (Brooks et al. 2018; Xu et al. 2018). The successful reports of in vivo use of senolytic compounds in the context of aging and age-related diseases open the possibility for a potential use as adjuvant therapy of cancer. In this sense, it is worth noting that some of these senolytics have been or are currently used in clinical trials for different neoplastic pathologies: ABT-263 was originally developed as an anti-cancer drug that failed to advance further due to the toxicity shown in patients during the early stages of clinical trials; Dasatinib, is a drug approved for clinical use in Chronic Myeloid Leukemia as an alternative to Imatinib; Cardiac Glycosides and Quercetin are currently being investigated as anti-cancer drugs in a number of neoplastic diseases in combination with other chemotherapeutic agents.

These potential treatments could be further improved by the specific targeting of senescent cells. Development of functionalized nanocarriers and formulations could increase the efficacy of the cytotoxic effect of senolytics while reducing their potential secondary effects. A nice example on this direction was provided recently with the encapsulation of senolytic compounds on mesoporous silica nanoparticles coated with galactose-based polymers to direct their delivery to senescent cells (Muñoz-Espín et al. 2018). This strategy also proved useful to label and monitor the fate of senescent cells in vivo.

5.3.1 A One-Two Punch Strategy Against Cancer

Since current chemotherapy for cancer has been shown to induce a potent senescence response in tumor cells in many cases, several laboratories reasoned that it should be possible to induce senolysis of the senescent tumor cells as a strategy to remove these damaged tumor cells to avoid tumor relapse and to reduce the undesired secondary effects that could potentially derive from the SASP. This led to the proposal of a one-two punch therapeutic strategy for cancer based on the use first of a compound inducing senescence in cancer cells followed by a senolytic drug that would specifically kill these cells (Fig. 5.1). With this idea in mind, several laboratories have addressed the issue using different systems. Initially, it was shown that treatment of cancer cells with aurora kinase inhibitors such as Alisertib and Barasertib, previously identified on genetic and compound screenings as potent senescence inducers, sensitizes the tumor cells in vitro to a BCL-2 family inhibitor and senolytic compound, ABT-263, leading to their specific killing (Wang et al. 2017). A similar strategy has been followed more recently after identifying the kinase CDC7 as a potential target to induce senescence of p53-deficient liver cancer cells using a genetic screening. The use of XL413, a specific inhibitor of CDC7 (Koltun et al. 2012), proved that targeting this kinase induces senescence in the tumor cells (Wang et al. 2019). However, this induction of senescence does not sensitize the cancer cells to known senolytics such as ABT-263 or Dasatinib. Further investigation on the molecular mechanisms involved in senescence induced by CDC7 inhibition revealed that blocking the mTOR pathway selectively triggered apoptosis in the XL413-treated cancer cells. The combination of the senogenic XL413 with the mTOR inhibitor AZD8055, acting as a specific senolytic agent, proved more effective than monotherapy in xenograft models of cancer growth as well as in a genetic model of hepatocellular carcinoma, providing an elegant example of the combined use of genetic and compound screenings to identify vulnerabilities of specific cancer cells (Wang et al. 2019).

Another good example of the potential use of this one-two punch strategy against cancer was reported recently, when the reevaluation of the effect caused by PARPi on ovarian cancer cells revealed that they can induce senescence (Fleury et al. 2019). Using Olaparib and other PARPi in combination with ABT-263 proved synthetic lethality in vivo leading to a more effective anticancer treatment. PARPi treatment of cancer cells however results in an incomplete cellular senescence response. Stopping

PARPi treatment allows the resumed proliferation of the cancer cells. Interestingly, this unstable senescent-like state provides a window of opportunity for therapeutic treatment with senolytics since this transient growth arrest is sufficient to sensitize the cancer cells to the effect of ABT-263.

Finally, Cardiac Glycosides (CGs) have been reported to have broad-spectrum senolytic activity causing cytotoxicity on a wide range of tumor cells after exposure to different senescence-inducing chemotherapeutic agents (Guerrero et al. 2019; Triana-Martínez et al. 2019). Interestingly, CGs had been previously proposed as cooperating antitumor compounds in combination with classic chemotherapeutic agents. It is tempting to speculate that perhaps the repeatedly observed antitumor activity of CGs could be, at least in part, the result of their senolytic activity.

5.3.2 Alleviation of Secondary Effects by Targeting Senescence Cells

As already exposed, chemotherapy of cancer can trigger the induction of senescence not only of the tumor cells but also of the normal surrounding tissue. Accumulation of these senescent cells actively secreting SASP factors could have detrimental consequences for the correct function of treated tissues (Demaria et al. 2017). The efficient elimination of these senescent cells could result on less toxic therapies by removing these damaged cells (Fig. 5.1). This possibility has been elegantly tested on a recent report in which they eliminated senescent cells produced by chemotherapy using a transgenic mouse model to target these cells or by treating animals with the senolytic ABT-263 (Chang et al. 2016; Zhu et al. 2016; Yosef et al. 2016). In both cases, removing normal senescent cells after treatment considerably decreased different parameters traditionally linked with secondary effects of chemotherapy, such as bone marrow suppression, cardiac disfunction, cancer relapse, and physical activity and strength. Furthermore, examining a senescence marker in T cells from cancer patients prior to chemotherapy showed that higher levels of the senescence marker correlated with increased risk of chemotherapy-induced fatigue.

Radiotherapy of cancer can also lead to secondary effects which have also been attributed to the induction of cellular senescence. For example, radiation-induced loss of salivary gland function or lung fibrosis after thoracic radiation have been shown to occur as a result of induction and accumulation of normal cells induced to senescence by radiation. In the case of the salivary glands it was possible to point to the SASP as responsible for this detrimental effect since IL6 modulation prevented the damage (Marmary et al. 2016). In the case of lung fibrosis, treatment of irradiated mice with ABT-263 after development of fibrosis reduced the number of senescent cells and reversed the disease (Pan et al. 2017).

5.4 Modulating the SASP to Improve Senotherapy

The composition and dynamics of the SASP are highly complex. Different molecules are released from senescent cells depending on the nature of the inducer, the cell type and the microenvironment (Coppé et al. 2010). Understanding this complexity is required to tease apart the different activities of SASP factors. The identification of key regulatory elements in the production of SASP factors could give us clues on how to modulate its activity to maximize the positive effects while avoiding the detrimental activities. This could lead to develop novel therapeutic compounds, known as senomorphics, that could be used to obtain therapeutic benefit by controlling the SASP in the context of cancer therapy (Fuhrmann-Stroissnigg et al. 2017) (Fig. 5.1).

There are now a number of reports showing how it is possible to alter the SASP without affecting the senescent cell cycle arrest. The first SASP regulator to be studied was NFkB (Chien et al. 2011). Chromatin proteomics identified p65 (also known as RelA), a member of the NFkB family, as an abundant protein associated with chromatin and regulating multiple genes during senescence, and among them, prominently SASP factors. When p65 is suppressed in vitro, senescent cells escape from immune recognition by natural killer cells and, when combined with p53 inactivation, senescence is bypassed. In vivo, using a mouse model of lymphoma development, inactivation of p65 leads to resistance to chemotherapy-induced senescence and reduced survival. These results demonstrate that NFkB controls cell intrinsic as well as cell extrinsic aspects of senescence and at least for some of these activities, it does so through its control of SASP production.

The SASP is also controlled by the inflammosome-mediated IL1 signaling (Acosta et al. 2013). Activation of this pathway during senescence leads to the production of SASP factors that can trigger senescence in a paracrine manner, extending the senescence response to neighboring cells, and activates and recruits the immune system to sites of senescence induction. In this manner, the SASP contributes to tumor suppression, as demonstrated in vivo using IL-1a inhibitors on a model of oncogene activation in liver. Blocking the SASP in this context leads to liver tumor progression in the presence of a defective immune surveillance. In contrast, other authors have reported that IL-1a inactivation impairs tumor progression by decreasing the SASP without affecting the senescent cell cycle arrest (Lau et al. 2019).

Another crucial regulator of the SASP is mTOR. Inhibition of mTOR or its effectors blocks the SASP in the context of cancer, revealing both its tumor suppressive and tumor promoting activity (Herranz et al. 2015). In particular, reducing a specific subset of SASP factors by mTOR inhibition decreases EMT induction and the invasive properties of cancer cells elicited by conditioned media derived from senescent cells. In vivo, co-injection of tumor cells with senescent cells enhances tumor growth but this pro-tumorigenic activity is lost when senescent cells express an shRNA targeted against mTOR. In contrast, the paracrine induction of senescence mediated by SASP factors present in conditioned media is clearly reduced when mTOR signaling is blocked in senescent cells. At the same time, senescence induction by oncogene expression in the liver of mice treated with rapamycin, an mTOR inhibitor, shows a defective immune surveillance activity that results in decreased tumor suppressive function of senescence. Livers of mice treated with the mTOR inhibitor showed reduced infiltration of T cells, B cells, NK cells and macrophages, immune cells in charge of the clearance of senescent cells and contributing to the tumor suppressor activity of senescence.

Finally, another important player regulating the SASP is the Bromodomain-containing 4 (BRD4), a member of the BET (bromodomain and extra terminal domain) family of transcriptional regulators. Inhibition of BRD4 in senescent cells does not result in bypass of senescence but decreases the SASP resulting in poor paracrine senescence activity, impaired macrophage polarization, and reduced NK cytotoxic activity (Tasdemir et al. 2016). As a result, BRD4 inhibition leads to a disruption in immune surveillance and elimination of senescent cells in vitro and in vivo, a crucial tumor suppressive activity of senescent cells mediated by the SASP.

5.5 Concluding Remarks and Future Perspectives

The recent development of compounds with specific cell killing activity against senescent cells, known as senolytics, has provided strong evidence of the possibility of removing senescent cells with therapeutic consequences. This idea has been initially applied to aging and age-related diseases. Selectively eliminating senescent cells improves the healthspan of premature and naturally aged mice while providing the beneficial tumor suppressive function of senescence (Childs et al. 2017). This has led many to consider the possibility of using this strategy in the context of cancer therapy. Chemotherapy-induced senescence of cancer encountered opposition when the potentially harmful effects of the SASP were first described. Besides, leaving damaged tumor cells hanging around in cancer patients represent a potential risk of relapse due to escape of senescent tumor cells from the growth arrest. Developing strategies to modulate the SASP, through senomorphics, and combining senescence-inducing drugs with senolytics could represent novel approaches to treat cancer more effectively. The induction of a senescent phenotype in cancer cells, even a partial one, might provide a therapeutic window of opportunity. As part of the process of senescence, cancer cells might acquire novel vulnerabilities that could make them susceptible to the cytotoxic activity of senolytics. On top of that, the possibility of removing non-tumor senescent cells generated by the unspecific targeting of normal cells by chemotherapy or radiotherapy offers the promise of reduced secondary effects. There are now a number of known chemotherapeutic drugs that can induce senescence in tumor cells. We need to identify more and to decipher their mode of action to potentially use them in the clinic. In addition, we also need to identify more powerful and specific senolytic compounds and try to understand the molecular basis of their action.

Understanding the molecular basis of the process that make us grow old could remarkably provide us with the key to develop more effective and less toxic treatments against cancer.

Acknowledgements M.C. is a "Miguel Servet II" investigator (CPII16/00015). Work at the laboratory of M.C. is funded by grant RTI2018-095818-B-100 (MCIU/AEI/FEDER, UE). P.P-R. receives support from a fellowship by the Fundación Instituto de Investigación Sanitaria de Santiago de Compostela, FIDIS.

Conflict of Interest Authors declare no conflict of interest.

References

Achuthan S, Santhoshkumar TR, Prabhakar J, Nair SA, Pillai MR (2011) Drug-induced senescence generates chemoresistant stemlike cells with low reactive oxygen species. J Biol Chem 286(43):37813–29

Acosta JC, O'Loghlen A, Banito A, Guijarro MV, Augert A, Raguz S et al (2008) Chemokine signaling via the CXCR2 receptor reinforces senescence. Cell 133(6):1006–18

Acosta JC, Banito A, Wuestefeld T, Georgilis A, Janich P, Morton JP et al (2013) A complex secretory program orchestrated by the inflammasome controls paracrine senescence. Nat Cell Biol 15(8):978–90

Angelini PD, Fluck MFZ, Pedersen K, Parra-Palau JL, Guiu M, Bernadó Morales C et al (2013) Constitutive HER2 signaling promotes breast cancer metastasis through cellular senescence. Cancer Res 73(1):450–8

Braig M, Lee S, Loddenkemper C, Rudolph C, Peters AHFM, Schlegelberger B et al (2005) Oncogene-induced senescence as an initial barrier in lymphoma development. Nature 436(7051):660–5

Brooks RW, Robbins PD (2018) Treating age-related diseases with somatic stem cells. Adv Exp Med Biol 1056:29-45

Chan ASL, Narita M (2019) Short-term gain, long-term pain: the senescence life cycle and cancer. Genes Dev 33(3–4):127–43

Chang J, Wang Y, Shao L, Laberge R-M, Demaria M, Campisi J et al (2016) Clearance of senescent cells by ABT263 rejuvenates aged hematopoietic stem cells in mice. Nat Med 22(1):78–83

Chen Z, Trotman LC, Shaffer D, Lin H-KK, Dotan ZA, Niki M et al (2005) Crucial role of p 53-dependent cellular senescence in suppression of Pten-deficient tumorigenesis. Nature 436(7051):725–30

Chien Y, Scuoppo C, Wang X, Fang X, Balgley B, Bolden JE et al (2011) Control of the senescence-associated secretory phenotype by NF-κB promotes senescence and enhances chemosensitivity. Genes Dev 25(20):2125–36

Childs BG, Baker DJ, Kirkland JL, Campisi J, van Deursen JM (2014) Senescence and apoptosis: dueling or complementary cell fates? EMBO Rep 15(11):1139–53

Childs BG, Gluscevic M, Baker DJ, Laberge RM, Marquess D, Dananberg J et al (2017) Senescent cells: an emerging target for diseases of ageing. Nat Rev Drug Discov 16(10):718–735

Collado M, Serrano M (2006) The power and the promise of oncogene-induced senescence markers. Nat Rev Cancer 6(6):472–6

Collado M, Gil J, Efeyan A, Guerra C, Schuhmacher AJ, Barradas M et al (2005) Tumour biology: senescence in premalignant tumours. Nature 436(7051):642

Coppé J-P, Patil CK, Rodier F, Sun Y, Muñoz DP, Goldstein J et al (2008) Senescence-associated secretory phenotypes reveal cell-nonautonomous functions of oncogenic RAS and the p53 tumor suppressor. PLoS Biol 6(12):e301

Coppé J-P, Desprez P-Y, Krtolica A, Campisi J (2010) The senescence-associated secretory phenotype: the dark side of tumor suppression. Annu Rev Pathol Mech Dis 5(1):99–118

Demaria M, O'Leary MN, Chang J, Shao L, Liu S, Alimirah F et al (2017) Cellular senescence promotes adverse effects of chemotherapy and cancer relapse. Cancer Discov 7(2):165–76

Ferreirós A, Pedrosa P, Da Silva-Álvarez S, Triana-Martínez F, Vilas JM, Picallos-Rabina P et al (2019) Context-dependent impact of RAS oncogene expression on cellular reprogramming to pluripotency. Stem Cell Rep 12(5):1099–112

Fleury H, Malaquin N, Tu V, Gilbert S, Martinez A, Olivier MA et al (2019) Exploiting interconnected synthetic lethal interactions between PARP inhibition and cancer cell reversible senescence. Nat Commun 10(1):2556

Fuhrmann-Stroissnigg H, Ling YY, Zhao J, McGowan SJ, Zhu Y, Brooks RW et al (2017) Identification of HSP90 inhibitors as a novel class of senolytics. Nat Commun 8(1):422

Gong X, Litchfield LM, Webster Y, Chio L-CC, Wong SS, Stewart TR et al (2017) Genomic aberrations that activate D-type cyclins are associated with enhanced sensitivity to the CDK4 and CDK6 inhibitor abemaciclib. Cancer Cell 32(6):761–776.e6

Guerrero A, Herranz N, Sun B, Wagner V, Gallage S, Guiho R et al (2019) Cardiac glycosides are broad-spectrum senolytics. Nat Metab 1(11):1074–88

Hahn WC, Weinberg RA (2002) Rules for making human tumor cells. N Engl J Med 347(20):1593–603

Hanahan D, Weinberg RA (2000) The hallmarks of cancer. Cell 100(1):57–70

Haugstetter AM, Loddenkemper C, Lenze D, Gröne J, Standfuss C, Petersen I et al (2010) Cellular senescence predicts treatment outcome in metastasised colorectal cancer. Br J Cancer 103(4):505–9

Hayflick L, Moorhead PS (1961) The serial cultivation of human diploid cell strains. Exp Cell Res 25(3):585–621

He Y, Thummuri D, Zheng G, Okunieff P, Citrin DE, Vujaskovic Z et al (2019) Cellular senescence and radiation-induced pulmonary fibrosis. Transl Res 209:14–21

Herranz N, Gallage S, Mellone M, Wuestefeld T, Klotz S, Hanley CJ et al (2015) mTOR regulates MAPKAPK2 translation to control the senescence-associated secretory phenotype. Nat Cell Biol 17(9):1205–17

Ito Y, Hoare M, Narita M (2017) Spatial and temporal control of senescence. Trends Cell Biol 27(11):820–32

Koltun ES, Tsuhako AL, Brown DS, Aay N, Arcalas A, Chan V et al (2012) Discovery of XL413, a potent and selective CDC7 inhibitor. Bioorg Med Chem Lett 22(11):3727–31

Krtolica A, Parrinello S, Lockett S, Desprez P-Y, Campisi J (2001) Senescent fibroblasts promote epithelial cell growth and tumorigenesis: a link between cancer and aging. Proc Natl Acad Sci 98(21):12072–7

Lau L, Porciuncula A, Yu A, Iwakura Y, David G (2019) Uncoupling the senescence-associated secretory phenotype from cell cycle exit via Interleukin-1 inactivation unveils its protumorigenic role. Mol Cell Biol 39(12):e00586-18

Lazzerini Denchi E, Attwooll C, Pasini D, Helin K (2005) Deregulated E2F activity induces hyperplasia and senescence-like features in the mouse pituitary gland. Mol Cell Biol 25(7):2660–72

Liu P, Lu Z, Wu Y, Shang D, Zhao Z, Shen Y et al (2019) Cellular senescence-inducing small molecules for cancer treatment. Curr Cancer Drug Targets 19(2):109–19

Llovet JM, Molina-Sánchez P, Ruiz de Galarreta M, Bian CB, Nguyen CB, Miguela V et al (2016) Palbociclib (PD-0332991), a selective CDK4/6 inhibitor, restricts tumour growth in preclinical models of hepatocellular carcinoma. Gut 66(7):1286–96

Luo X, Fu Y, Loza AJ, Murali B, Leahy KM, Ruhland MK et al (2016) Stromal-initiated changes in the bone promote metastatic niche development. Cell Rep 14(1):82–92

Marmary Y, Adar R, Gaska S, Wygoda A, Maly A, Cohen J et al (2016) Radiation-induced loss of salivary gland function is driven by cellular senescence and prevented by IL6 modulation. Cancer Res 76(5):1170–80

McHugh D, Gil J (2018) Senescence and aging: causes, consequences, and therapeutic avenues. J Cell Biol 217(1):65–77

Michaloglou C, Vredeveld LCW, Soengas MS, Denoyelle C, Kuilman T, van der Horst CMAM et al (2005) BRAFE600-associated senescence-like cell cycle arrest of human naevi. Nature 436(7051):720–4

Milanovic M, Fan DNY, Belenki D, Däbritz JHM, Zhao Z, Yu Y et al (2018) Senescence-associated reprogramming promotes cancer stemness. Nature 553(7686):96–100
Mosteiro L, Pantoja C, Alcazar N, Marión RM, Chondronasiou D, Rovira M et al (2016) Tissue damage and senescence provide critical signals for cellular reprogramming in vivo. Science 354(6315):pii: aaf4445
Muñoz-Espín D, Rovira M, Galiana I, Giménez C, Lozano-Torres B, Paez-Ribes M et al (2018) A versatile drug delivery system targeting senescent cells. EMBO Mol Med 10(9):e9355
Pan J, Li D, Xu Y, Zhang J, Wang Y, Chen M et al (2017) Inhibition of Bcl-2/xl With ABT-263 selectively kills senescent type II pneumocytes and reverses persistent pulmonary fibrosis induced by ionizing radiation in mice. Int J Radiat Oncol 99(2):353–61
Rhinn M, Ritschka B, Keyes WM (2019) Cellular senescence in development, regeneration and disease. Development 146(20):dev151837
Roberson RS, Kussick SJ, Vallieres E, Chen S-YJ, Wu DY (2005) Escape from therapy-induced accelerated cellular senescence in p53-null lung cancer cells and in human lung cancers. Cancer Res 65(7):2795–803
Sanoff HK, Deal AM, Krishnamurthy J, Torrice C, Dillon P, Sorrentino J et al (2014) Effect of cytotoxic chemotherapy on markers of molecular age in patients with breast cancer. J Natl Cancer Inst 106(4):dju057
Schmitt CA, Fridman JS, Yang M, Lee S, Baranov E, Hoffman RM et al (2002) A senescence program controlled by p53 and p16INK4a contributes to the outcome of cancer therapy. Cell 109(3):335–46
Serrano M (2007) Cancer regression by senescence. N Engl J Med 356(19):1996–7
Serrano M, Lin AW, McCurrach ME, Beach D, Lowe SW (1997) Oncogenic ras provokes premature cell senescence associated with accumulation of p53 and p16(INK4a). Cell 88(5):593–602
Tasdemir N, Banito A, Roe J-S, Alonso-Curbelo D, Camiolo M, Tschaharganeh DF et al (2016) BRD4 connects enhancer remodeling to senescence immune surveillance. Cancer Discov 6(6):612–29
te Poele RH, Okorokov AL, Jardine L, Cummings J, Joel SP (2002) DNA damage is able to induce senescence in tumor cells in vitro and in vivo. Cancer Res 62(6):1876–83
Torres-Guzmán R, Calsina B, Hermoso A, Baquero C, Alvarez B, Amat J et al (2017) Preclinical characterization of abemaciclib in hormone receptor positive breast cancer. Oncotarget 8(41):69493–507
Triana-Martínez F, Picallos-Rabina P, Da Silva-Álvarez S, Pietrocola F, Llanos S, Rodilla V et al (2019) Identification and characterization of cardiac glycosides as senolytic compounds. Nat Commun 10(1):1–12
Ventura A, Kirsch DG, McLaughlin ME, Tuveson DA, Grimm J, Lintault L et al (2007) Restoration of p53 function leads to tumour regression in vivo. Nature 445(7128):661–5
Wang E (1995) Senescent human fibroblasts resist programmed cell death, and failure to suppress bcl2 is involved. Cancer Res 55(11):2284–92
Wang L, Leite de Oliveira R, Wang C, Fernandes Neto JM, Mainardi S, Evers B et al (2017) High-throughput functional genetic and compound screens identify targets for senescence induction in cancer. Cell Rep 21(3):773–83
Wang C, Vegna S, Jin H, Benedict B, Lieftink C, Ramirez C et al (2019) Inducing and exploiting vulnerabilities for the treatment of liver cancer. Nature 574(7777):268–72
Wu C-H, van Riggelen J, Yetil A, Fan AC, Bachireddy P, Felsher DW (2007) Cellular senescence is an important mechanism of tumor regression upon c-Myc inactivation. Proc Natl Acad Sci 104(32):13028–33
Xu M, Pirtskhalava T, Farr JN, Weigand BM, Palmer AK, Weivoda MM et al (2018) Senolytics improve physical function and increase lifespan in old age. Nat Med 24(8):1246–56
Xue W, Zender L, Miething C, Dickins RA, Hernando E, Krizhanovsky V et al (2007) Senescence and tumour clearance is triggered by p53 restoration in murine liver carcinomas. Nature 445(7128):656–60

Yosef R, Pilpel N, Tokarsky-Amiel R, Biran A, Ovadya Y, Cohen S et al (2016) Directed elimination of senescent cells by inhibition of BCL-W and BCL-XL. Nat Commun 7(1):1–11

Zhu Y, Tchkonia T, Pirtskhalava T, Gower AC, Ding H, Giorgadze N et al (2015) The Achilles' heel of senescent cells: from transcriptome to senolytic drugs. Aging Cell 14(4):644–58

Zhu Y, Tchkonia T, Fuhrmann-Stroissnigg H, Dai HM, Ling YY, Stout MB et al (2016) Identification of a novel senolytic agent, navitoclax, targeting the Bcl-2 family of anti-apoptotic factors. Aging Cell 15(3):428–35

Part III
Understanding and Targeting the Senescent Surfaceome

Chapter 6
Targeted Senolytic Strategies Based on the Senescent Surfaceome

Akang E. Ekpenyong-Akiba, Marta Poblocka, and Salvador Macip

Abstract The clearance of senescent cells has the potential to become a therapeutic strategy that could be used in many pathologies, ranging from pulmonary fibrosis and diabetes to ageing itself. The initial genetic experiments performed in mouse models are now being recapitulated using chemical compounds that preferentially kill senescent cells, known as senolytics. Senolytic drugs hold an immense clinical potential but their lack of specificity could lead to side effects that would limit their use, especially when treating otherwise healthy aged individuals. Thus, it would be convenient to develop more targeted approaches to the elimination of senescent cells. Here, we discuss the ongoing efforts to design targeted senolytics, with special focus on the utilization of the extracellular epitopes displayed by the senescent surfaceome, and we summarize the avenues of research that have shown the most promising results so far: nanotechnology, cellular therapies and antibody-based drug delivery.

Keywords Senolytics · Senescent cell clearing · Surfaceome · Nanoparticles · Antibody-drug conjugates

6.1 The Road to Targeted Senolytics

According to the current theories of ageing, therapeutic clearance of senescent cells could prevent comorbidities and prolong lifespan in mammals (Naylor et al. 2013; de Magalhães and Passos 2017; Demaria 2017). This is because the progressive accumulation of senescent cells in tissues is thought to play an active role in ageing and in the reduction of healthspan (Deursen 2014; de Keizer 2016). This hypothesis is supported by the fact that genetic or chemical interventions to remove senescent cells

A. E. Ekpenyong-Akiba · M. Poblocka · S. Macip (✉)
Mechanisms of Cancer and Aging Laboratory, Department of Molecular and Cell Biology, University of Leicester, Leicester, UK
e-mail: sm460@le.ac.uk

S. Macip
Faculty of Health Sciences, Universitat Oberta de Catalunya, Rambla del Poblenou 156, Barcelona, Spain

D. Muñoz-Espin and M. Demaria (eds.), *Senolytics in Disease, Ageing and Longevity*, Healthy Ageing and Longevity 11,
https://doi.org/10.1007/978-3-030-44903-2_6

have been recently shown to improve organismal health in several models (Kirkland and Tchkonia 2015; Moreno-Blas et al. 2018).

The hallmark study of Baker and colleagues (Baker et al. 2011), which used an elegant transgenic mouse model to prevent the build-up of senescent cells, was the first to demonstrate this possibility in vivo. As shown in this paper, the apoptosis induced in p16^{Ink4a}-positive cells from muscle, fat and the lens of a *BubR1*$^{H/H}$ progeroid mice promoted normal functioning of these tissues, delayed the onset of age-related diseases, slowed down the progression of already established age-related disorders and extended animal lifespan. The same group later confirmed that clearance of p16 positive cells from normally aged mice had similar effects (Baker et al. 2016). Other mouse models followed, including one in which the clearance of p19^{ARF}-expressing cells from mice was achieved using a toxin receptor-mediated cell knockout system, which ameliorated age related decline in lung function (Hashimoto et al. 2016). Also, the elimination of senescent cells by a suicide gene-meditated ablation of p16^{Ink4a}-expressing senescent cells in the *INK-ATTAC* mouse model was shown to reduce overall hepatic steatosis (Ogrodnik et al. 2017). These ground-breaking experiments started a race to translate these observations made in transgenics into a clinically applicable strategy based on chemical compounds.

The first preliminary results were provided by a new class of chemicals, which were named 'senolytics' for their selective ability to induce death in senescent cells. For instance, a pharmacological approach with drugs that inhibit Bcl-2 and activate caspase 3/7 to partially eliminate senescent cells in vivo, was shown to decrease age-related phenotypes in wild type C57BL/6 mice and prolonged their lifespan (Zhu et al. 2015, 2016). The senolytic drug ABT263 was found to rejuvenate haematopoietic and muscle stem cells, and also improve healthspan in mice (Chang et al. 2016). Additionally, using a suicide gene as well as senolytic compounds, senescent cell clearance was found to prevent ageing-associated bone loss in mice (Farr et al. 2017). Senolytics have been shown to clear senescent type II pneumocytes and alveolar epithelial cells, suggesting and application in the treatment of idiopathic pulmonary fibrosis (Lehmann et al. 2017; Pan et al. 2017) and have also been proposed to be useful in Alzheimer's disease (Zhang et al. 2019) and tau-dependent pathologies (Bussian et al. 2018).

Many other senolytic compounds are currently being developed and tested (Fuhrmann-Stroissnigg et al. 2017), including some derived from natural products (Li et al. 2019), but so far they all share the issue of a lack of specificity. The action of these drugs are not exclusively limited to senescent cells and the potential side effects could be significant, although preliminary experiments in mouse models of ageing seem promising (van Deursen 2019; Xu et al. 2018). Moreover, the first in human clinical trial of senolytics using dasatinib plus quercetin have already shown positive results in alleviating physical dysfunction in IPF patients (Justice et al. 2019). In this context, it would now be convenient to develop more efficient strategies to clear senescent cells as the field of senolytics moves towards the first clinical applications (Justice et al. 2019). This is where the concept of targeted senolytics comes to relevance.

6.2 Recognizing Senescent Cells

The first step in the design of targeted senolytics approaches, and probably the most important, would be to devise the best possible system to identify the cells to be eliminated. Several characteristics that distinguish senescent cells from normal proliferating cells could be used for this purpose. As it was originally observed (Hayflick and Moorehead 1961), the main feature of these cells is being growth inhibited and not able to respond to growth factors or mitogens. Moreover, they express unique characteristic features, including morphological changes, senescence-associated β-galactosidase (SA β-Gal) positivity, increased expression of certain proteins (such as p21, p53, p16, p15, p27 and ARF), presence of senescence-associated heterochromatin foci (SAHF), enlarged and prominent nucleoli and a senescence-associated secretory phenotype (SASP) (Naylor et al. 2013; Hayflick 1965; Narita et al. 2003; Campisi 2011; Rodier and Campisi 2011; Rufini et al. 2013; Munoz-Espin and Serrano 2014; Pérez-Mancera et al. 2014).

To date, several of these markers have been extensively used to detect senescent cells in vitro and in vivo, alone or in combination. The fact that none of them is exclusive to the senescent state (Rodier and Campisi 2011; Munoz-Espin and Serrano 2014; Sikora et al. 2011) highlights the need to search for novel and more reliable substitutes for clinical applications. A universal senescence marker would need to be very robust and exclusive to the senescent state (Lawless et al. 2010; Matjusaitis et al. 2016). Also, a suitable biomarker should be able to distinguish between senescent and non-senescent cells in vitro and in vivo, its expression should be detectable regardless of sample preparation and it should be minimal in non-senescence states. Considering the limitations faced so far, it seems unlikely that such a marker will ever be found, although it would be expected that many that at least fulfil some of those criteria will be eventually identified.

The senescence program harbours transcriptional heterogeneity with a dynamic phenotype, which changes at variable intervals (Sharpless and Sherr 2015; Hernandez-Segura et al. 2017). Thus, the expression of thousands of genes is significantly altered in cellular senescence, and this could be the basis of new screens for markers. However, these changes are largely conserved within individual cell types and are specific to these cells only (Coppe et al. 2010). Of all the remarkable changes in gene expression observed during senescence, very few of them are stably expressed, specific and of great enough magnitude to be considered proper markers (Sharpless and Sherr 2015; Wang et al. 2009).

The fact that none of the markers has so far been proven to be sufficiently specific, having many false positives and negatives as well as cell type- and tissue-dependent expression, would be the main limiting factor when it comes to designing targeted senolytic strategies. Thus, a single strategy to clear all senescent cells from the body without noticeable side effects is a goal that may prove to be too ambitious. It will be more likely that a multi-faceted approach would be the most successful when translated to clinical use. We will first review some of the most used markers before discussing the surfaceome as a source of novel clinical opportunities.

6.2.1 *Staining of Senescent Cells*

The most used protocol to identify senescent cells ex vivo is the classic senescence associated β-galactosidase [SA-β-Gal] assay, in which senescent cells are stained blue at a pH of 6.0 (Dimri et al. 1995). Several human cells express high levels of the enzyme β-galactosidase upon becoming senescent, which is detectable cytologically or histochemically in freshly fixed cells or tissues at pH 6.0 (Campisi 2011; Munoz-Espin and Serrano 2014; Dimri et al. 1995). SA-β-Gal was originally observed in senescent fibroblasts and keratinocytes but not in terminally differentiated keratinocytes, quiescent fibroblasts or immortal cells, and an age-dependent increase was described in dermal fibroblasts and epidermal keratinocytes in human skin samples of different ages (Dimri et al. 1995).

Although the functional implications are not clear, the increased levels of lysosomal β-galactosidase activity in senescent cells is thought to be as a result of increased content and mass of the lysosomes (Kurz et al. 2000a). However, it is worthy of note that almost all cells show endogenous lysosomal β-galactosidase activity at pH 4.0 and certain cell types, such as human epithelial cells and mouse fibroblasts, stain with some intensity for SA-β-Gal at pH 6.0 (Itahana et al. 2013). Moreover, when cells are maintained at confluency for long periods, a false positive density-induced SA-β-Gal activity can be sometimes detected, and serum starvation is also known to produce false positives (Itahana et al. 2007; Campisi et al. 2009; Evangelou et al. 2017). Despite this, SA-β-gal is still considered to be one of the best senescence markers available (Kurz et al. 2000b) and it has been used as the basis for the first nanoparticle-based targeted senolytics (see below).

Senescent cells also stain positive for the histochemical Sudan Black-B (SBB), which stains the lysosomal aggregate and age-pigment lipofuscin (Terman and Brunk 2004; Georgakopoulou et al. 2013). SBB stains senescent cells in tissues regardless of sample preparation and can be used on formalin-fixed, paraffin-embedded tissues, giving it an advantage over SA-β-Gal, which requires the use of fresh cells or tissue samples (Munoz-Espin and Serrano 2014; Georgakopoulou et al. 2013).

6.2.2 *The Senescence-Associated Secretory Phenotype*

Senescent cells secrete factors collectively known as the senescence associated secretory phenotype (SASP), which includes inflammatory cytokines, chemokines, matrix remodelling proteins, damage-associated molecular pattern proteins (DAMPs), growth factors (such as interleukins 6, 7, and 8), Macrophage Inflammatory Protein 3α (MIP-3α), Growth Regulated Oncogene alpha (GROα), Monocyte Chemoattractant Proteins 1 and 2 (MCP-1 and MCP-2), Insulin-like Growth Factor Binding Protein (IGFBP) and Hepatocyte Growth Factor (HGF), among others (Kuilman and Peeper 2009; Young et al. 2013; Salama et al. 2014; Laberge et al. 2015; Maciel-Barón et al. 2016; Kirkland and Tchkonia 2017; McHugh and Gil

2018). It is thought that these molecules are able to disrupt tissue structure and function, being then responsible in part for chronic age-related diseases and cancer progression (Cahu 2013).

The mechanisms involved in the generation of the SASP are still being investigated. It has been found that the production of SASP is not tied to characteristics of senescence such as enlarged cell morphology and SA-β-Gal expression or even cell cycle arrest (Laberge et al. 2015; Herranz et al. 2015; Wang et al. 2017). Also, ectopic expression of $p16^{INK4a}$ and $p21^{WAF1/Cip1/Sdi1}$ can induce senescence without production of SASP (Coppe et al. 2011). It has been suggested that the SASP is not an essential feature of senescent cells and can develop independently of $p16^{INK4a}$ status (Coppe et al. 2011). The SASP may thus be a result of severe DNA damage, which can activate secretion independently of p53 (Coppe et al. 2008). The SASP is also regulated by microRNAs, the mTOR, cGAS/STING and JAK/STAT signalling pathways, the cytokine receptor for interleukin–1 [IL-1] and IL-8 chemokine receptor CXCR2, as well as transcription factors such as NF-kB (Coppe et al. 2011). It has also recently been reported that NOTCH1 plays a vital role as both a regulator of the composition SASP and a regulator of juxtacrine signalling within the context of oncogene-induced senescence (OIS) (Hoare and Narita 2017; Ito et al. 2017; Hoare et al. 2016).

The SASP elements could be good markers to identify senescent cells, given the fact that they constitute a protein profile not observed in normal cells. Pro-inflammatory cytokines such as IL-6 and IL-8 seem to be the most conserved aspects of the secretome, and they play key roles in the maintenance of the SASP (Lasry and Ben-Neriah 2015; Kuilman et al. 2008). However, the SASP varies significantly between tissues and different senescence triggers, and the molecular regulation of the SASP program is complex and multifaceted in space and time (Wang et al. 2017; Malaquin et al. 2016). Moreover, it has been shown that mitochondrial dysfunction can lead to a type of cellular senescence, known as mitochondrial dysfunction associated senescence (MiDAS) (Wiley et al. 2016; Gallage and Gil 2016; Herranz and Gil 2016; Wiley and Campisi 2016), which produces a distinct secretory phenotype (Wiley et al. 2016). Also, using models of stress induced premature senescence (SIPS), proteasome inhibition induced premature senescence (PIIPS) and replicative senescence (RS), it was shown that the SASP due to PIIPS was substantially different, with significantly lower amounts of cytokines and chemokines when compared to the SASP due to RS or SIPS (Maciel-Barón et al. 2016). Thus, markers based on the SASP are unlikely to be specific enough for targeted senolytics, although a profile of secreted factors specific to certain type of senescent cells could alternatively be used in certain situations.

6.2.3 Nuclear Features

Cellular senescence is associated with an altered chromatin assembly, forming easy to visualize heterochromatin regions called senescence associated heterochromatic foci

(SAHF) (Narita et al. 2003; Cichowski and Hahn 2008). These are highly condensed regions of chromatin characterised by a build-up of histone H3 that is tri-methylated at lysine 9 (K9M-H3), as well as heterochromatin proteins, such as high-mobility group A (HMGA) proteins, macroH2A and heterochromatin protein 1 (HP1) (Salama et al. 2014; Adams 2007; Kuilman et al. 2010). In proliferating and quiescent cells, the euchromatin markers K9Ac-H3 (histone H3 acetylated on lysine 9) and K4M-H3 (histone H3 methylated on lysine 4) homogeneously stain the DNA. The SAHF usually lack lysine 9 acetylated H3 (K9Ac-H3) and lysine 4 methylated H3 (K4M-H3) but rather are enriched in K9M-H3 (Narita et al. 2003).

$p16^{INK4A}$, Rb, p53, interleukin-6 and C/EBPβ are thought to be vital for the formation of SAHF, which can be prevented by interfering with the signalling of these pathways (Narita et al. 2003; Cichowski and Hahn 2008; Kuilman et al. 2010). It is likely that SAHF contribute to the collective changes in gene expression observed in senescent cells (Young et al. 2013; Mduff and Turner 2011). The SAHF are mainly found only in human cells, and even taking this in consideration, they are not consistently expressed in all models (Narita et al. 2003; Swanson et al. 2013), which limits their use as markers.

Senescent cells also present a senescent-associated distension of satellites (SADS), a display of constitutive peri/centromeric satellite heterochromatin decondensation, which are not exclusive to either the p53–p21 or the p16 ^{INK4A}/Rb pathways. These SADS occur prior to and independently of SAHF formation (Ogrodnik et al. 2017; Swanson et al. 2013).

6.2.4 Molecular Markers

Although it has been over half a century since it was first described (Hayflick and Moorhead 1961), the molecular pathways involved in senescence are yet to be fully understood. Senescence can be signalled through various routes, many of which activate p53 and/or Rb and the cyclin dependent kinase [CDK] inhibitors $p16^{INK4A}$, $p15^{INK4B}$, $p21^{WAF1/Cip1/Sdi1}$ and p27 (Munoz-Espin and Serrano 2014; Campisi 2005). Upregulation of these and other genes has been used as markers of senescence, although none of them is specific of the senescent phenotype. Nevertheless, p16 is currently the most widely used genetic marker of senescence and many in vivo experiments have equated p16-expresing cells to senescent cells (Baker et al. 2011; Baker et al. 2016; Palmer et al. 2019).

The changes in gene expression patterns of senescent cells also affect cell cycle regulatory genes, extra-cellular matrix remodelling genes, as well as genes involved in cytokine signalling and inflammation (Mduff and Turner 2011). These various senescence-associated gene expression changes are specific to and conserved within individual cell types (Coppe et al. 2010). It has also been observed that mTOR activation is necessary for cellular senescence to take place, as mTOR enables transition from cell cycle arrest and quiescence to senescence (Blagosklonny 2014; Leontieva and Blagosklonny 2017). Rapamycin, an inhibitor of the mTOR pathway, was found to slow down cellular senescence and prevent the accompanying irreversible loss of proliferative capacity (Demidenko et al. 2010; Xu et al. 2014).

It is also interesting to consider that ontologic and gene expression analysis of skin from Caucasian females between the ages of 20 years and 70 years revealed changes in the expression of thousands of genes with age (Kimball et al. 2018). Studies have shown that the expression of cellular damage related genes such as inflammatory or stress response genes increase with age while biosynthetic and metabolic genes expression decrease with age (Kimball et al. 2018; Edwards et al. 2007).

In order to facilitate research on the mechanisms of gene regulation in cellular senescence, a database of senescence associated genes is greatly needed. To this end, Dong and colleagues have established the Human Cellular Senescence Gene Database (HCSGD), using a combination of data from published literature sources, gene expression profiling as well as protein-protein interaction networks (Dong et al. 2017). Profiles of gene expression taken from databases such as HCSGD could provide useful alternatives to single-gene markers, which are the ones mostly used now and heavily rely on the p16/Rb and the p53/p21 pathways.

6.2.5 The P16/Rb Axis

Rb is an important gatekeeper during cell cycle progression through the G1 phase and its activity is tightly controlled by several post-translational modifications, such as phosphorylation, acetylation and ubiquitination (Campisi 2005; Takahashi et al. 2007). $p21^{Cip1/Waf1/Sdi1}$ and $p16^{INK4A}$ inhibit the kinases that phosphorylate Rb, leading to the accumulation of its active, hypo-phosphorylated form (Xu et al. 2014; Herbig et al. 2004; di Fagagna and Campisi 2007). The two products of the *INK4a/ARF* locus, $p16^{Ink4a}$ and $p19^{Arf}$ [$p14^{ARF}$ in humans], are key tumour suppressors which regulate the activities of p53 and Rb and are expressed from partly overlapping nucleotide sequences read in alternative reading frames (Lowe and Sherr 2003). $p14^{ARF}$ increases the growth inhibitory functions of p53 by sequestering its negative regulator, Mdm2. Both $p16^{INK4a}$ and $p19^{ARF}$ could be used as markers, since they have been found to accumulate in many senescent cells and their overexpression also promotes senescence (Lundberg et al. 2000). Mutations affecting INK4a or ARF can compromise senescence on various levels depending on the cell type and species (Narita et al. 2003; Lowe and Sherr 2003; Park and Sin 2014).

As previously mentioned, $p16^{INK4A}$ is considered a leading marker for indicating the presence of senescent cells, as most senescent cells express it (Campisi 2011; Taniguchi et al. 1999; Simboeck and Di Croce 2013). The levels of $p16^{INK4A}$ have even been used as a biomarker of ageing in humans as it is seen to increase with ageing—up to 7 fold in some human tissues and up to 30 fold in mouse tissues (Wang et al. 2009; Krishnamurthy et al. 2004; Collado et al. 2007; Romagosa et al. 2011; Sherr 2012). However, p16 does not seem to be a key regulator of development, as it is not expressed during embryogenesis, even though senescence is known to play an important role during these stages (Guney and Sedivy 2006). Instead, this is usually mediated by p21 in a p53-independent manner (Storer et al. 2013; Munoz-Espin et al. 2013).

Immunostaining of benign tumours show a positive p16^{INK4A} expression and a negative or very low Ki67 index (a marker of proliferation), while malignant tumours often stain positive for Ki67 but negative for p16^{INK4A} (Romagosa et al. 2011; Collado et al. 2005). However, in high grade malignant tumours, where there are alterations in the p16^{INK4A}/Rb pathway, a high p16^{INK4A} as well as a high Ki67 immunostaining has been observed. In human tumours, the overexpression of p16 could either be a result of OIS, as seen in benign or pre-malignant lesions, or a result of an Rb pathway failure, as seen in some malignant lesions. Hence, p16^{INK4A} is often used as a prognostic marker for several cancers in order to grade or distinguish between benign and malignant lesions (Romagosa et al. 2011). Staining for p16 in vivo has been quite challenging and requires robust positive and negative controls.

6.2.6 The P53/P21 Axis

p53 is a cell cycle regulator and guardian of the genome, and an important tumour-suppressor protein involved in senescence (Munoz-Espin and Serrano 2014). p53 induction after genotoxic stress and DNA damage can lead to either DNA repair, transient cell cycle arrest, senescence or apoptosis, depending on various factors (Sionov and Haupt 1999; Macip et al. 2002; Bieging et al. 2014). The activity of p53 as well as levels of its downstream effector p21 increase during late passage of cells, leading to a slowdown in proliferation (Murray-Zmijewski et al. 2008; Kastan et al. 1991; Lane 1992; Levine 1997). It has been proposed that p53 and p21 are not required for the maintenance of senescence, even though they are important in inducing the senescence-like proliferation arrest (Chang et al. 1999). Linked to this, p16 expression has been proposed to be a second barrier to cell proliferation during senescence to ensure irreversibility of the arrest (Macip et al. 2002; Beausejour et al. 2003).

p21$^{Waf1/Cip1/Sdi1}$ is a well characterised transcriptional target of p53 and links the p53 and Rb pathways to enhance tumour suppression (Takahashi et al. 2007). p21 is a dual inhibitor of proliferating cell nuclear antigen (PCNA) and cyclin dependent kinases, both crucial in the cell cycle, and is significantly overexpressed during cellular senescence (El-Deiry et al. 1993; Waldman et al. 1995; Brugarolas et al. 1995). In late passage human diploid fibroblasts, an increase in p21$^{Waf1/Cip1/Sdi1}$ expression was observed, and targeting the deregulation of p21$^{Waf1/Cip1/Sdi1}$ in these cells increased their lifespan. In p21-null mice, p53 is unable to cause G1 cell cycle arrest upon DNA damage (Shiohara et al. 1994; Abbas and Dutta 2009; Romanov and Rudolph 2016). Expression of p21 can also induce apoptosis and this is thought to be mediated by reactive oxygen species (Masgras et al. 2012). Genomic and phenotypic analyses using p21-inducible, p53-null, malignant as well as near-normal cellular models has shown that a subpopulation of proliferating cells emerged after an initial senescence-like phase, which expressed p21 and exhibited increased genomic instability, aggressiveness and resistance to chemotherapeutic agents (Galanos et al. 2016; Georgakilas et al. 2017).

The Bruton's tyrosine kinase (BTK) is a dual-specificity non-receptor tyrosine kinase belonging to the Tec family that is vital for B cell maturation (Mohamed et al. 2009) and it has recently been reported to be a part of the p53 senescent pathway, phosphorylating p53 and modulating its responses (Althubiti et al. 2016). BTK was found to be expressed in response to damage and induces phosphorylation of p53 at serine 15 at the N-terminus, increasing its protein levels and activity. In addition, it was found that BTK binds to and phosphorylates MDM2, facilitating its loss of ubiquitination activity and further stabilising p53 and also plays a role in the p73 pathway (Rada et al. 2017, 2018). Inhibition of BTK reduced the expression of p53 and interfered with the upregulation of p53 target genes, leading to an impairment in the induction of p53-mediated senescence. Thus, BTK has been proposed to be useful as a novel marker of p53-induced senescence (Rada et al. 2018).

6.3 The Senescent Surfaceome

In the search for better markers of senescence, the profile of proteins present on the surface of these cells, known as the senescent surfaceome, has emerged as a potentially interesting alternative (Table 6.1). In general, the surfaceome could be defined as the panel of all plasma membrane proteins of a given cell type that exhibit at

Table 6.1 Main validated members of the senescent surfaceome

Surfaceome protein	References
CD264	Collado et al. (2005), Madsen et al. (2017)
NOTCH1	Hoare et al. (2016)
NOTCH3	Cui et al. (2013), Althubiti et al. (2014)
DEP1	Althubiti et al. (2014)
NTAL	Althubiti et al. (2014)
EBP50	Althubiti et al. (2014)
STX4	Althubiti et al. (2014)
VAMP3	Althubiti et al. (2014)
ARMCX3	Althubiti et al. (2014)
B2M	Althubiti et al. (2014)
LANCL1	Althubiti et al. (2014)
VPS26A	Althubiti et al. (2014)
PLD3	Althubiti et al. (2014)
DPP4	Kim et al. (2017)
SCAMP4	Kim et al. (2018)

least one amino acid residue at the extracellular space. The majority of the human surfaceome consist of all types of transmembrane proteins, single and or polytopic α-helical or β-sheet transmembrane proteins (Bausch-Fluck et al. 2018). Like other senescent markers, those in the surfaceome are expressed varyingly but consistently across different tissues and conditions and are unlikely to be universal. One of the most interesting features of the novel markers of senescence present in the senescent surfaceome are their extracellular epitopes, which could be used for targeted therapies, as discussed below.

The cell surface is a hub that coordinates information transmission to and from the outside environment. The proteins present at the cell surface are crucial factors connecting cellular signaling networks and determining the cell's interaction and communication with its neighbourhood. Thus, the surfaceome consists of a broad range of receptors, transporters, channels and enzymes, and each of them could be a valuable diagnostic marker of cellular senescence, as well as a potential therapeutic target.

6.3.1 Current Known Members of the Senescent Surfaceome

CD264 (also known as TRAIL-R4, DcR2, TRUNDD and TNFRSF10D) was one of the first proteins preferentially expressed on the surface of senescent cells identified, and was accordingly proposed as a biomarker of cellular aging (Madsen et al. 2017). CD264 is a decoy receptor for tumour necrosis factor-related apoptosis- inducing ligand (TRAIL) (Falschlehner et al. 2007). Due to the presence of a truncated intercellular death domain, CD264 competes for TRAIL binding and prevents the formation of a death-inducing signalling complex (Pan et al. 1998; Lee et al. 2005). CD264 upregulation was observed in senescent premalignant tumours induced by *ras*, whereby there was downregulation of CD264 in the malignant *ras*-induced tumours that had escaped senescence (Collado et al. 2005). CD264 was also further detected in stress-induced senescence of non-malignant smooth muscle cells (Zhu et al. 2011) and, in contrast, showed to be not upregulated in replicative senescence (Collado et al. 2005).

NOTCH1 was one of the first membrane-associated proteins shown to play a role in senescence, specifically as a key determinant of SASP expression (Hoare et al. 2016). NOTCH3, a member of the Notch family receptors, which also reported to have tumour suppressor functions and is overexpressed in senescent fibroblasts (Cui et al. 2013; Althubiti et al. 2014), is another classic member of the senescent surfaceome that has been known for a while.

A mass spectrometry screening of the membrane fraction of senescent cell lysates revealed a new subset of proteins that were preferentially expressed in the membranes of senescent cells over their proliferative counterparts (Althubiti et al. 2014), in what, to the best of our knowledge, was the first systematic screening of the senescent surfaceome. Later, gene expression profiling analysis of G protein-coupled receptor kinase (GRK) 4-induced senescent HEK293 cells showed a further differential expression of a new set of 17 senescence-related genes compared to control cells (Xiao et al. 2017). Also, a more comprehensive screen of membrane-associated proteins, this time focusing on oncogene-induced senescence, was later published as part as the characterization of the involvement of Notch signalling in SASP (Hoare et al. 2016). In this study, 521 proteins were identified on the surface, 32 increasing in senescence and 135 being downregulated instead. Most of these candidates were not found in the first screen described above, but some were present in both (such as NTAL). This confirms that different cell types and methods of induction will lead to completely different surfaceomes, although some elements may still be common in more than one model.

It has to be taken in consideration that the plasma membrane proteome is only a subset of all membrane proteins within the cells. Thus, in any proteomic screen of the senescent surfaceome, it would be challenging to distinguish between intercellular membrane proteins (i.e. endoplasmic reticulum, Golgi), the ones bound to the intracellular part of the plasma membrane (such as BTK (Althubiti et al. 2016)) and the ones present in the plasma membrane that actually cross over the cell surface. Thus, a biochemical validation and characterization of the individual hits of the proteomic screen would always be necessary.

One of the approaches to identifying proteins uniquely expressed on the surface of senescent cells mentioned above was done in our lab and focused on previously described genetic models of senescence (Althubiti et al. 2014). EJp16 and EJp21 are derivatives of a bladder carcinoma cell line (EJ) with a tetracycline regulatable expression system for p16 and p21, respectively, that entered senescence after 3–4 days of induction (Macip et al. 2002; Chang et al. 1999). EJ-based models allow for the establishment of a genetically induced senescent state due to the prolonged expression of core genes from the main pathways responsible for the induction and maintenance of the phenotype, without any additional effects caused by the trigger of senescence itself, as seen in more physiological models. Thus, the proposed panel of surfaceome proteins presented in this study can be linked to the activation of either the p53/p21 or the pRb/p16 pathways and could help determine which markers are activated preferentially in response to each of them. In this regard, 107 proteins were reported to be exclusively present on the plasma membrane in EJp21 senescent cells and 132 in EJp16 were identified.

Ten proteins that had not been previously linked with cellular senescence were validated using immunofluorescence microscopy and Western blots: DEP1, NTAL, EBP50, STX4, VAMP3, ARMCX3, B2M, LANCL1, VPS26A and PLD3. The biological role of the proposed markers is heterogeneous. Some, like DEP1 or EBP50, have been identified as important players in the regulation of cancer progression (Liu et al. 2018; Iuliano et al. 2010). Others, including VPS26A and PLD3, have been found to be involved in Alzheimer's disease, providing a direct link to an age-associate disorder (Tan et al. 2019; Choi et al. 2018). Some of them, like VPS26A, VAMP3, PLD3 and STX4, may have a role in vesicle trafficking and therefore could potentially contribute to some aspect of the SASP (Veale et al. 2011; Caceres et al. 2016; Bugarcic et al. 2011). All of them showed a tissue- and pathway-specific profile. For instance, in tissue sections collected from a ^{V600E}BRAF lung adenoma mouse model, previously showed to have a high percentage of senescent cells (Mercer et al. 2005; Carragher et al. 2010), DEP1, STX4, B2M and NTAL were preferentially detected. Analysis of human naevi also revealed positive staining for the same markers.

An alternative approach to these three screens described, focused only on replicative senescence, using proliferating passages of WI-38 human diploid fibroblasts [population doubling level (PDL) = 23) compared with their senescent counterparts, in which senescence was induced by extensive cell culture (PDL = 59) (Kim et al. 2017). Following the fractionation of membrane-associated proteins, the samples were analysed using mass-spectrometry as well. The results revealed 2 groups of proteins: those abundant in intracellular membranes [i.e. endoplasmic reticulum, mitochondria] and those in the cell surface. There were 118 proteins at the overlap of these two groups and this list was narrowed down to 15. Out of these, the leading candidate marker to be further studied was DPP4 (CD26). DPP4 is responsible for deactivation of the glucose-dependent insulinotropic peptide (GIP) and glucagon-like peptide-1 (GLP-1) (Zhong et al. 2013). GIP and GLP-1 hormones are beneficial in diabetes as they induce the pancreatic β cells to release insulin after a meal, which results with the suppression of the glucose level in the blood (Wu et al. 2016). However, the role of the elevation of DPP4 in senescent cells remains unclear and further tests will need to be done to confirm its impact on biological ageing.

The expression of DPP4 in the membrane fraction of senescent but not proliferating cells was further validated by Western blot and RT-qPCR. Validation experiments included also different cell models (WI-38, IMR-90, HUVEC, HAEC) subjected to ionizing radiation (IR). The expression of DPP4 was confirmed in IR models, although with more modest effects than in replicative senescence. The expression level of DPP4 was next validated with other triggers of stress-induced senescence, including doxorubicin and OIS in MEFs expressing the oncogene $HRAS^{G12V}$. DPP4 was upregulated in all of them, demonstrating that is a widely expressed marker.

Besides DPP4, SCAMP4 has also been proposed by the same group to be preferentially expressed on the surface of senescent cells (Kim et al. 2018). They confirmed its membrane localization and its abundancy in different senescent models (Lee et al. 2005; Kim et al. 2018). Western blot analysis revealed increased level of SCAMP4 in replicative senescence, together with the upregulation of other family proteins (SCAMP1-3). The SCAMP4 upregulation was also successfully determined in models of SIPS, including doxorubicin-induced and oncogene-induced senescence or HAECs and HUVECs cells subjected to IR. However, SCAMP4 mRNA levels remained unchanged in all the tested models, suggesting a post-translational effect. Testing its half-life in WI-38 fibroblasts after treatment with the translational inhibitor cycloheximide showed a SCAMP4 decline in proliferating cells, without the changes in the protein stability in senescent cells. After proteasome inhibition, SCAMP4 protein was found to accumulate in proliferating but not in senescent cells. In turn, autophagy inhibition revealed SCAMP4 accumulation in both proliferating and senescent cells, suggesting proteasome but not autophagy involvement in the control of protein levels. This was further supported by previous findings on SCAMP4 ubiquitination in Lys4 and Lys185 in lung cancer (Wu et al. 2015).

It is worth mentioning that the level of SASP also increased after proteasome inhibition and that SCAMP4 protein stabilization occurred before the induction of SASP. On the other hand, the secretion of many SASP factors decreased after SCAMP4 silencing (IL6, IL8, GDF-15, IL1B, MIF and CCL2), together with the reduced level of core senescence markers (p53, p21 and SA-β-gal activity). These findings were further confirmed by short-term and stable overexpression of SCAMP4 in proliferating cells, in which there was increased mRNA and protein level of several senescent and SASP markers (p16, IL1A, IL1B, IL6, and IL8). SCAMP4 overexpression resulted also in increased SA-β-gal activity and reduced [3H]-thymidine incorporation.

6.3.2 Clinical Relevance

The potential clinical implications for the senescence surfaceome have already started to be discussed in the literature. For instance, CD264 has been proposed to be specific for the detection of senescent mesenchymal stem cells (MSC) population. The regenerative capacity and therapeutic applications of MSC are well documented (Barrilleaux et al. 2006; Caplan 2007). However, the main limiting aspect that impedes the standardization and optimization of MSC therapies for clinical use is their cellular heterogeneity, which is considered the main cause of strong variability in treatment outcomes. The immunolabelling of surface antigens can be an effective approach for standardization of MSC composition, for instance before transplantation. The identification of aging cells among the MSC population would facilitate the enrichment of multipotent progenitors by negative selection, slowing cellular aging during MSC expansion by the improvement of culture conditions, and allowing for selective elimination of aging MSC. The proliferation potential of MSC cultures has been shown to be inversely related to the CD264 cellular content. The expression of CD264 was also tested in replicative senescence by serial passage of MSCs, and the number of $CD264^+$ cells increased from 30% at the passage 3–7 to 90% by passage 14.

After these initial studies, more detail on the senescent surfaceome will no doubt appear in the future. As the profiles of extracellular epitopes of different types of senescent cells become clearer, new markers will emerge, which could be used for diagnostic and therapeutic purposes. Apart from that, the surfaceome will also generate information on novel effectors and modulators of senescence, like the first reports have shown. Thus, the proteomic screen of the membrane of senescent cells will not only provide a list of extracellular epitopes that could be used for detection and targeting of the cells, but will likely uncover new important members of the senescent pathways.

6.4 Targeted Senolytics as a New Anti-senescence Approach

Targeted therapies have been developed for the management of various cancers and they have proven over the years to be more specific and more effective than the traditional approaches. Unlike classic antineoplastic drugs, which are highly cytotoxic, targeted therapies can inhibit specific molecular targets of interests without acting in a non-specific cytotoxic manner. Although not without adverse effects, targeted therapies are often better tolerated and generally more efficient (Walter and Ahmed 2017; Ke 2017). This principle can be applied to other diseases as well (Florence and Lee 2011; Daste et al. 2016; Pauliah et al. 2018). Indeed, targeted therapies tailored to individual patients, with new approaches towards disease assessment and dosing, is likely to be the choice for most treatments in the near future, provided that the technical and financial hurdles are resolved.

Senescence-related conditions, including ageing, could benefit from targeted drugs as well. In this instance, the goal would be to eliminate senescent cells in a more specific way than the current available senolytics, which could be compared to blunt treatments, such as classic chemotherapy for cancer. Several avenues for these novel targeted senolytics are currently being developed, some of them based on epitopes present on the senescent surfaceome, some against intracellular targets (Fig. 6.1). The three main categories of potential targeted therapies for clearance of senescent cells that have been most investigated are based on nanoparticles, the immune system and monoclonal antibodies.

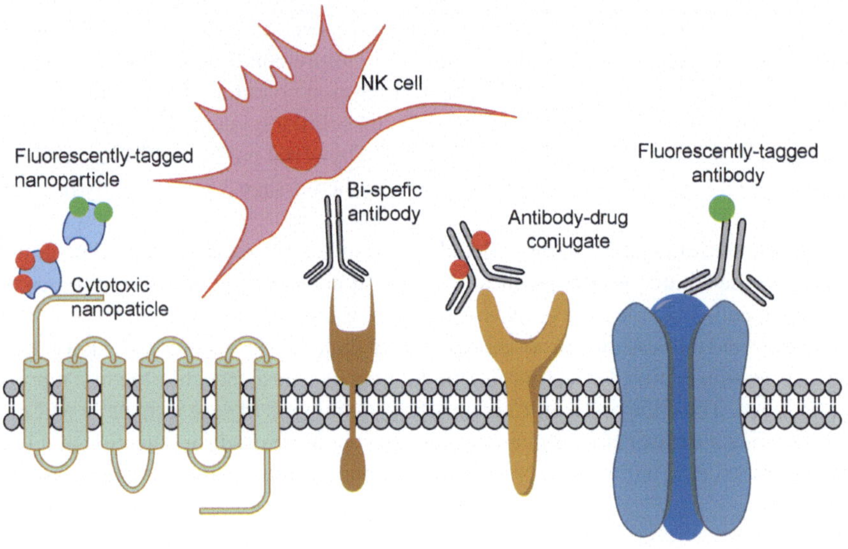

Fig. 6.1 Strategies for targeted senolysis using the senescent surfaceome. Schematic representation of the type of proteins present in the plasma membrane of senescent cells, together with the different approaches to targeted senolysis described in this chapter

6.4.1 Nanoparticles

Unlike cancer, senescence lacks the concept of an oncogene-like abnormal protein that triggers the phenotype. It is, instead, the result of the interaction of different networks of physiological signals that respond to damage. Thus, it is unlikely that drugs that inhibit a single protein to induce apoptosis only in senescent cells will ever be identified. One alternative approach for targeted senolytics could be to design a specific mechanism of toxic drug delivery that would preferentially reach senescent cells. As we will discuss below, this kind of tools have been successfully used in cancer. Nanotechnology has already provided proof of principle examples that this could also be a relevant avenue for senescence.

The first published report of nanoparticles applied to targeting senescent cells used a galactosidase-encapsulated drug (Munoz-Espin et al. 2018). Sugar-coated beads of ~100 nm diameter produced using MCM-41, a silica porous scaffold, had been previously shown to be internalized by endocytosis and digested in the lysosomes. This approach was modified using a 6-mer galacto-oligosaccharide to coat the silica scaffold and containing in their core drugs such as the cytotoxic doxorubicin or the senolytic navitoclax. Due to their high levels of lysosomal β-galactosidase, these nanoparticles only released the drug in senescent cells, which were able to lyse the galactosidase coat, while the cargo was protected in other cells. The same principle was modified to detect senescent cells, using a fluorescent cargo for the nanoparticles instead of drugs.

Importantly, it was shown that this drug delivery system worked in vivo. Toxic or senolytic drugs were specifically released in xenografts of cancer cells after senescence-inducing treatments, which importantly reduced the size of the tumours formed in the mice. Also, a mouse model of bleomyicin-induced pulmonary fibrosis responded to the nanoparticles with a severe decrease of the associated respiratory symptomatology and a reduction of damage in the tissues, consistent with the clearance of the senescent cells known to aggravate this disease (Munoz-Espin and Serrano 2014).

β-galactosidase was also used as a senolytic target by a later study that showed that enzyme-instructed peptide self-assembly (EISA) can lead to the formation of nanofibers and hydrogels specifically in senescent cells (Xu et al. 2019). These nanofibers inhibit the expression of senescence-driving proteins [p53, p21 and p16] and eventually cause death by apoptosis. This is the first example of the use of peptide-based supramolecular nanomaterials directed by the high levels of an enzyme (β-galactosidase in this case) to eliminate senescent cells.

Other delivery systems used porous calcium carbonate nanoparticles ($CaCO_3$) loaded with the mTOR inhibitor rapamycin, which then blocked SASP (Thapa et al. 2017). Similarly to the previous examples, these nanoparticles were coated with a conjugate of lactose to increase specificity based on β-galactosidase enzymatic activity. On top of that, another layer of specificity was added by tagging them to a monoclonal antibody against CD9, which some senescent cells express. These nanoparticles were shown to be senolytic in human fibroblasts.

Another alternative to these senolytic strategies that bypass the need for an antibody would be the use Molecularly Imprinted NanoPolymers (nanoMIPs), which are synthetic antibody-like nanoparticles with high stability, cost effectiveness, ease of preparation and adaptation (Sellergren and Allender 2005; Piletsky et al. 2006; Vasapollo et al. 2011; Canfarotta et al. 2016a, b) to deliver drugs only to cells that express a certain surface marker. NanoMIPs are synthesized from methacrylic or acrylic monomers polymerized by creating a matrix around a template target molecule, which is later removed, leaving behind cavities complementary to its shape and functional groups. The produced nanoMIPs are thereafter able to selectively rebind to their target molecule (Canfarotta et al. 2016a; Poma A, Turner APF, Piletsky SA. Advances in the manufacture of MIP nanoparticles 2010).

The advantage of this approach is that it relies on high affinity and specificity of the final product toward their targets, and the binding site affinities exhibit homogenous distribution (Canfarotta et al. 2016c). These "plastic antibodies" have recently found application in the detection of different cancers (Sengupta and Sasisekharan 2007; Voigt et al. 2014; Tyagi et al. 2016; Cecchini et al. 2017), in sustained drug release and drug delivery systems using small molecules (Sellergren and Allender 2005; Cunliffe et al. 2005; Puoci et al. 2011; Tieppo et al. 2012; Kempe et al. 2015; Luliński 2017), and also for transport across an in vitro model of the blood-brain barrier (Dadparvar et al. 2011). This suggests that nanoMIPs could be used as a system to deliver drugs to senescent cells, providing an extracellular marker as an epitope for the nanoparticles to bind.

NanoMIPs were created in our lab to specifically recognize one of the previously identified members of the senescent surfaceome, β2-Microglobulin (B2M) (Ekpenyong-Akiba et al. 2019). These MIPs, fluorescently-tagged, were tested in vitro and the selective detection of senescent cells using FACS was confirmed. Furthermore, it was observed that the nanoMIPs preferentially accumulated on the membrane surface of these cells, consistent with their selective detection of the extracellular epitope. The internalization of nanoMIPs was observed after binding to the marker, together with its accumulation in the cytosol forming perinuclear aggregates.

In order to analyse their potential for the detection of senescent cells in vivo, the B2M nanoMIPs were conjugated with a DyLight 800 NHS ester and injected intravenously into mice. A strong fluorescent signal was detectable in old animals, suggesting that the nanoMIPs were selectively binding to senescent cells, specifically in the gastrointestinal tract. These proof-of-principle experiments demonstrate that B2M nanoMIPs accumulate in certain senescent cells in vitro and in vivo and supports the hypothesis that nanoMIPs could be a promising approach for targeted drug delivery into senescent cells. To test this, B2M nanoMIPs were loaded with different drugs, including dasatinib, to minimize the off-target toxicity and impact on other cells. The dasatinib-containing nanoMIPs significantly reduced the senescent cells viability comparing to the proliferating cells and the effect was significantly stronger than the activity of dasatinib as a drug alone. Of note, nanoMIPs had no cytotoxic effects on their own, as the long-term exposure of the B2M targeted nanoparticles did not affect the viability and survival of senescent and control cells. This had been previously shown for other nanoMIPs (Canfarotta et al. 2016b; Hoshino et al. 2010).

Also, the administration of B2M targeted nanoMIPs to mice did not significantly affect the health of the animals during 14-days follow-up after treatment despite the route of injection (oral gavage, intraperitoneally and intravenously).

These experiments show that nanoparticles, in different formats, are a stable and inexpensive tool for the detection and targeting of senescent cells. The senolytic nanoparticles tested so far have been shown to be biocompatible, non-toxic to cells in the long-term and non-toxic to animals in the short-term. However, not much is currently known about the bio-distribution and clearance of nanoparticles from the body when applied in vivo. For instance it has been proposed that nanoMIPs are cleared from the animal's blood by the liver's mononuclear phagocytes (Hoshino et al. 2010). Moreover, technical and economic limitations may occur when nanoparticle production needs to be scaled for clinical use. Further studies to study these and other parameters related to biosafety will be needed before nanoparticles could be used in humans, but these tools can already provide useful pre-clinical information on potential targets for targeted clearance of senescent cells.

6.4.2 Antibody-Drug Conjugates

A more readily translatable approach to the same principle would be to use antibodies instead on the nanoparticles for drug delivery. Antibody-drug conjugates (ADCs) are monoclonal antibodies to which cytotoxic drugs are bound through a chemical linker to reduce systemic toxicity and increase the therapeutic benefit for patients (Strohl and Strohl 2012; Casi and Neri 2012; Perez et al. 2014; Gébleux and Casi 2016; Kumar et al. 2017). The concept of ADCs was first introduced over a century ago by Paul Ehrlich, a German physician who proposed the use of a targeting agent to selectively deliver a cytotoxic drug to a tumour (Perez et al. 2014; Ehrlich 1906). From the introduction of the idea by Ehrlich to date, advances have been made in the development of ADCs for cancer therapy (Trail et al. 1993; Sievers and Senter 2013; Sau et al. 2017; Donnell et al. 2017). So far, they have only been used to kill cancer cells, but their therapeutic potential could be exploited for other targets, such as senescent cells.

Several ADCs have entered clinical trials that have been developed to target breast cancer (Trail et al. 1993; Kolodych et al. 2017; Trail et al. 2018), ovarian cancer (Jiang et al. 2016), lung and colon cancers (Trail et al. 1993) and hematologic malignancies (Sievers and Senter 2013), among others (Birrer et al. 2019). The US Food and Drug Administration (FDA) has so far approved three ADCs: Mylotarg® and Adcetris® for the treatment of haematological cancers, and Kadcyla® for the treatment of HER2 positive breast cancer. Of note, Mylotarg® was withdrawn from the market a decade after its approval due to poor overall survival of patients (Perez et al. 2014; Gébleux and Casi 2016; Sau et al. 2017).

An ADC typically consists of three parts namely, an antibody, a linker and a cytotoxic drug or "payload" (Kumar et al. 2017; Thomas et al. 2016). In the design of ADCs, the antibody's specificity, stability of the linker, potency of the payload, the drug to antibody ratio (DAR) as well as rate of internalization are important determinants of efficacy (Perez et al. 2014; Gébleux and Casi 2016; Sievers and Senter 2013; Kolodych et al. 2017; Trail et al. 2018). ADCs are eventually cleared from circulation either via the renal or hepatobiliary route and studies have been carried out to evaluate the toxicity, higher-than-normal tissue exposure and to improve the localization and clearance of ADCs from the body (Casi and Neri 2012; Sievers and Senter 2013).

An anti-senescent ADC would have to be targeted against an epitope on the surfaceome and based on a monoclonal antibody that was easily internalized after binding. Several drugs could be chosen as payloads, from the most non-specific (we have observed that senescent cells are sensitive to different concentrations of doxorubicin and duocarmycin, for instance) to actual senolytics (like we used in some of the mentioned nanoparticles). The linker should be degraded once the antibody has been internalized, in order to free the payload. Also, the DAR should be high enough to elicit a strong effect. Preliminary data from our laboratory shows that an ADC against B2M with an average DAR of 2 (in line with currently used ADCs (Birrer et al. 2019)), generated using a commercially available monoclonal antibody linked to duocarmycin, can selectively kill senescent cells in culture, as seen with the targeted nanoMIPs. This suggests that ADCs could be used in humans to specifically clear senescent cells. However, further in vitro and in vivo tests would be necessary to understand the action of these treatments and evaluate their safety. More importantly, it has to be taken in consideration that the choice of the right surface marker would determine their specific biological effect.

6.4.3 Cellular Therapies

Antibodies can also be used to elicit cytotoxic effects independently of drug delivery. New approaches have already been developed for other diseases, such as antibody-dependent cell-mediated cytotoxicity (ADCC), which guides natural killer (NK) cells to selectively destroy a target (Lo Nigro et al. 2019). This has already been tested in senescence using a protein from the surfaceome. In this previously mentioned study (Kim et al. 2017), an anti-DPP4 antibody was added to both proliferating and senescent WI-38 cells. NK cells were isolated from human peripheral blood mononuclear cells and added to the conditioned media of WI-38 previously treated with a specific anti-DPP4 antibody. Results showed up to 40% reduction in cell viability relative to proliferating cells, proving that the antibody attracted the NK cells to the DPP4-expressing senescent targets. One variant of these immune-based therapies that does not require the use of antibodies would be to engineer NK cells to be attracted to SASP molecules such as IL-6 (Qudrat et al. 2017).

6.5 Conclusions: The Road Ahead

Targeted senolytics represent an attractive mode of therapy that could help speed up the translational applications of senescent cell clearance. The increased effectivity and, more importantly, reduced side effects could allow a widespread use of senolytics in senescence-related pathologies, including ageing. However, this is an area of research still in its infancy and many issues will have to be resolved before the first clinical trials can be designed. Given that the cellular events and molecular pathways that lead to senescence vary, it would be necessary to define clearly the pharmacodynamics of each potential anti-senescence therapeutic candidate not only to minimise off-target effects, ensure sensitivity and specificity, but to also identify potential synergistic effects or even contraindications when they are used in combination with other medications.

Identifying specific targets is key to the selectivity of these targeted therapies, since all studies so far seem to point to the fact that each tissue and model will have a particular set of senescent markers preferentially expressed. Thus, knowing the profile of senescent proteins present in each particular condition will allow the development of the right drug. In this context, the senescent surfaceome has the potential to provide several novel candidates, with the added value of determining extracellular epitopes for nanoparticle- and antibody-based strategies. Many proteins overexpressed on the surface of senescent cells have already been described in several labs, and more are likely to be published in the near future. However, like all known senescent markers, their expression is not universal, showing both tissue and disease specificity.

Since none of the current markers of senescence is exclusively specific to the senescent state, there is still the need for more reliable alternatives. One could be to use a combination of markers to increase specificity. Following the examples described above, a cytotoxic mechanism split in two portions (such as those used in certain mouse models (Baker et al. 2011)) could be delivered into senescent cells by separate ADCs or nanoparticles, thus killing only the cells that express both markers simultaneously. This and other approaches could help solve the issue of lacking a distinct and unique marker of senescence.

In parallel, better and more exhaustive screens would need to be developed. Mass spectrometry is still the best option for an unbiased approached. However, it would be interesting to compare the results of different pathways of induction of senescence (OIS, RS, MiDAS, radiotherapy, etc.) on the same cellular model in the same experiments, in order to provide a more balanced comparison of the surfaceomes. Manual validation of the main targets will still be essential. Therefore, high throughput techniques will need to be implemented to confirm the hits of the proteomic screens, which it has not been done in the published studies so far.

In light of recent findings (Zhang et al. 2019; van Deursen 2019; Xu et al. 2018; Roos et al. 2016), there is no doubt that senolytics are going to become a new class of drugs that will be intensively studied for their wide range of clinical applications. Their potential impact in the healthcare of an ever-growing aged population is immense. Targeted senolytics could be the most useful form of these drugs and efforts should be made to propose options that could be tested in specific diseases as soon as possible. In order to do this, better markers of senescence need to be identified, while the first generation of targeted senolytics are generated against the ones that are currently available. In following years, important advances are likely to be made in these areas and is to be expected that senolytics and targeted senolytics will routinely reach the patients.

References

Abbas T, Dutta A (2009) p21 in cancer: intricate networks and multiple activities. Nat Rev Cancer 9(6):400–414

Adams PD (2007) Remodeling of chromatin structure in senescent cells and its potential impact on tumor suppression and aging 84–93

Althubiti M, Lezina L, Carrera S, Jukes-Jones R, Giblett SM, Antonov A et al (2014a) Characterization of novel markers of senescence and their prognostic potential in cancer. Cell Death Dis 5(11):e1528

Althubiti M, Lezina L, Carrera S, Jukes-Jones R, Giblett SM, Antonov A et al (2014b) Characterization of novel markers of senescence and their prognostic potential in cancer. Cell Death Dis 5:e1528

Althubiti M, Rada M, Samuel J, Escorsa JM, Najeeb H, Lee KG et al (2016) BTK modulates p53 activity to enhance apoptotic and senescent responses. Can Res 76(18):5405–5414

Baker DJ, Wijshake T, Tchkonia T, LeBrasseur NK, Childs BG, van de Sluis B et al (2011) Clearance of p16Ink4a-positive senescent cells delays ageing-associated disorders. Nature 479(7372):232–236

Baker DJ, Childs BG, Durik M, Wijers ME, Sieben CJ, Zhong J et al (2016) Naturally occurring p16(Ink4a)-positive cells shorten healthy lifespan. Nature 530(7589):184–189

Barrilleaux B, Phinney DG, Prockop DJ, O'Connor KC (2006) Review: ex vivo engineering of living tissues with adult stem cells. Tissue Eng 12(11):3007–3019

Bausch-Fluck D, Goldmann U, Muller S, van Oostrum M, Muller M, Schubert OT et al (2018) The in silico human surfaceome. Proc Natl Acad Sci USA 115(46):E10988–E10997

Beausejour CM, Krtolica A, Galimi F, Narita M, Lowe SW, Yaswen P et al (2003) Reversal of human cellular senescence: roles of the p53 and p16 pathways. EMBO J 22(16):4212–4222

Bieging KT, Mello SS, Attardi LD (2014) Unravelling mechanisms of p53-mediated tumour suppression. Nat Rev Cancer 14:359

Birrer MJ, Moore KN, Betella I, Bates RC (2019) Antibody-Drug conjugate-based therapeutics: state of the science. J Natl Cancer Inst

Blagosklonny MV (2014) Geroconversion: irreversible step to cellular senescence. Cell Cycle (Georgetown, Tex) 13(23):3628–3635

Brugarolas J, Chandrasekaran C, Gordon JI, Beach D, Jacks T, Hannon GJ (1995) Radiation-induced cell cycle arrest compromised by p21 deficiency. Nature 377(6549):552–557

Bugarcic A, Zhe Y, Kerr MC, Griffin J, Collins BM, Teasdale RD (2011) Vps26A and Vps26B subunits define distinct retromer complexes. Traffic 12(12):1759–1773

Bussian TJ, Aziz A, Meyer CF, Swenson BL, van Deursen JM, Baker DJ (2018) Clearance of senescent glial cells prevents tau-dependent pathology and cognitive decline. Nature 562(7728):578–582

Caceres PS, Mendez M, Haque MZ, Ortiz PA (2016) Vesicle-associated membrane protein 3 (VAMP3) mediates constitutive trafficking of the renal co-transporter NKCC2 in thick

ascending limbs: ROLE IN RENAL FUNCTION AND BLOOD PRESSURE. J Biol Chem 291(42):22063–22073
Cahu J (2013) SASP: roadblock for tissue re-organization. Aging 5(9):641–642
Campisi J (2005) Senescent cells, tumor suppression, and organismal aging: good citizens. Bad Neighb Cell 120(4):513–522
Campisi J (2011) Cellular senescence: putting the paradoxes in perspective. Curr Opin Genet Dev 21(1):107–112
Campisi J, Debacq-Chainiaux F, Erusalimsky JD, Toussaint O (2009) Protocols to detect senescence-associated beta-galactosidase (SA-βgal) activity, a biomarker of senescent cells in culture and in vivo. Nat Protoc 4(12):1798–1806
Canfarotta F, Poma A, Guerreiro A, Piletsky S (2016a) Solid-phase synthesis of molecularly imprinted nanoparticles. Nat Prot 11(3):443–455
Canfarotta F, Waters A, Sadler R, McGill P, Guerreiro A, Papkovsky D et al (2016b) Biocompatibility and internalization of molecularly imprinted nanoparticles. Nano Res 9(11):3463–3477
Canfarotta F, Poma A, Guerreiro A, Piletsky S (2016c) Solid-phase synthesis of molecularly imprinted nanoparticles. Nat Prot 11:443
Caplan AI (2007) Adult mesenchymal stem cells for tissue engineering versus regenerative medicine. J Cell Physiol 213(2):341–347
Carragher LA, Snell KR, Giblett SM, Aldridge VS, Patel B, Cook SJ et al (2010) V600EBraf induces gastrointestinal crypt senescence and promotes tumour progression through enhanced CpG methylation of p16INK4a. EMBO Mol Med 2(11):458–471
Casi G, Neri D (2012) Antibody–drug conjugates: basic concepts, examples and future perspectives 422–428
Cecchini A, Raffa V, Canfarotta F, Signore G, Piletsky S, MacDonald MP et al (2017) In Vivo recognition of human vascular endothelial growth factor by molecularly imprinted polymers. Nano Lett
Chang BD, Xuan Y, Broude EV, Zhu H, Schott B, Fang J et al (1999) Role of p53 and p21waf1/cip1 in senescence-like terminal proliferation arrest induced in human tumor cells by chemotherapeutic drugs. Oncogene 18(34):4808–4818
Chang J, Wang Y, Shao L, Laberge RM, Demaria M, Campisi J et al (2016) Clearance of senescent cells by ABT263 rejuvenates aged hematopoietic stem cells in mice. Nat Med 22(1):78–83
Choi SA, Kim YH, Park YH, Yang HJ, Jeong PS, Cha JJ et al (2018) Novel crosstalk between Vps26a and Nox4 signaling during neurogenesis. Cell Death Differ
Cichowski K, Hahn WC (2008) Unexpected pieces to the senescence puzzle 958–961
Collado M, Gil J, Efeyan A, Guerra C, Schuhmacher AJ, Barradas M et al (2005) Tumour biology: senescence in premalignant tumours. Nature 436(7051):642
Collado M, Blasco MA, Serrano M (2007) Cellular senescence in cancer and aging. Cell 130(2):223–233
Coppe JP, Patil CK, Rodier F, Sun Y, Munoz DP, Goldstein J et al (2008) Senescence-associated secretory phenotypes reveal cell-nonautonomous functions of oncogenic RAS and the p53 tumor suppressor. PLoS Biol 6(12):2853–2868
Coppe JP, Desprez PY, Krtolica A, Campisi J (2010) The senescence-associated secretory phenotype: the dark side of tumor suppression. Ann Rev Pathol 5:99–118
Coppe JP, Rodier F, Patil CK, Freund A, Desprez PY, Campisi J (2011) Tumor suppressor and aging biomarker p16(INK4a) induces cellular senescence without the associated inflammatory secretory phenotype. J Biol Chem 286(42):36396–36403
Cui H, Kong Y, Xu M, Zhang H (2013) Notch3 functions as a tumor suppressor by controlling cellular senescence. Cancer Res 73(11):3451
Cunliffe D, Kirby A, Alexander C (2005) Molecularly imprinted drug delivery systems 1836–1853
Dadparvar M, Wagner S, Wien S, Kufleitner J, Worek F, von Briesen H et al (2011) HI 6 human serum albumin nanoparticles—development and transport over an in vitro blood–brain barrier model 60–66

Daste A, Chakiba C, Domblides C, Gross-goupil M, Quivy A, Ravaud A et al (2016) Targeted therapy and elderly people: a review 199–215
de Keizer PL (2016) The fountain of youth by targeting senescent cells? Trends Mol Med
de Magalhães JP, Passos JF (2017) Stress, cell senescence and organismal ageing
Demaria M (2017) Senescent cells: New target for an old treatment? Mol Cell Oncol 4(3):e1299666
Demidenko ZN, Korotchkina LG, Gudkov AV, Blagosklonny MV (2010) Paradoxical suppression of cellular senescence by p53. Proc Natl Acad Sci USA 107(21):9660–9664
Deursen JMAV (2014) The role of senescent cells in ageing 509(7501):439–446
di Fagagna FDA, Campisi J (2007) Cellular senescence: when bad things happen to good cells. Nat Rev Mol Cell Biol 8(9):729–740
Dimri GP, Lee X, Basile G, Acosta M, Scott G, Roskelley C et al (1995) A biomarker that identifies senescent human cells in culture and in aging skin in vivo. Proc Natl Acad Sci USA 92(20):9363–9367
Dong Q, Han H, Liu X, Wei L, Zhang W, Zhao Z et al (2017) HCSGD: an integrated database of human cellular senescence genes 227–234
Donnell AF, Zhang Y, Stang EM, Wei DD, Tebben AJ, Perez HL et al (2017) Macrocyclic pyrrolobenzodiazepine dimers as antibody-drug conjugate payloads 5267–5271
Edwards MG, Anderson RM, Yuan M, Kendziorski CM, Weindruch R, Prolla TA (2007) Gene expression profiling of aging reveals activation of a p53-mediated transcriptional program. BMC Genom 8:80
Ehrlich P (1906) The relationship existing between chemical constitution, distribution and pharmacological action. Collect Stud Immun 441–450
Ekpenyong-Akiba AE, Canfarotta F, Abd HB, Poblocka M, Casulleras M, Castilla-Vallmanya L et al (2019) Detecting and targeting senescent cells using molecularly imprinted nanoparticles. Nanoscale Horiz. 4(3):757–768
El-Deiry WS, Tokino T, Velculescu VE, Levy DB, Parsons R, Trent JM et al (1993) WAF1, a potential mediator of p 53 tumor suppression. Cell 75(4):817–825
Evangelou K, Lougiakis N, Rizou SV, Kotsinas A, Kletsas D, Munoz-Espin D et al (2017) Robust, universal biomarker assay to detect senescent cells in biological specimens. Aging Cell 16(1):192–197
Falschlehner C, Emmerich CH, Gerlach B, Walczak H (2007) TRAIL signalling: Decisions between life and death. Int J Biochem Cell Biol 39(7):1462–1475
Farr JN, Xu M, Weivoda MM, Monroe DG, Fraser DG, Onken JL et al (2017) Targeting cellular senescence prevents age-related bone loss in mice. Nat Med 23(9):1072–1079
Florence AT, Lee VHL (2011) Personalised medicines: More tailored drugs, more tailored delivery 29–33
Fuhrmann-Stroissnigg H, Ling YY, Zhao J, McGowan SJ, Zhu Y, Brooks RW et al (2017) Identification of HSP90 inhibitors as a novel class of senolytics. Nat Commun 8(1):z
Galanos P, Vougas K, Walter D, Polyzos A, Maya-Mendoza A, Haagensen EJ et al (2016) Chronic p53-independent p21 expression causes genomic instability by deregulating replication licensing. Nat Cell Biol 18:777
Gallage S, Gil J (2016) Mitochondrial dysfunction meets senescence. Trends Biochem Sci 41(3):207–209
Gébleux R, Casi G (2016) Antibody-drug conjugates: current status and future perspectives 48–59
Georgakilas AG, Martin OA, Bonner WM (2017) p 21: a two-faced genome guardian 310–319
Georgakopoulou EA, Tsimaratou K, Evangelou K, Fernandez Marcos PJ, Zoumpourlis V, Trougakos IP et al (2013) Specific lipofuscin staining as a novel biomarker to detect replicative and stress-induced senescence. A method applicable in cryo-preserved and archival tissues. Aging. 5(1):37–50
Guney I, Sedivy JM (2006) Cellular senescence, epigenetic switches and c-Myc. Cell Cycle (Georgetown, Tex) 5(20):2319–2323
Hashimoto M, Asai A, Kawagishi H, Mikawa R, Iwashita Y, Kanayama K et al (2016) Elimination of p19(ARF)-expressing cells enhances pulmonary function in mice. JCI insight. 1(12):e87732

Hayflick L (1965) The limited in vitro lifetime of human diploid cell strains. Exp Cell Res 37(3):614–636

Hayflick L, Moorehead P (1961) The serial cultivation of human diploid strains. Exp Cell Res 25:585–621

Hayflick L, Moorhead PS (1961) The serial cultivation of human diploid cell strains. Exp Cell Res 25(3):585–621

Herbig U, Jobling WA, Chen BP, Chen DJ, Sedivy JM (2004) Telomere shortening triggers senescence of human cells through a pathway involving ATM, p53, and p21(CIP1), but not p16(INK4a). Mol Cell 14(4):501–513

Hernandez-Segura A, de Jong TV, Melov S, Guryev V, Campisi J, Demaria M (2017) Unmasking transcriptional heterogeneity in senescent cells. Curr Biol: CB

Herranz N, Gil J (2016) Mitochondria and senescence: new actors for an old play. The EMBO J 35(7):701–702

Herranz N, Gallage S, Mellone M, Wuestefeld T, Klotz S, Hanley CJ et al (2015) mTOR regulates MAPKAPK2 translation to control the senescence-associated secretory phenotype. Nat Cell Biol 17:1205

Hoare M, Narita M (2017) NOTCH and the 2 SASPs of senescence. Cell Cycle 16(3):239–240

Hoare M, Ito Y, Kang TW, Weekes MP, Matheson NJ, Patten DA et al (2016) NOTCH1 mediates a switch between two distinct secretomes during senescence. Nat Cell Biol 18(9):979–992

Hoshino Y, Koide H, Urakami T, Kanazawa H, Kodama T, Oku N et al (2010) Recognition, neutralization and clearance of target peptides in the blood stream of living mice by molecular imprinted polymer nanoparticles: a plastic antibody. J Am Chem Soc 132(19):6644–6645

Itahana K, Campisi J, Dimri GP (2007) Methods to detect biomarkers of cellular senescence: the senescence-associated beta-galactosidase assay. Methods Mol Biol (Clifton, NJ) 371:21

Itahana K, Itahana Y, Dimri GP (2013) Colorimetric detection of senescence-associated beta galactosidase. Methods Mol Biol 965:143–156

Ito Y, Hoare M, Narita M (2017) Spatial and temporal control of senescence 820–832

Iuliano R, Palmieri D, He H, Iervolino A, Borbone E, Pallante P et al (2010) Role of PTPRJ genotype in papillary thyroid carcinoma risk. Endocr Relat Cancer 17(4):1001–1006

Jiang J, Dong L, Wang L, Wang L, Zhang J, Chen F et al (2016) HER2-targeted antibody drug conjugates for ovarian cancer therapy 274–286

Justice JN, Nambiar AM, Tchkonia T, LeBrasseur NK, Pascual R, Hashmi SK et al (2019) Senolytics in idiopathic pulmonary fibrosis: results from a first-in-human, open-label, pilot study. EBioMedicine 40:554–563

Kastan MB, Onyekwere O, Sidransky D, Vogelstein B, Craig RW (1991) Participation of p 53 protein in the cellular response to DNA damage. Cancer Res 51(23 Part 1):6304–6311

Ke X (2017) Shen L. The progress and future prospect, Molecular targeted therapy of cancer, pp 69–75

Kempe H, Parareda Pujolràs A, Kempe M (2015) Molecularly imprinted polymer nanocarriers for sustained release of erythromycin. Pharm Res 32(2):375–388

Kim KM, Noh JH, Bodogai M, Martindale JL, Yang X, Indig FE et al (2017) Identification of senescent cell surface targetable protein DPP4. Genes Dev 31(15):1529–1534

Kim KM, Noh JH, Bodogai M, Martindale JL, Pandey PR, Yang X et al (2018) SCAMP4 enhances the senescent cell secretome. Genes Dev 32(13–14):909–914

Kimball AB, Alora-Palli MB, Tamura M, Mullins LA, Soh C, Binder RL et al (2018) Age-induced and photoinduced changes in gene expression profiles in facial skin of Caucasian females across 6 decades of age 39.e7

Kirkland JL, Tchkonia T (2015) Clinical strategies and animal models for developing senolytic agents. Exp Gerontol 68:19–25

Kirkland JL, Tchkonia T (2017) Cellular senescence: a translational perspective. EBioMedicine 21:21–28

Kolodych S, Michel C, Delacroix S, Koniev O, Ehkirch A, Eberova J, et al (2017) Development and evaluation of β-galactosidase-sensitive antibody-drug conjugates 376–382

Krishnamurthy J, Torrice C, Ramsey MR, Kovalev GI, Al-Regaiey K, Su L et al (2004) Ink4a/Arf expression is a biomarker of aging. J Clin Investig 114(9):1299–1307
Kuilman T, Peeper DS (2009) Senescence-messaging secretome: SMS-ing cellular stress. Nat Rev Cancer 9(2):81–94
Kuilman T, Michaloglou C, Vredeveld LC, Douma S, van Doorn R, Desmet CJ et al (2008) Oncogene-induced senescence relayed by an interleukin-dependent inflammatory network. Cell 133(6):1019–1031
Kuilman T, Michaloglou C, Mooi WJ, Peeper DS (2010) The essence of senescence. Genes Dev 24(22):2463–2479
Kumar A, White J, James Christie R, Dimasi N, Gao C (2017) Chapter twelve–antibody-drug conjugates. In: Goodnow RA (ed). Academic Press, pp 441–480
Kurz DJ, Decary S, Hong Y, Erusalimsky JD (2000a) Senescence-associated (beta)-galactosidase reflects an increase in lysosomal mass during replicative ageing of human endothelial cells. J Cell Sci 113(Pt 20)(Pt 20):3613–3622
Kurz DJ, Decary S, Hong Y, Erusalimsky JD (2000b) Senescence-associated beta-galactosidase reflects an increase in lysosomal mass during replicative ageing of human endothelial cells. J Cell Sci 113(Pt):3613–3622
Laberge R-M, Sun Y, Orjalo AV, Patil CK, Freund A, Zhou L et al (2015) MTOR regulates the pro-tumorigenic senescence-associated secretory phenotype by promoting IL1A translation. Nat Cell Biol 17:1049
Lane DP (1992) p53, guardian of the genome. Nature 358:15
Lasry A, Ben-Neriah Y (2015) Senescence-associated inflammatory responses: aging and cancer perspectives. Trends Immunol 36(4):217–228
Lawless C, Wang C, Jurk D, Merz A, Zglinicki Tv, Passos JF (2010) Quantitative assessment of markers for cell senescence, pp 772–778
Lee HW, Lee SH, Lee HW, Ryu YW, Kwon MH, Kim YS (2005) Homomeric and heteromeric interactions of the extracellular domains of death receptors and death decoy receptors. Biochem Biophys Res Commun 330(4):1205–1212
Lehmann M, Korfei M, Mutze K, Klee S, Skronska-Wasek W, Alsafadi HN et al (2017) Senolytic drugs target alveolar epithelial cell function and attenuate experimental lung fibrosis ex vivo. Eur Respir J 50(2):2016. Print 7 Aug
Leontieva OV, Blagosklonny MV (2017) While reinforcing cell cycle arrest, rapamycin and Torins suppress senescence in UVA-irradiated fibroblasts. Oncotarget 8(65):109848–109856
Levine AJ (1997) p 53, the cellular gatekeeper for growth and division 323–331
Li W, Qin L, Feng R, Hu G, Sun H, He Y et al (2019) Emerging senolytic agents derived from natural products. Mech Ageing Dev 181:1–6
Liu H, Zhao WL, Wang JP, Xin BM, Shao RG (2018) EBP50 suppresses the proliferation of MCF-7 human breast cancer cells via promoting Beclin-1/p62-mediated lysosomal degradation of c-Myc. Acta Pharmacol Sin 39(8):1347–1358
Lo Nigro C, Macagno M, Sangiolo D, Bertolaccini L, Aglietta M, Merlano MC (2019) NK-mediated antibody-dependent cell-mediated cytotoxicity in solid tumors: biological evidence and clinical perspectives. Ann Transl Med 7(5):105
Lowe SW, Sherr CJ (2003) Tumor suppression by Ink4a–Arf: progress and puzzles 77–83
Luliński P (2017) Molecularly imprinted polymers based drug delivery devices: a way to application in modern pharmacotherapy. A Rev 1344–1353
Lundberg AS, Hahn WC, Gupta P, Weinberg RA (2000) Genes involved in senescence and immortalization 705–709
Maciel-Barón LA, Morales-Rosales S, Aquino-Cruz A, Triana-Martínez F, Galván-Arzate S, Luna-López A et al (2016) Senescence associated secretory phenotype profile from primary lung mice fibroblasts depends on the senescence induction stimuli. Age 38(1):26
Macip S, Igarashi M, Fang L, Chen A, Pan ZQ, Lee SW et al (2002) Inhibition of p21-mediated ROS accumulation can rescue p21-induced senescence. EMBO J 21(9):2180–2188

Madsen SD, Russell KC, Tucker HA, Glowacki J, Bunnell BA, O'Connor KC (2017) Decoy TRAIL receptor CD264: a cell surface marker of cellular aging for human bone marrow-derived mesenchymal stem cells. Stem Cell Res Ther 8(1):201

Malaquin N, Martinez A, Rodier F (2016) Keeping the senescence secretome under control: molecular reins on the senescence-associated secretory phenotype 39–49

Masgras I, Carrera S, de Verdier PJ, Brennan P, Majid A, Makhtar W et al (2012) Reactive oxygen species and mitochondrial sensitivity to oxidative stress determine induction of cancer cell death by p21. J Biol Chem 287(13):9845–9854

Matjusaitis M, Chin G, Sarnoski EA, Stolzing A (2016) Biomarkers to identify and isolate senescent cells, pp 1–12

McHugh D, Gil J (2018) Senescence and aging: Causes, consequences, and therapeutic avenues. J Cell Biol 217(1):65

Mduff FKE, Turner SD (2011) Jailbreak: oncogene-induced senescence and its evasion 6–13

Mercer K, Giblett S, Green S, Lloyd D, DaRocha Dias S, Plumb M et al (2005) Expression of endogenous oncogenic V600EB-raf induces proliferation and developmental defects in mice and transformation of primary fibroblasts. Can Res 65(24):11493–11500

Mohamed AJ, Yu L, Backesjo CM, Vargas L, Faryal R, Aints A et al (2009) Bruton's tyrosine kinase (Btk): function, regulation, and transformation with special emphasis on the PH domain. Immunol Rev 228(1):58–73

Moreno-Blas D, Gorostieta-Salas E, Castro-Obregón S (2018) Connecting chaperone-mediated autophagy dysfunction to cellular senescence 34–41

Munoz-Espin D, Serrano M (2014) Cellular senescence: from physiology to pathology. Nat Rev Mol Cell Biol 15(7):482–496

Munoz-Espin D, Canamero M, Maraver A, Gomez-Lopez G, Contreras J, Murillo-Cuesta S et al (2013) Programmed cell senescence during mammalian embryonic development. Cell 155(5):1104–1118

Munoz-Espin D, Rovira M, Galiana I, Gimenez C, Lozano-Torres B, Paez-Ribes M et al (2018) A versatile drug delivery system targeting senescent cells. EMBO Mol Med

Murray-Zmijewski F, Slee EA, Lu X (2008) A complex barcode underlies the heterogeneous response of p53 to stress. Nat Rev Mol Cell Biol 9(9):702–712

Narita M, Nunez S, Heard E, Narita M, Lin AW, Hearn SA et al (2003) Rb-mediated heterochromatin formation and silencing of E2F target genes during cellular senescence. Cell 113(6):703–716

Naylor RM, Deursen JMAV, Baker DJ (2013) Senescent cells: a novel therapeutic target for aging and age-related diseases 93(1):105–116

Ogrodnik M, Miwa S, Tchkonia T, Tiniakos D, Wilson CL, Lahat A et al (2017) Cellular senescence drives age-dependent hepatic steatosis. Nat Commun 8:15691

Palmer AK, Xu M, Zhu Y, Pirtskhalava T, Weivoda MM, Hachfeld CM et al (2019) Targeting senescent cells alleviates obesity-induced metabolic dysfunction. Aging Cell 18(3):e12950

Pan G, Ni J, Yu G-l, Wei Y-F, Dixit VM (1998) TRUNDD, a new member of the TRAIL receptor family that antagonizes TRAIL signalling. FEBS Lett 424(1):41–45

Pan J, Li D, Xu Y, Zhang J, Wang Y, Chen M et al (2017) Inhibition of Bcl-2/xl with ABT-263 selectively kills senescent type II pneumocytes and reverses persistent pulmonary fibrosis induced by ionizing radiation in Mice, pp 353–361

Park HY, Sin DD (2014) Chapter 16—stress-induced premature senescence: another mechanism involved in the process of accelerated aging in chronic obstructive pulmonary disease. In: Rahman I, Bagchi D (eds) Inflammation, advancing age and nutrition. Academic Press, San Diego, pp 193–202

Pauliah M, Zanganeh S, Erfanzadeh M, Ho JQ (2018) Chapter 10—tumor-targeted therapy. In: Mahmoudi M, Laurent S (eds). Elsevier; pp 273–290

Perez HL, Cardarelli PM, Deshpande S, Gangwar S, Schroeder GM, Vite GD (2014) Antibody–drug conjugates: current status and future directions 869–881

Pérez-Mancera PA, Young ARJ, Narita M (2014) Inside and out: the activities of senescence in cancer. Nat Rev Cancer 14(8):547

Piletsky SA, Turner NW, Laitenberger P (2006) Molecularly imprinted polymers in clinical diagnostics–future potential and existing problems. Med Eng Phys 28(10):971–977

Poma A, Turner APF, Piletsky SA. Advances in the manufacture of MIP nanoparticles. 2010. p. 629–37

Puoci F, Cirillo G, Curcio M, Parisi OI, Iemma F, Picci N (2011) Molecularly imprinted polymers in drug delivery: state of art and future perspectives. Expert Opin Drug Deliv 8(10):1379–1393

Qudrat A, Wong J, Truong K (2017) Engineering mammalian cells to seek senescence associated secretory phenotypes. J Cell Sci

Rada M, Althubiti M, Ekpenyong-Akiba AE, Lee KG, Lam KP, Fedorova O et al (2017) BTK blocks the inhibitory effects of MDM2 on p53 activity. Oncotarget 8(63):106639–106647

Rada M, Barlev N, Macip S (2018a) BTK modulates p73 activity to induce apoptosis independently of p53. Cell Death Discov 4:30

Rada M, Barlev N, Macip S (2018b) BTK: a two-faced effector in cancer and tumour suppression. Cell Death Dis 9(11):1064

Rodier F, Campisi J (2011) Four faces of cellular senescence. J cell Biol 192(4):547–556

Romagosa C, Simonetti S, López-Vicente L, Mazo A, Lleonart ME, Castellvi J et al (2011) p16Ink4a overexpression in cancer: a tumor suppressor gene associated with senescence and high-grade tumors. Oncogene 30(18):2087–2097

Romanov VS, Rudolph KL (2016) p21 shapes cancer evolution. Nat Cell Biol 18:722

Roos CM, Zhang B, Palmer AK, Ogrodnik MB, Pirtskhalava T, Thalji NM et al (2016) Chronic senolytic treatment alleviates established vasomotor dysfunction in aged or atherosclerotic mice. Aging Cell 15(5):973–977

Rufini A, Tucci P, Celardo I, Melino G (2013) Senescence and aging: the critical roles of p53. Oncogene 32(43):5129

Salama R, Sadaie M, Hoare M, Narita M (2014) Cellular senescence and its effector programs. Genes Dev 28(2):99–114

Sau S, Alsaab HO, Kashaw SK, Tatiparti K, Iyer AK (2017) Advances in antibody–drug conjugates: a new era of targeted cancer therapy 1547–1556

Sellergren B, Allender CJ (2005) Molecularly imprinted polymers: a bridge to advanced drug delivery. Adv Drug Deliv Rev 57(12):1733–1741

Sengupta S, Sasisekharan R (2007) Exploiting nanotechnology to target cancer. Br J Cancer 96(9):1315–1319

Sharpless NE, Sherr CJ (2015) Forging a signature of in vivo senescence. Nat Rev Cancer 15(7):397–408

Sherr CJ (2012) Ink4-Arf locus in cancer and aging. Wiley Interdiscip Rev Dev Biol 1(5):731–741

Shiohara M, El-Deiry WS, Wada M, Nakamaki T, Takeuchi S, Yang R et al (1994) Absence of WAF1 mutations in a variety of human malignancies. Blood 84(11):3781

Sievers EL, Senter PD (2013) Antibody-drug conjugates in cancer therapy. Annu Rev Med 64:15–29

Sikora E, Arendt T, Bennett M, Narita M (2011) Impact of cellular senescence signature on ageing research, pp 146–152

Simboeck E, Di Croce L (2013) p16INK4a in cellular senescence. Aging 5(8):590

Sionov RV, Haupt Y (1999) The cellular response to p53: the decision between life and death. Oncogene 18(45):6145–6157

Storer M, Mas A, Robert-Moreno A, Pecoraro M, Ortells MC, Di Giacomo V et al (2013) Senescence is a developmental mechanism that contributes to embryonic growth and patterning. Cell 155(5):1119–1130

Strohl WR, Strohl LM (2012) 15—Antibody-drug conjugates. In: Strohl WR, Strohl LM (eds) Therapeutic antibody engineering. Woodhead Publishing Series in Biomedicine, Woodhead Publishing; pp 345–595

Swanson EC, Manning B, Zhang H, Lawrence JB (2013) Higher-order unfolding of satellite heterochromatin is a consistent and early event in cell senescence. J Cell Biol 203(6):929–942

Takahashi A, Ohtani N, Hara E (2007) Irreversibility of cellular senescence: dual roles of p16INK4a/Rb-pathway in cell cycle control. Cell Div 2:10

Tan M, Li J, Ma F, Zhang X, Zhao Q, Cao X (2019) PLD3 rare variants identified in late-onset alzheimer's disease affect amyloid-beta levels in cellular model. Front Neurosci 13:116
Taniguchi K, Kohsaka H, Inoue N, Terada Y, Ito H, Hirokawa K et al (1999) Induction of the p16INK4a senescence gene as a new therapeutic strategy for the treatment of rheumatoid arthritis. Nat Med 5(7):760
Terman A, Brunk UT (2004) Lipofuscin 1400–1404
Thapa RK, Nguyen HT, Jeong JH, Kim JR, Choi HG, Yong CS et al (2017) Progressive slowdown/prevention of cellular senescence by CD9-targeted delivery of rapamycin using lactose-wrapped calcium carbonate nanoparticles. Sci Rep 7:43299
Thomas A, Teicher BA, Hassan R (2016) Antibody–drug conjugates for cancer therapy e262
Tieppo A, White CJ, Paine AC, Voyles ML, McBride MK, Byrne ME (2012) Sustained in vivo release from imprinted therapeutic contact lenses 391–397
Trail PA, Willner D, Lasch SJ, Henderson AJ, Hofstead S, Casazza AM et al (1993) Cure of xenografted human carcinomas by BR96-doxorubicin immunoconjugates. Sci (New York, NY) 261(5118):212–215
Trail PA, Dubowchik GM, Lowinger TB (2018) Antibody drug conjugates for treatment of breast cancer: novel targets and diverse approaches in ADC design 126–142
Tyagi N, Arora S, Deshmukh SK, Singh S, Marimuthu S, Singh AP (2016) Exploiting nanotechnology for the development of MicroRNA-Based cancer therapeutics. J Biomed Nanotechnol 12(1):28–42
van Deursen JM (2019) Senolytic therapies for healthy longevity. Science 364(6441):636–637
Vasapollo G, Sole RD, Mergola L, Lazzoi MR, Scardino A, Scorrano S et al (2011) Molecularly imprinted polymers: present and future prospective. Int J Mol Sci 12(9):5908–5945
Veale KJ, Offenhauser C, Lei N, Stanley AC, Stow JL, Murray RZ (2011) VAMP3 regulates podosome organisation in macrophages and together with Stx4/SNAP23 mediates adhesion, cell spreading and persistent migration. Exp Cell Res 317(13):1817–1829
Voigt J, Christensen J, Shastri VP (2014) Differential uptake of nanoparticles by endothelial cells through polyelectrolytes with affinity for caveolae. Proc Natl Acad Sci USA 111(8):2942–2947
Waldman T, Kinzler KW, Vogelstein B (1995) p21 is necessary for the p53-mediated G1 arrest in human cancer cells. Can Res 55(22):5187–5190
Walter HS, Ahmed S (2017) Targeted therapies in cancer. Surgery (Oxford)
Wang C, Jurk D, Maddick M, Nelson G, Martin-Ruiz C, von Zglinicki T (2009) DNA damage response and cellular senescence in tissues of aging mice. Aging Cell 8(3):311–323
Wang R, Sunchu B, Perez VI (2017) Rapamycin and the inhibition of the secretory phenotype 89–92
Wiley CD, Campisi J (2016) From ancient pathways to aging cells-connecting metabolism and cellular senescence. Cell Metab 23(6):1013–1021
Wiley CD, Velarde MC, Lecot P, Liu S, Sarnoski EA, Freund A et al (2016) Mitochondrial dysfunction induces senescence with a distinct secretory phenotype 303–314
Wu Q, Cheng Z, Zhu J, Xu W, Peng X, Chen C et al (2015) Suberoylanilide hydroxamic acid treatment reveals crosstalks among proteome, ubiquitylome and acetylome in non-small cell lung cancer A549 cell line. Sci Rep 5:9520
Wu T, Rayner CK, Horowitz M (2016) Incretins. In: Herzig S (ed) Metabolic control. Springer International Publishing, Cham, pp 137–171
Xiao P, Huang X, Huang L, Yang J, Li A, Shen K et al (2017) G protein-coupled receptor kinase 4-induced cellular senescence and its senescence-associated gene expression profiling 273–280
Xu Y, Li N, Xiang R, Sun P (2014) Emerging roles of the p 38 MAPK and PI3K/AKT/mTOR pathways in oncogene-induced senescence 268–276
Xu M, Pirtskhalava T, Farr JN, Weigand BM, Palmer AK, Weivoda MM et al (2018) Senolytics improve physical function and increase lifespan in old age. Nat Med 24(8):1246–1256
Xu T, Cai Y, Zhong X, Zhang L, Zheng D, Gao Z et al (2019) Beta-Galactosidase instructed supramolecular hydrogelation for selective identification and removal of senescent cells. Chem Commun (Camb)

Young ARJ, Narita M, Narita M (2013) Cell senescence as both a dynamic and a static phenotype. In: Galluzzi L, Vitale I, Kepp O, Kroemer G (eds) Cell senescence: methods and protocols. Humana Press, Totowa, NJ, pp 1–13

Zhang P, Kishimoto Y, Grammatikakis I, Gottimukkala K, Cutler RG, Zhang S et al (2019) Senolytic therapy alleviates Abeta-associated oligodendrocyte progenitor cell senescence and cognitive deficits in an Alzheimer's disease model. Nat Neurosci 22(5):719–728

Zhong J, Rao X, Deiuliis J, Braunstein Z, Narula V, Hazey J et al (2013) A potential role for dendritic cell/macrophage-expressing DPP4 in obesity-induced visceral inflammation. Diabetes 62(1):149

Zhu C, Zhang L, Zheng Y, Xu J, Song J, Rolfe BE et al (2011) Effects of estrogen on stress-induced premature senescence of vascular smooth muscle cells: a novel mechanism for the "time window theory" of menopausal hormone therapy. Atherosclerosis 215(2):294–300

Zhu Y, Tchkonia T, Pirtskhalava T, Gower AC, Ding H, Giorgadze N et al (2015) The Achilles' heel of senescent cells: from transcriptome to senolytic drugs. Aging Cell 14(4):644–658

Zhu Y, Tchkonia T, Fuhrmann-Stroissnigg H, Dai HM, Ling YY, Stout MB et al (2016) Identification of a novel senolytic agent, navitoclax, targeting the Bcl-2 family of anti-apoptotic factors. Aging Cell 15(3):428–435

Chapter 7
Senolysis and Senostasis Through the Plasma Membrane

Kyoung Mi Kim, Ji Heon Noh, and Myriam Gorospe

Abstract With increasing evidence that senescent cells are detrimental towards a range of age-associated diseases and physiologic declines, there is rising urgency to develop interventions to suppress their adverse effects. Most *senolytic* approaches aim to eliminate senescent cells by rendering them vulnerable to apoptosis, while *senostatic (senomorphic)* approaches do not destroy the cell and instead suppress a specific senescent trait. In both senolysis and senostasis, the major goals include reducing the senescence-associated secretory phenotype (SASP) and to enhance the immunogenicity of the senescent compartment. These therapeutic aims are best elicited from the plasma membrane, although efforts to identify plasma membrane targets are only now beginning. We discuss several plasma membrane proteins expressed preferentially in senescent cells and their roles in neutralizing senescent cells by immune-mediated senolysis (as reported for DPP4, VIM, and NKFB2 ligands) and by suppressing the SASP (as reported for SCAMP4 and CD36). We identify the advantages and challenges of developing therapeutic approaches directed at the plasma membrane of senescent cells.

Keywords Senolytic therapy · Senostatic therapy · Senomorphic therapy · Senescent surfaceome · Plasma membrane markers

K. M. Kim · J. H. Noh · M. Gorospe
Laboratory of Genetics and Genomics, National Institute on Aging Intramural Research Program, National Institutes of Health, Baltimore, MD 21224, USA

J. H. Noh
Department of Biochemistry, Chungnam National University, Daejeon 34134, Republic of Korea

K. M. Kim (✉)
Department of Biological Sciences, Chungnam National University, Daejeon 34134, Republic of Korea
e-mail: km.kim@cnu.ac.kr

D. Muñoz-Espin and M. Demaria (eds.), *Senolytics in Disease, Ageing and Longevity*, Healthy Ageing and Longevity 11,
https://doi.org/10.1007/978-3-030-44903-2_7

7.1 Introduction

7.1.1 Cellular Senescence in Physiology and Disease

Cellular senescence is a specific response to cell damage that leads to long-term, generally irreversible growth arrest. Senescence can be triggered by stresses such as telomere attrition, DNA damage, oncogene activation, oxidative injury, and inflammation (von Zglinicki 2002; Guerra et al. 2003; Herbig et al. 2004; van Deursen 2014; Gorgoulis et al. 2019). In young persons, senescence is considered to be tumor-suppressive as it prevents the propagation of potentially transformed cells (Lowe et al. 2004; Collado and Serrano 2010; Muñoz-Espín and Serrano 2014; Chen et al. 2005; Kirkland and Tchkonia 2017). However, in older organisms, an excessive accumulation of senescent cells can become detrimental to tissues and organs, leading to organ dysfunction and promoting pathologies such as neurodegeneration, cardiovascular disease, diabetes, cancer, arthritis, liver disease, and sarcopenia (van Deursen 2014; Campisi 2005). Accordingly, in young persons, cellular senescence is recognized as a homeostatic response to sublethal damage, but in older persons it promotes aging-associated decline and disease (López-Otín et al. 2013). Additionally, senescence improves tissue repair by promoting wound healing (Demaria et al. 2014; Krizhanovsky et al. 2008).

Senescent cells display distinct features. Compared with proliferating cells, senescent cells are larger, contain cytoplasmic vacuoles, exhibit increased senescence-associated β-galactosidase (SA-βGal) activity, and have higher levels of reactive oxygen species (ROS) (Gorgoulis et al. 2019; Ben-Porath and Weinberg 2004). In addition, senescent cells express different subsets of RNAs, including noncoding transcripts such as *PURPL*, and proteins such as cyclin-dependent kinase (CDK) inhibitors p21 (CDKN1A) and p16 (CDKN2A) (Alcorta et al. 1996; Casella et al. 2019; Wiley et al. 2017). They also exhibit a trait known as the senescence-associated secretory phenotype (SASP), whereby they display increased secretion of distinct subsets of cytokines, growth factors, and matrix remodeling proteins, including interleukin (IL)6 and IL8, vascular endothelial growth factor (VEGF), tumor necrosis factor (TNF), and matrix metalloproteinase 9 (MMP9) (Coppe et al. 2008; Parrinello et al. 2003; Dimri et al. 1995; Narita et al. 2003).

Among the differentially expressed proteins, the suppressor of growth and tumorigenesis p16 [an activator of the protein retinoblastoma (RB)], has been considered one of the most promising triggers and biomarkers of senescence in tissues and organs, as well as in cultured cells (Alcorta et al. 1996; Krishnamurthy et al. 2004; Benanti and Galloway 2004; Zindy et al. 1997). Supporting this notion, prolonged p16 expression promoted cellular senescence and decreased regenerative capacity in aging tissues, including the hematopoietic, pancreatic, and neural stem cell compartments (Kim and Sharpless 2006). Recent studies also showed that the numbers of p16-positive cells correlate with biological (but not chronological) aging in different tissues (Liu et al. 2009; Baker et al. 2011; Waaijer et al. 2012) and served as a surrogate marker to evaluate clinical outcomes following therapeutic interventions (Nelson et al. 2012). However, p16 appears not to be a universal biomarker of senescence (reviewed in Hernandez-Segura et al. 2018) and its low abundance and intracellular location pose further limitations on its usefulness as a marker. Thus, efforts have intensified in recent years to identify senescent markers that (1) improve the detection of senescent cells including live cells, (2) identify all senescent cells universally or subsets of senescent cells, and (3) provide a means to isolate senescent cells for elimination or for further analysis. Such markers are particularly valuable in therapeutic efforts aimed at protecting or destroying senescent cells.

7.2 Senolytics

7.2.1 Strategies to Reduce the Burden of Senescent Cells During Age

'Senolysis', or the elimination of senescent cells, is a strategy that has been actively pursued in recent years. A key landmark in this effort was a report by Baker and colleagues of a genetic mouse model (INK-ATTAC) in which activation of caspase 8 in p16-expressing cells eliminated senescent cells and ameliorated age-related losses in strength, adiposity, and vision (Baker et al. 2011). Recapitulating this genetic approach and similar mouse models (Demaria et al. 2014), a handful of chemical compounds named 'senolytics' have been identified that selectively eliminate senescent cells (Hernandez-Segura et al. 2018). Typically, senolytic drugs potently block pathways that protect senescent cells, thereby enabling their destruction (Ovadya and Krizhanovsky 2018). For example, various studies suggested that antiapoptotic protein members of the BCL2 family are significantly increased during senescence; accordingly, inhibition of those proteins in senescent cells activates programmed cell death (Yosef et al. 2016). Following similar strategies, approaches to augment apoptotic signals have been effective in achieving senolysis. Inhibitors of PI3K (phosphotidyl inositol 3 kinase) such as quercetin, inhibitors of tyrosine kinases including EPHA2 (the ephryn receptor A2) such as dasatinib, inhibitors of glucose homeostasis, such as 2-deoxyglucose (2-DG), as well as the aforementioned BCL2 inhibitors (ABT263, ABT737) were found to remove senescent cells selectively from heterogeneous cell populations (Czabotar et al. 2014; Zhu et al. 2016; Chang et al. 2016; Tse et al. 2008).

Given the detrimental impact of senescent cells in many aspects of aging pathophysiology, as mentioned above, efforts to identify specific and effective senolytics have intensified in aging research (López-Otín et al. 2013; Childs et al. 2014; Newgard and Sharpless 2013). However, although senolytics show promise in cultured cells and in some animal models, their broad application in age-related pathologies faces important obstacles. Top among them is the fact that since most senolytic compounds are strong inducers of apoptosis, they are often toxic also for proliferating healthy cells and can delay wound healing (Demaria et al. 2014; Cang et al. 2015; Zhu et al. 2017). Thus, it is critical that advanced approaches be developed in which senescent cells are targeted more specifically. An earlier study suggested alternative drug therapies targeting multiple cellular pathways simultaneously (Zhu et al. 2015). Along these lines, combination therapies such as administration of dasatinib + quercetin (D + Q) have shown synergistic cytotoxicity on senescent cells in different age-related pathologic conditions (Farr et al. 2017; Lehmann et al. 2017; Roos et al. 2016; Ogrodnik et al. 2017). However, senolytic combinations do not appear sufficient to solve the problem of specificity, and the mechanisms of action of senolytics remain poorly understood at present. A number of alternative approaches are being explored to thwart the damage of senescent cells.

'Senostasis' (or 'senomorphism') refers to the suppression of a senescent trait to reduce the detrimental impact of senescent cells, as proposed in the case of lung senescence by Birch et al. (Birch et al. 2018). Initiatives to identify senostatic approaches focused on two properties of senescent cells have emerged in recent times. First, as mentioned above, senescent cells secrete high amounts of proinflammatory factors (the SASP, as mentioned above) and extracellular matrix proteases (Coppe et al. 2008; Kuilman and Peeper 2009). Given that SASP factors are capable of remodeling tissue and promoting a pro-inflammatory phenotype, they can exacerbate age-related pathology (Coppé et al. 2010). In a paracrine fashion, SASP factors enhance senescence in neighboring cells. Therefore, neutralizing SASP factors or preventing their expression has been proposed as a possible strategy to reduce the levels and deleterious actions of senescent cells in culture and in vivo (Moiseeva et al. 2013). In this regard, mTOR inhibition decreased the secretion of IL1-dependent cytokines, including IL6 (Laberge et al. 2015; Herranz et al. 2015; Noh et al. 2019), and the use of neutralizing antibodies against the IL1 receptor (IL1R) and soluble protein IL1A was sufficient to reduce the downstream activity of transcription factor NF-kB, thereby abolishing the transcription of SASP factors (Orjalo et al. 2009). Second, senescent cells are inherently immunogenic and are therefore targets of immunological surveillance (Sagiv and Krizhanovsky 2013; Sagiv et al. 2016; Biran et al. 2015), pointing to possible immune-associated interventions in senostatic therapy.

Importantly, the above-mentioned features of senescent cells are centered on the plasma membrane. In the following section, we review several membrane proteins that were identified as distinguishing senescent from non-senescent cells and we address the advantages of alternative senolytic and senostatic strategies aiming at cell surface proteins. We discuss their impact as well as future challenges to transform proof-of-concept evidence into successful therapies.

7.3 Targeting the Membrane of Senescent Cells

7.3.1 Biomarkers on the Plasma Membrane of Senescent Cells

Efforts are increasing to find novel senescence biomarkers associated with the outside of the plasma membrane because intracellular (nuclear or cytoplasmic) markers are believed to be less useful candidates for therapeutic interventions, particularly those directed at live cells. Plasma membrane markers are especially desirable for strategies aimed at removing senescent cells through the immune system (Krizhanovsky et al. 2008; Xue et al. 2007).

A number of proteins were found to be elevated on the plasma membrane of senescent cells, including DCR2, ICAM1, and NOTCH3 (Collado et al. 2005; Gorgoulis et al. 2005; Cui et al. 2013) (Table 7.1). Althubiti and coworkers (Althubiti et al. 2014) subsequently identified a number of plasma membrane-associated proteins including DEP1 and B2MG as potential biomarkers of senescence in cell culture and tissues. In a recent screen for senescent biomarkers using mice immunized with

Table 7.1 Reported protein markers on the plasma membrane of senescent cells

Surface protein	Function reported	Comments	References
DCR2	Biomarker		Collado et al. (2005)
ICAM1	Biomarker		Gorgoulis et al. (2005)
NOTCH3	Biomarker		Cui et al. (2013)
DEP1	Biomarker		Althubiti et al. (2014)
B2MG	Biomarker		Althubiti et al. (2014)
VIM (modified)	Biomarker		Frescas et al. (2017)
MICA/B	NKG2D ligands	Implicated in immunosurveillance	Sagiv et al. (2016), Spear et al. (2013), Muñoz et al. (2019)
ULBP2	NKG2D ligands	Implicated in immunosurveillance	Sagiv et al. (2016), Spear et al. (2013), Muñoz et al. (2019)
DPP4	Immune target Peptidase	Implicated in glucose metabolism, prostate cancer Membrane-bound and secreted Inhibited by gliptins	Kim et al. (2017)
SCAMP4	Cytokine secretion	Implicated in SASP	Kim et al. (2018)
CD36	Signaling Transcription by NF-κB	Implicated in SASP Inhibited by Sulfo-N-succinimidyl oleate (SSO)	Chong et al. (2018)

senescent lung fibroblasts, Frescas et al. (2017) found antibodies that recognized senescence-associated cell surface antigens. Among them, IgM clone 9H4 specifically detected a posttranslationally modified form of the protein vimentin, present in high levels on the plasma membrane of senescent cells (Frescas et al. 2017). The fact that the modified vimentin was also detectable in the plasma of aged mice (SAMP8, senescence-accelerated mouse prone 8) compared to wild-type (C57BL6) suggested that this protein could both accumulate on the cell surface and be secreted into the extracellular space (Frescas et al. 2017).

7.3.2 Cell Surface Proteins Enhancing Immune Surveillance

Besides the studies mentioned above, there have been other recent attempts to identify and remove senescent cells by targeting cell surface proteins specifically expressed in senescent cells. Sagiv et al. (2016) discovered that the expression of two ligands of NKG2D, MICA and ULBP2, increased greatly in cells rendered senescent following DNA damage, expression of oncogenes, or replicative exhaustion. It was already known that the two NKG2D ligands were more highly expressed in response to cellular stresses (Spear et al. 2013), and their accumulation on the surface of senescent fibroblasts facilitated NK cell-mediated cytotoxicity through a mechanism previously described (Sagiv et al. 2016; Soriani et al. 2009; Iannello et al. 2013; Schmiedel and Mandelboim 2018). The introduction of antibodies directed at MICA or ULBP2 as well as the ablation of *Nkg2d* genes in mice suppressed the elimination of senescent cells by NK cells, demonstrating that cell surface antigens can drive the removal of senescent cells by the immune system (Sagiv et al. 2016). In a follow-up study, shedding of NKG2D ligands from the cell surface was found to enable severely damaged senescent cells to escape immunosurveillance (Muñoz et al. 2019).

Kim et al. (2017) reported evidence that senescent cells could be selectively eliminated by exploiting the presence of another plasma membrane protein, DPP4 (dipeptidyl peptidase 4, also known as CD26). A screen of cell surface markers uncovered > 100 proteins more highly expressed on the plasma membrane of senescent fibroblasts than proliferating fibroblasts; among them, DPP4 was most highly increased in several senescent cells and was studied as a potential biomarker (Kim et al. 2017). DPP4 had been studied extensively in the context of glucose metabolism and type 2 diabetes (T2D) (Mentlein and Dipeptidyl-peptidase 1999; Röhrborn et al. 2015), as DPP4 cleaves and thereby inactivates glucagon-like peptide-1 (GLP-1), a protein that stimulates insulin production and helps to remove intestinal glucose. Members of the gliptin family of DPP4 inhibitors (e.g., sitagliptin, saxagliptin, and vildagliptin) have been used in combination with insulin for T2D therapy (Gomez-Peralta et al. 2018). Although they are effective for T2D therapy and are generally safe (Scheen 2018), their impact on senescence has not been examined at present. However, as DPP4 was highly abundant on the plasma membrane of senescent cells but not proliferating cells, antibody-dependent cell-mediated cytotoxicity (ADCC) was used to eliminate senescent cells selectively. A humanized anti-DPP4 antibody was employed to tag senescent (DPP4-positive) fibroblasts, and these were subsequently eliminated by using NK cells, while proliferating cells (DPP4-negative) were preserved (Kim et al. 2017).

7.3.3 Membrane-Associated Proteins Modulating the SASP

Among the unique features of senescence, the rise in SASP is particularly detrimental, as the ensuing proinflammatory phenotype and debilitated extracellular matrix compromise tissue and organ function and exacerbate age-related pathology (Hernandez-Segura et al. 2018; Campisi 2011). Thus, the SASP trait itself represents an attractive therapeutic target, as suppressing SASP can ameliorate the harmful impact of senescent cells. Along these lines, compounds such as rapamycin, metformin, or inhibitors of NFKB1 (NF-κB) or p38 have been developed to suppress SASP in senescent cells (Xie et al. 2016; Nacarelli and Sell 2017). However, as mentioned earlier, the long-term use of chemical compounds has inherent side effects.

Studies highlighting the role of plasma membrane proteins in promoting the SASP are beginning to emerge. The levels of the plasma membrane protein CD36 were found to increase strongly in response to different senescent stimuli in a range of cell types (Chong et al. 2018). The interaction of several ligands, including Aβ42 and oxidized low-density lipoprotein (oxLDL), with CD36 triggered signaling through SRC and p38 that culminated with the transcription of NF-κB-driven SASP factors. In this manner, CD36 was found to promote the SASP trait and to foment senescence (Chong et al. 2018).

An earlier report showed that IL1A (IL-1α) accumulated on the senescent cell surface and that the levels of major proinflammatory factors IL6 and IL8 decreased when IL1A was blocked by the IL1 receptor antagonist (Orjalo et al. 2009). This study suggested that IL1 was a major upstream regulator of the SASP program, and further indicated that regulation was elicited from the cell membrane (or the extracellular region). In agreement with this form of regulation, a surfaceome analysis of proteins on the plasma membrane by Kim et al. (2018) led the authors to propose a novel function for the membrane-associated protein SCAMP4 (secreted carrier membrane protein 4) in the SASP. As reported earlier for the related member SCAMP5, which promotes the calcium-regulated exocytosis of signal peptide-containing cytokines such as CCL5 in carcinoma cells and monocytes (Han et al. 2009), SCAMP4 was also found to be capable of enhancing the secretion of SASP factors in senescent cells (Kim et al. 2018), although the specific mechanism was not elucidated. At the molecular level, SCAMP4 protein accumulated on the surface of senescent fibroblasts due to the greater stability of SCAMP4 in senescent cells, opening a window for possible therapeutic actions. The authors concluded that SCAMP4 promotes the secretion of SASP factors including IL6, IL8, CCL2, CXCL1 and MIF, with a possible autocrine influence on spreading and further strengthening the senescent phenotype (Kim et al. 2018).

7.4 Future Directions

7.4.1 Future Directions in Senolysis and Senostasis Focused on the Plasma Membrane

The initial examples presented here provide proof of concept that plasma membrane proteins are suitable targets of senolytic interventions, aimed at killing senescent cells (as with DPP4, modified VIM, and NKG2D ligands) as well as senostatic interventions, aimed at preventing a detrimental senescent phenotype (as with SCAMP4 and CD36) (Fig. 7.1).

Universal markers of senescent on the plasma membrane appear particularly attractive for both senolytic and senostatic interventions. Many of the membrane markers listed in Table 7.1 were found elevated primarily in fibroblasts and endothelial cells that were rendered senescent via different mechanisms. However, other senescent cells residing in liver, muscle, pancreas, lung, epithelial tissues, immune system, etc., should also be investigated in order to identify truly ubiquitous plasma membrane targets. At the same time, plasma membrane markers of senescence that are specific to given cell types or to specific senescence programs may also be desirable for interventions aimed at select senescent cell populations. Whether universal or specific, more comprehensive knowledge of the plasma membrane proteome

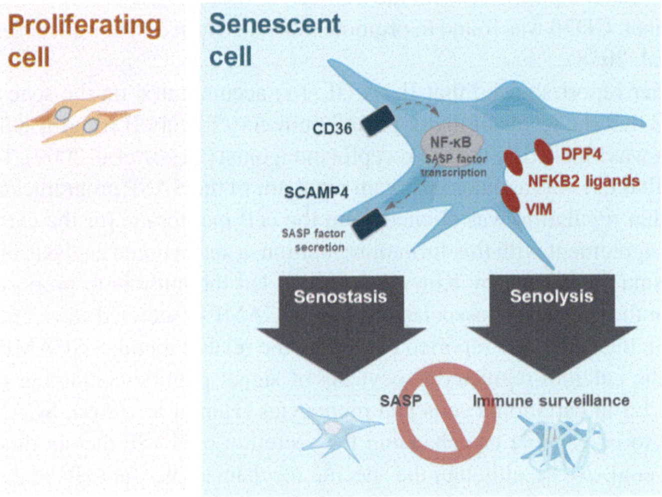

Fig. 7.1 Proteins present on the surface of senescent cells present opportunities for the elimination of senescent cells (senolysis) or the inhibition of senescent cell function (senostasis). Proof-of-principle examples include several plasma membrane proteins recently shown to be drastically elevated in senescent cells. DPP4, NFKB2 ligands, and modified VIM were found to serve as antigens that triggered strong immunosurveillance, while suppression of SCAMP4 and CD36 potently reduced the senescence-associated secretory trait

in a wide range of senescent cell types and senescence inducers is needed. Moreover, in principle other plasma membrane-associated molecules (e.g., lipids, carbohydrates, and nucleic acids) may also be suitable targets of universal or specific senescence-modulatory interventions.

Additional attention should be paid to develop improved approaches (molecular, biochemical, pharmacological, etc.) for senolysis and senostasis. An immediate advance over ADCC is the technology known as antibody-drug conjugate (ADC), through which specific cell surface markers can be identified by an antibody linked to a highly cytotoxic drug (Sievers and Senter 2013). Developed for cancer therapy, ADC antibodies have been shown capability to deliver drugs that cause cell death, generally by eliciting DNA damage, but could be repurposed to deliver senolytic or senostatic agents.

In specific cases in which the plasma membrane markers of senescence are enzymes, select drugs capable of modulating their function could be exploited. The peptidase DPP4 is a putative example of this approach, as numerous safe and effective inhibitors of DPP4 are already in the market for T2D and warrant direct attention. With increased knowledge of senescence-associated plasma membrane markers, other proteins may also emerge that can be modulated pharmacologically.

In addition, superior models in which to test the therapeutic benefit of senostatic and senolytic interventions are needed. Mouse models have provided helpful information thus far, but improved human models, e.g., organoids and tissue grafts, will help to establish the experimental specifics of efficacy of such approaches.

Finally, methods to deliver therapeutic agents directed at factors on the plasma membrane of senescent cells, whether broad-spectrum of selective, represent an ongoing challenge. Senescent cells directly accessible via the circulatory system (immune cells, endothelial cells, etc.) may be readily targeted. However, strategies to reach senescent cells located deeper in organs are necessary in order to gain full benefit from interventions to modulate the senescent cell compartment and achieve favorable outcomes.

Acknowledgements This work was supported by the NIA IRP, NIH, and by the Chungnam National University research fund.

References

Alcorta DA, Xiong Y, Phelps D, Hannon G, Beach D, Barrett JC (1996) Involvement of the cyclin-dependent kinase inhibitor p16 (INK4a) in replicative senescence of normal human fibroblasts. Proc Natl Acad Sci USA 93(24):13742–13747

Althubiti M, Lezina L, Carrera S, Jukes-Jones R, Giblett SM, Antonov A et al (2014) Characterization of novel markers of senescence and their prognostic potential in cancer. Cell Death Dis 5(11):e1528

Baker DJ, Wijshake T, Tchkonia T, LeBrasseur NK, Childs BG, van de Sluis B et al (2011) Clearance of p16Ink4a-positive senescent cells delays ageing-associated disorders. Nature 479:232–236

Benanti JA, Galloway DA (2004) Normal human fibroblasts are resistant to RAS-induced senescence. Mol Cell Biol 24:2842–2852

Ben-Porath I, Weinberg RA (2004) When cells get stressed: an integrative view of cellular senescence. J Clin Invest 113(1):8–13

Biran A, Perelmutter M, Gal H, Burton DG, Ovadya Y, Vadai E et al (2015) Senescent cells communicate via intercellular protein transfer. Genes Dev 29(8):791–802

Birch J, Barnes PJ, Passos JF (2018) Mitochondria, telomeres and cell senescence: implications for lung ageing and disease. Pharmacol Ther 183:34–49

Campisi J (2005) Senescent cells, tumor suppression, and organismal aging: good citizens, bad neighbors. Cell 120(4):513–522

Campisi J (2011) Cellular senescence: putting the paradoxes in perspective. Curr Opin Genet Dev 21(1):107–112

Cang S, Iragavarapu C, Savooji J, Song Y, Liu D (2015) ABT-199 (venetoclax) and BCL-2 inhibitors in clinical development. J Hematol Oncol 8:129

Casella G, Munk R, Kim KM, Piao Y, De S, Abdelmohsen K et al (2019) Transcriptome signature of cellular senescence. Nucl Acids Res 47(14):7294–7305

Chang J, Wang Y, Shao L, Laberge RM, Demaria M, Campisi J et al (2016) Clearance of senescent cells by ABT263 rejuvenates aged hematopoietic stem cells in mice. Nat Med 22(1):78–83

Chen Z, Trotman LC, Shaffer D, Lin HK, Dotan ZA, Niki M et al (2005) Crucial role of p53-dependent cellular senescence in suppression of Pten-deficient tumorigenesis. Nature 436(7051):725–730

Childs BG, Baker DJ, Kirkland JL, Campisi J, van Deursen JM (2014) Senescence and apoptosis: dueling or complementary cell fates? EMBO Rep 15(11):1139–1153

Chong M, Yin T, Chen R, Xiang H, Yuan L, Ding Y et al (2018) CD36 initiates the secretory phenotype during the establishment of cellular senescence. EMBO Rep 19(6):pii: e45274

Collado M, Serrano M (2010) Senescence in tumours: evidence from mice and humans. Nat Rev Cancer 10(1):51–57

Collado M, Gil J, Efeyan A, Guerra C, Schuhmacher AJ, Barradas M et al (2005) Tumour biology: senescence in premalignant tumours. Nature 436:642

Coppe JP, Patil CK, Rodier F, Sun Y, Munoz DP, Goldstein J et al (2008) Senescence-associated secretory phenotypes reveal cell-nonautonomous functions of oncogenic RAS and the p53 tumor suppressor. PLoS Biol 6:2853–2868

Coppé JP, Desprez PY, Krtolica A, Campisi J (2010) The senescence-associated secretory phenotype: the dark side of tumor suppression. Annu Rev Pathol 5:99–118

Cui H, Kong Y, Xu M, Zhang H (2013) Notch3 functions as a tumor suppressor by controlling cellular senescence. Cancer Res 73:3451–3459

Czabotar PE, Lessene G, Strasser A, Adams JM (2014) Control of apoptosis by the BCL-2 protein family: implications for physiology and therapy. Nat Rev Mol Cell Biol 15(1):49–63

Demaria M, Ohtani N, Youssef SA, Rodier F, Toussaint W, Mitchell JR et al (2014) An essential role for senescent cells in optimal wound healing through secretion of PDGF-AA. Dev Cell 31(6):722–733

Dimri GP, Lee X, Basile G, Acosta M, Scott G, Roskelley C et al (1995) A biomarker that identifies senescent human cells in culture and in aging skin in vivo. Proc Natl Acad Sci USA 92:9363–9367

Farr JN, Xu M, Weivoda MM, Monroe DG, Fraser DG, Onken JL et al (2017) Targeting cellular senescence prevents age-related bone loss in mice. Nat Med 23(9):1072–1079

Frescas D, Roux CM, Aygun-Sunar S, Gleiberman AS, Krasnov P, Kurnasov OV et al (2017) Senescent cells expose and secrete an oxidized form of membrane-bound vimentin as revealed by a natural polyreactive antibody. Proc Natl Acad Sci USA 114(9):E1668–E1677

Gomez-Peralta F, Abreu C, Gomez-Rodriguez S, Barranco RJ, Umpierrez GE (2018) Safety and efficacy of DPP4 inhibitor and basal insulin in type 2 diabetes: an updated review and challenging clinical scenarios. Diabetes Ther 9(5):1775–1789

Gorgoulis VG, Pratsinis H, Zacharatos P, Demoliou C, Sigala F, Asimacopoulos PJ et al (2005) p53-dependent ICAM-1 overexpression in senescent human cells identified in atherosclerotic lesions. Lab Invest 85:502–511

Gorgoulis V, Adams PD, Alimonti A, Bennett DC, Bischof O, Bishop C et al (2019) Cellular senescence: defining a path forward. Cell 179(4):813–827

Guerra C, Mijimolle N, Dhawahir A, Dubus P, Barradas M, Serrano M et al (2003) Tumor induction by an endogenous K-ras oncogene is highly dependent on cellular context. Cancer Cell 4:111–120

Han C, Chen T, Yang M, Li N, Liu H, Cao X (2009) Human SCAMP5, a novel secretory carrier membrane protein, facilitates calcium-triggered cytokine secretion by interaction with SNARE machinery. J Immunol 182(5):2986–2996

Herbig U, Jobling WA, Chen BP, Chen DJ, Sedivy JM (2004) Telomere shortening triggers senescence of human cells through a pathway involving ATM, p53, and p21(CIP1), but not p16(INK4a). Mol Cell 14(4):501–513

Hernandez-Segura A, Nehme J, Demaria M (2018) Hallmarks of Cellular Senescence. Trends Cell Biol 28(6):436–453

Herranz N, Gallage S, Mellone M, Wuestefeld T, Klotz S, Hanley CJ et al (2015) mTOR regulates MAPKAPK2 translation to control the senescence-associated secretory phenotype. Nat Cell Biol 17(9):1205–1217

Iannello A, Thompson TW, Ardolino M, Lowe SW, Raulet DH (2013) p53-dependent chemokine production by senescent tumor cells supports NKG2D-dependent tumor elimination by natural killer cells. J Exp Med 210(10):2057–2069

Kim WY, Sharpless NE (2006) The regulation of INK4/ARF in cancer and aging. Cell 127(2):265–275

Kim KM, Noh JH, Bodogai M, Martindale JL, Yang X, Indig FE et al (2017) Identification of senescent cell surface targetable protein DPP4. Genes Dev 31(15):1529–1534

Kim KM, Noh JH, Bodogai M, Martindale JL, Pandey PR, Yang X et al (2018) SCAMP4 enhances the senescent cell secretome. Genes Dev Jul 32(13–14):909–914

Kirkland JL, Tchkonia T (2017) Cellular senescence: a translational perspective. EBioMedicine 21:21–28

Krishnamurthy J, Torrice C, Ramsey MR, Kovalev GI, Al-Regaiey K, Su L et al (2004) Ink4a/Arf expression is a biomarker of aging. J Clin Invest 114:1299–1307

Krizhanovsky V, Yon M, Dickins RA, Hearn S, Simon J, Miething C et al (2008) Senescence of activated stellate cells limits liver fibrosis. Cell 134(4):657–667

Kuilman T, Peeper DS (2009) Senescence-messaging secretome: SMS-ing cellular stress. Nat Rev Cancer 9(2):81–94

Laberge RM, Sun Y, Orjalo AV, Patil CK, Freund A, Zhou L et al (2015) MTOR regulates the pro-tumorigenic senescence-associated secretory phenotype by promoting IL1A translation. Nat Cell Biol 17(8):1049–1061

Lehmann M, Korfei M, Mutze K, Klee S, Skronska-Wasek W et al (2017) Senolytic drugs target alveolar epithelial cell function and attenuate experimental lung fibrosis ex vivo. Eur Respir J 50(2):pii: 1602367

Liu Y, Sanoff HK, Cho H, Burd CE, Torrice C, Ibrahim JG et al (2009) Expression of p16(INK4a) in peripheral blood T-cells is a biomarker of human aging. Aging Cell 8:439–448

López-Otín C, Blasco MA, Partridge L, Serrano M, Kroemer G (2013) The hallmarks of aging. Cell 153(6):1194–1217

Lowe SW, Cepero E, Evan G (2004) Intrinsic tumour suppression. Nature 432(7015):307–315

Mentlein R, Dipeptidyl-peptidasc IV (1999) (CD26)–role in the inactivation of regulatory peptides. Regul Pept 85(1):9–24

Moiseeva O, Deschênes-Simard X, St-Germain E, Igelmann S, Huot G, Cadar AE et al (2013) Metformin inhibits the senescence-associated secretory phenotype by interfering with IKK/NF-κB activation. Aging Cell 12(3):489–498

Muñoz DP, Yannone SM, Daemen A, Sun Y, Vakar-Lopez F, Kawahara M et al (2019) Targetable mechanisms driving immunoevasion of persistent senescent cells link chemotherapy-resistant cancer to aging. JCI Insight, pii: 124716
Muñoz-Espín D, Serrano M (2014) Cellular senescence: from physiology to pathology. Nat Rev Mol Cell Biol 15:482–496
Nacarelli T, Sell C (2017) Targeting metabolism in cellular senescence, a role for intervention. Mol Cell Endocrinol 455:83–92
Narita M, Nunez S, Heard E, Narita M, Lin AW, Hearn SA et al (2003) Rb-mediated heterochromatin formation and silencing of E2F target genes during cellular senescence. Cell Jun 113:703–716
Nelson JA, Krishnamurthy J, Menezes P, Liu Y, Hudgens MG, Sharpless NE et al (2012) Expression of p16(INK4a) as a biomarker of T-cell aging in HIV-infected patients prior to and during antiretroviral therapy. Aging Cell 11:916–918
Newgard CB, Sharpless NE (2013) Coming of age: molecular drivers of aging and therapeutic opportunities. J Clin Invest 123(3):946–950
Noh JH, Kim KM, Pandey PR, Noren Hooten N, Munk R, Kundu G et al (2019) Loss of RNA-binding protein GRSF1 activates mTOR to elicit a proinflammatory transcriptional program. Nucl Acids Res 47(5):2472–2486
Ogrodnik M, Miwa S, Tchkonia T, Tiniakos D, Wilson CL, Lahat A, Jurk D et al (2017) Cellular senescence drives age-dependent hepatic steatosis. Nat Commun 8:15691
Orjalo AV, Bhaumik D, Gengler BK, Scott GK, Campisi J (2009) Cell surface-bound IL-1alpha is an upstream regulator of the senescence-associated IL-6/IL-8 cytokine network. Proc Natl Acad Sci USA 106(40):17031–17036
Ovadya Y, Krizhanovsky V (2018) Strategies targeting cellular senescence. J Clin Invest 128(4):1247–1254
Parrinello S, Samper E, Krtolica A, Goldstein J, Melov S, Campisi J (2003) Oxygen sensitivity severely limits the replicative lifespan of murine fibroblasts. Nat Cell Biol 5:741–747
Röhrborn D, Wronkowitz N, Eckel J (2015) DPP4 in Diabetes. Front Immunol 6:386
Roos CM, Zhang B, Palmer AK, Ogrodnik MB, Pirtskhalava T, Thalji NM et al (2016) Chronic senolytic treatment alleviates established vasomotor dysfunction in aged or atherosclerotic mice. Aging Cell 15(5):973–977
Sagiv A, Krizhanovsky V (2013) Immunosurveillance of senescent cells: the bright side of the senescence program. Biogerontology 14(6):617–628
Sagiv A, Burton DG, Moshayev Z, Vadai E, Wensveen F, Ben-Dor S et al (2016) NKG2D ligands mediate immunosurveillance of senescent cells. Aging 8(2):328–344
Scheen AJ (2018) The safety of gliptins: updated data in 2018. Expert Opin Drug Saf 17(4):387–405
Schmiedel D, Mandelboim O (2018) NKG2D ligands-critical targets for cancer immune escape and therapy. Front Immunol 9:2040
Sievers EL, Senter PD (2013) Antibody-drug conjugates in cancer therapy. Annu Rev Med 64:15–29
Soriani A, Zingoni A, Cerboni C, Iannitto ML, Ricciardi MR, Di Gialleonardo V et al (2009) ATM-ATR-dependent up-regulation of DNAM-1 and NKG2D ligands on multiple myeloma cells by therapeutic agents results in enhanced NK-cell susceptibility and is associated with a senescent phenotype. Blood 113(15):3503–3511
Spear P, Wu MR, Sentman ML, Sentman CL (2013) NKG2D ligands as therapeutic targets. Cancer Immun 13:8
Tse C, Shoemaker AR, Adickes J, Anderson MG, Chen J, Jin S et al (2008) ABT-263: a potent and orally bioavailable Bcl-2 family inhibitor. Cancer Res 68(9):3421–3428
van Deursen JM (2014) The role of senescent cells in ageing. Nature 509(7501):439–446
von Zglinicki T (2002) Oxidative stress shortens telomeres. Trends Biochem Sci 27(7):339–344
Waaijer ME, Parish WE, Strongitharm BH, van Heemst D, Slagboom PE, de Craen AJ et al (2012) The number of p16INK4a positive cells in human skin reflects biological age. Aging Cell 11(4):722–725

Wiley CD, Flynn JM, Morrissey C, Lebofsky R, Shuga J, Dong X et al (2017) Analysis of individual cells identifies cell-to-cell variability following induction of cellular senescence. Aging Cell 16(5):1043–1050

Xie J, Wang X, Proud CG (2016) mTOR inhibitors in cancer therapy, F1000Res

Xue W, Zender L, Miething C, Dickins RA, Hernando E, Krizhanovsky V et al (2007) Senescence and tumour clearance is triggered by p53 restoration in murine liver carcinomas. Nature 445(7128):656–660

Yosef R, Pilpel N, Tokarsky-Amiel R, Biran A, Ovadya Y, Cohen S et al (2016) Directed elimination of senescent cells by inhibition of BCL-W and BCL-XL. Nat Commun 7:11190

Zhu Y, Tchkonia T, Pirtskhalava T, Gower AC, Ding H, Giorgadze N et al (2015) The Achilles' heel of senescent cells: from transcriptome to senolytic drugs. Aging Cell 14(4):644–658

Zhu Y, Tchkonia T, Fuhrmann-Stroissnigg H, Dai HM, Ling YY, Stout MB et al (2016) Identification of a novel senolytic agent, navitoclax, targeting the Bcl-2 family of anti-apoptotic factors. Aging Cell 15(3):428–435

Zhu Y, Doornebal EJ, Pirtskhalava T, Giorgadze N, Wentworth M, Fuhrmann-Stroissnigg H et al (2017) New agents that target senescent cells: the flavone, fisetin, and the BCL-XL inhibitors, A1331852 and A1155463. Aging 9(3):955–963

Zindy F, Quelle DE, Roussel MF, Sherr CJ (1997) Expression of the p16INK4a tumor suppressor versus other INK4 family members during mouse development and aging. Oncogene 15:203–211

Part IV
Novel Senescence-Associated Markers and Targets

Chapter 8
In Situ Detection of miRNAs in Senescent Cells in Archival Material

Ioannis S. Pateras, Sofia D. P. Theodorou, Niki V. Chouliari, Dimitris Veroutis, Panagiotis-Georgios Passias, Konstantinos Evangelou, and Vassilis G. Gorgoulis

Abstract The detection of senescent cells has been challenging. We have recently developed a novel hybrid histo-/immunohisto-chemical method that can bypass several constraints of detection of senescent cells. It is based on the development of a novel reagent called GL13 (SenTraGor™) that binds lipofuscin, a non-degradable metabolic by-product that is known as a hallmark of senescence. This chapter provides a unique approach to detect formalin fixed paraffin embedded tissues miRNAs in senescent GL13-reactive cells in routine. Given the significant role of miRNAs in senescent programs, this approach enables for the first time to monitor miRNAs in the context of senescence in situ in archival material. Notably, this assay improves our capacity to detect in vivo senescent cells which favors rationalization of senotherapeutic drugs. Although these agents are in early clinical trials, their introduction in routine practice will transform healthcare as we know it, bringing to the fore the necessity for precise detection of senescent cells in tissues.

Keywords Senescence · Co-detection · miRNAs · GL13 (SenTraGor™) · Senotherapeutics

---

I. S. Pateras · S. D. P. Theodorou · N. V. Chouliari · D. Veroutis · P.-G. Passias · K. Evangelou · V. G. Gorgoulis (✉)
Molecular Carcinogenesis Group, Department of Histology and Embryology, Medical School, National and Kapodistrian University of Athens, Athens, Greece
e-mail: vgorg@med.uoa.gr; vgorgoulis@gmail.com

V. G. Gorgoulis
Center for New Biotechnologies and Precision Medicine, Medical School, National and Kapodistrian University of Athens, Athens, Greece

Biomedical Research Foundation, Academy of Athens, Athens, Greece

Faculty Institute for Cancer Sciences, Manchester Academic Health Sciences Centre, University of Manchester, Manchester, UK

Manchester Centre for Cellular Metabolism, University of Manchester, Manchester Academic Health Science Centre, Manchester, UK

D. Muñoz-Espin and M. Demaria (eds.), *Senolytics in Disease, Ageing and Longevity*, Healthy Ageing and Longevity 11,
https://doi.org/10.1007/978-3-030-44903-2_8

8.1 Introduction

Hayflick and Moorhead in 1961 introduced the term cellular senescence by stating that "*A consideration of the cause of the eventual degeneration of these strains leads to the hypothesis that non-cumulative external factors are excluded and that the phenomenon is attributable to intrinsic factors which are expressed as senescence at the cellular level*" in order to describe the limited in vitro lifetime of diploid fibroblasts (Hayflick and Moorhead 1961). The latter was later linked to telomere attrition and this type of senescence was termed replicative senescence (Levy et al. 1992; Allsopp et al. 1995; Campisi 1997). Over the years it has been recognized that other forms of senescence exist that are independent of telomere shortening. They are collectively known as stress-induced premature senescence (SIPS) and can be induced by various stressogenic stimuli (Debacq-Chainiaux et al. 2016; Gorgoulis and Halazonetis 2010; Gorgoulis et al. 2019; Myrianthopoulos et al. 2019). Hence senescence can be considered as a stress response mechanism that preserves homeostasis. It has been linked with tumor suppression, although it seems that it is an imperfect anti-tumor barrier (Gorgoulis and Halazonetis 2010; Childs et al. 2014).

One of the major challenges to study cellular senescence is the accurate detection of senescent cells in clinical material. Precise recognition of senescent cells in vivo is essential not only to understand the role of senescence, but also to monitor the effect of ongoing therapeutic modalities in various clinical settings. Currently available assays exhibit several disadvantages for in vivo detection of senescent cells (Muñoz-Espín and Serrano 2014). To overcome this issue, we recently synthesized a lipophilic, biotin-linked Sudan Black-B (SBB) analogue called GL13 (SenTraGor™) suitable for the detection of senescent cells in archival formalin-fixed paraffin-embedded tissues (FFPE) (Evangelou et al. 2017). FFPE tissues are the most common resource for the study of different biomarkers in pathological laboratories.

Here we demonstrate a novel approach that allows the detection of miRNAs in senescent cells in routinely processed material employing the new commercially available GL13 (SenTraGor™) compound. Given that numerous microRNAs are involved in senescent pathways, this approach enables for the first time the in situ detection of miRNAs in senescent cells in FFPE (Abdelmohsen and Gorospe 2015; Bischof and Martínez-Zamudio 2015). As a miR target we selected miR34c, since it is involved in senescence in various settings (He et al. 2007; Disayabutr et al. 2016; Komseli et al. 2018). Specifically, for in situ co-detection we proceed as follows: (a) visualization of miR34c employing the Chromogenic 3, 3'-Diaminobenzidine (DAB) system (brown), (b) hematoxylin counterstain (blue), (c) visualization of senescent cells utilizing Emerald chromogen (bluish-green) (Fig. 8.1). For ISH we utilized the Locked Nucleic Acid (LNA) technology that allows superior hybridization properties (Vester and Wengel 2004). To improve detection sensitivity we employed double-digoxigenin (DIG) labeled (at both 5′ and 3′ ends) probes as previously described (Komseli et al. 2018) (Table 8.1). Double-DIG labeled U6 small nuclear RNA (snRNA) and Scramble-miR served as positive and negative control of the

Step	Process	Time	Comments
1	**Deparaffinization - Rehydration**	70 min	
2	***In situ* Hybridization**		
	Proteinase K treatment	10 min	
	Pre-hybridization	15 min	optional
	Preparation of Hybridization mixtures - Denaturation of probes	4 min	
	Hybridization	60 min	
	Post-hybridization washes	30 min	
3	**Blocking**	30 min	
4	**Incubation with anti-DIG-POD**	60 min	
5	**Visualization of miR34c positivity with DAB chromogen**	1-5 min	miR34c positive cells stain with brown color
6	**GL13 (SenTraGor™) staining**		
	Permeabilization	3-7 min	
	Dehydration with Ethanol	10 min	
	Incubation with GL13 (SenTraGor™)	10 min	
	Post-GL13 staining washes	1-3 min	
7	**Blocking**	45 min	
8	**Incubation with mouse anti-biotin**	overnight	
9	**Incubation with mouse HRP polymer**	15 min	
10	**Hematoxylin counterstaining**	5-10 sec	Nuclei stain blue
11	**Visualization of GL13 (SenTraGor™) positivity with emerald chromogen**		
	Treatment with Emerald chromogen	5-10 min	GL13 (SenTraGorTM) positive cells stain with bluish-green color
	Air dry	2 min	
12	**Dehydration - Mounting**	2 min	

Fig. 8.1 Workflow of the protocol employed to detect micro-RNA in senescent cells in routinely processed FFPE tissues. Washing steps with PBS1x or PBS-T or tap water are not included in the overview

method. Notably, this assay can be employed in conjuction with the detection of proteins, enabling the co-detection of coding and non-coding RNAs on senescent cells.

8.2 Translational Aspect and Future Perspectives

The establishment of co-detection approaches that render feasible the analysis of multiple parameters at the same time in senescent context in archival material, opens a new exciting field. GL13 (SenTraGor^{TM}) provides a unique opportunity of co-visualization with additional markers, which has the following advantages: (a) improve the detection accuracy of senescent cells, (b) address how various biological markers are altered in different senescent settings, (c) reveal the existence of senescent cells subpopulations. Within this frame this assay allows for the first time monitoring of the spatiotemporal status of miRNAs involved in p53/p21$^{WAF1/CIP1}$,

Table 8.1 Troubleshooting tips

Assay	Problem	Possible cause	Solution—suggestion
ISH	Absent or low signal	Suboptimal treatment with Proteinase K	Increase duration of Proteinase K treatment
		Probe or target degradation	Avoid contamination by RNases
		Absence of target miRNA	Include a positive control along with LNA U6 snRNA probe
		Hybridization temperature higher than the proper one	Decrease the hybridization temperature (Tm)
		Increased post-hybridization washes	Optimize the post-hybridization washes by adjusting the following parameters: (a) temperature of incubation, (b) concentration of SCC buffer, (c) duration of incubation with SCC buffer
	High background staining	Hybridization temperature lower than the proper one	Increase the hybridization temperature (Tm)
		Increased concentration of probe applied	Reduce the concentration of the probe
		Insufficient post hybridization washes	Optimize the post-hybridization washes by adjusting the following parameters: (a) temperature of incubation, (b) concentration of SCC buffer, (c) duration of incubation with SCC buffer
		Insufficient incubation with blocking solution	Increase the duration of treatment with the blocking solution

(continued)

Table 8.1 (continued)

Assay	Problem	Possible cause	Solution—suggestion
		Increased concentration of the anti-DIG-POD	Decrease the concentration of the anti-DIG-POD
		Dry out of sections	Ensure that all treatments are employed in humidifying chambers
GL13 (SenTraGo^{rTM}) staining	Absent or low signal	Insufficient penetration into cells	Increase duration of treatment with Triton X-100 0.5%
		Low duration of GL13 (SenTraGor^{TM}) incubation	Increase the duration of incubation with GL13 (SenTraGor^{TM})
		Increased post—GL13 (SenTraGor^{TM}) staining washes	Decrease the duration of washing with Ethanol 50% and/or Triton X-100 0.5%. Monitor washing under light microscope
		Reduced mouse-anti-biotin incubation	Increase the duration of incubation with mouse anti-biotin
		Absence of lipofuscin	Choose a positive control
	High background staining	High duration of GL13 (SenTraGor^{TM}) incubation	Reduce GL13 (SenTraGor^{TM}) incubation in a step-wise manner
		Decreased post—GL13 (SenTraGor^{TM}) staining washes	Increase the duration of washing with Ethanol 50% and/or Triton X-100 0.5%. Monitor washing under light microscope

p16^{INK4A}/pRb senescent signaling pathways and Senescence Associated Secretory Phenotype (SASP) in archival material (Munk et al. 2018).

Additionally taking into consideration that several single in situ assays already drive modern oncological therapies (Cooks et al. 2019), precise detection of senescent cells in vivo will further encourage the precision medical approach. Within this context it has been shown that during aging senescent cells accumulate in humans

(Campisi et al. 2019). Besides, accumulating evidence support that senescent cells are implicated in numerous age-related pathologies including cancer (Myrianthopoulos et al. 2019; Campisi et al. 2019). We recently identified in vivo senescent cells in classical Hodgkin lymphoma (cHL) and Langerhans cell Histiocytosis (Myrianthopoulos et al. 2019). Notably, cHL cases with increased number of senescent Hodgkin Reed-Sternberg cells are associated with unresponsiveness to first-line and salvage treatment, revealing that GL13 positivity can be a biomarker with prognostic utility in a common hematological malignancy. Additionally, we detected senescent cells in archival material from cancer patients receiving radiotherapy (Myrianthopoulos et al. 2019). The effect of these cells on disease prognosis remains to be clarified.

Along this line senotherapeutic agents that specifically target and eliminate senescent cells by triggering apoptosis or suppress SASP have developed in recent years. Although, the safety and efficacy of these drugs in humans is tested, the introduction of effective agents that could target selectively senescent cells will transform clinical practice as we know it (Kirkland et al. 2017). However, the introduction of senotherapeutics renders necessary the accurate detection of senescent cells in archival material. Cumulatively from a translational angle the in situ co-detection of GL13 (SenTraGor™) and miRNAs in archival material could help us: (i) to determine the senescent burden from biopsies in age-related disorders prior and after senotherapeutic treatment, monitoring the efficacy of the therapeutic intervention, (ii) to know the "biological age" or "functional status" of different tissues during ageing (i.e. using the status of senescent cells as a biomarker of ageing), (iii) to determine the senescent burden in premalignant lesions or after treatment with common therapeutic modalities (i.e. chemotherapy and radiotherapy) with potential prognostic value. Overall, the method presented in this chapter first of all improves our ability to monitor senescent cells in different settings which in turn facilitates the rationalization of senotherapeutic approaches.

8.3 Materials

(1) **Equipment**

Coplin staining jars
Coverslips
Forceps
Glass trough
Gloves
Humidifying chamber
Hybridizer
Laboratory bench centrifuge
Laboratory oven
Light microscope
Microcentrifuge

Micropipettes
PCR cycler
Pipettes
Positively Charged Glass Slides
Soft paper
Staining/Slide racks
Sterile 1.5 and 0.5 ml micro-tubes (e.g., Sarstedt, Cat. no. 72.695.500)
Sterile pipette tips (e.g. Greiner bio-one, Cat. no.: 739290)
Surgical blade
Surgical blade handle
Syringes Q-Max Syringe Filters (Frisenette 13 mm filter, membrane 0.22 μm, Cat. no. 13CA022-100)

(2) **Reagents**

(2a) **In Situ Hybridization**

Xylene
Ethanol (EtOH)
RNase removal solution
Hydrogen peroxide 30% (Scharlau, Cat no: HI01351000)
Tween 20 (Sigma, Cat. no. P9416)
Triton X-100 (Sigma, Cat. no. T8787)
miRCURY LNA miRNA ISH Buffer Set (FFPE) (Qiagen, Cat. no 339450)
Proteinase K (Qiagen, Cat. no. 339450 or Macherey-Nagel, Cat. no. 740506)
PBS tablets (Gibco by life technologies, Cat. no.: 18912-014)
UltraPure 20X SSC Buffer (Invitrogen, Cat. no.: 15557-044)
Sheep serum (Jackson Immunoresearch, Cat. no. 013-000-121)
BSA (Serva, Cat. no. 11930)
Liquid Blocker Super Pan Pen or alternatively DAKO Pen

Antibody:
Anti-Digoxigenin-POD Fab fragments (Merck, Cat. no. 11207733910)

Probes:
hsa-miR-34c-3p (Exiqon, Cat. no.: 611870-360, now complemented by Qiagen)
Scramble-miR (Exiqon, Cat. no.: 699004-360, now complemented by Qiagen)
U6, hsammurno (Exiqon, Cat. no.: 699002-360, now complemented by Qiagen)

Detection Assay:
DAB Chromogen and DAB Substrate (included in the IHC Kit provided by Abcam, Cat. no. ab183286)

(2b) **GL13 staining**

GL13 (SenTraGor™) (Lab Supplies). Reconstitute it following manufacturer's instructions.
Sodium Chloride (NaCl).
Tris HCl.

Counterstaining:
Hematoxylin

Antibodies:
Anti-biotin antibody (Abcam, Hyb-8, ab201341)

Detection Assay:
Emerald chromogen (included in the IHC Kit provided by Abcam, Cat. no. ab183286)

Mounting:
Non-aqueous mounting medium (included in the IHC Kit provided by Abcam, Cat. no. ab183286)

(3) **Setup of Solutions**
Ethanol (EtOH) 50%, Ethanol (EtOH) 70%: dissolve Ethanol (EtOH) 100% with dH2O to the appropriate final concentration.

Proteinase K buffer (Qiagen, Cat no 339450): to prepare 20 mg/ml, add to lyophilized proteinase K, 900 ml RNase-free water, 5 ml of 1 M Tris-HCl (pH 7.4), 2 ml of 0.5 M EDTA and 0.2 ml of 5 M NaCl. Adjust final volume to 1000 ml. [based on the Instruction manual miRCURY® LNA® miRNA ISH Optimization Kits (FFPE) Handbook (provided by Qiagen, page 19)]. *Alternatively*, Proteinase K buffer (Macherey-Nagel, Cat no. 740506): to prepare 20 mg/ml, add to lyophilized proteinase K the appropriate quantity of stabilizing solution following the manufacturer's instructions.

Triton X-100 0.5% in PBS 1x: Dilute Triton X-100 30 to 0.5% with PBS 1x. (Sterilize).

Hydrogen peroxide (H2O2) 3% solution: Dilute Hydrogen peroxide 30 to 3% with dH2O.

PBS 1X buffer: Dilute 1 PBS tablet with 500 mL dH2O (sterilize).

PBS-T (0.1%) buffer: To 1L of PBS 1x add 1 ml Tween 20.

TBS-T: Dilute 50 nM Tris HCl and 150 nM NaCl along with Tween 20 0.05% in ddH2O. Adjust pH to 7.6.

1X miRCURY LNA miRNA ISH Buffer Set (FFPE): Dilute 2X miRCURY LNA miRNA ISH Buffer Set (FFPE, Qiagen) with equal amount of ddH2O based on the Instruction manual miRCURY® LNA® miRNA ISH Optimization Kits (FFPE) Handbook (provided by Qiagen, page 20).

5x SSC; 1x SSC; 0.2x SSC: Dilute 20x SSC buffer to 5x, 1x and 0.2x accordingly with ddH2O.

Blocking and Dilutant reagent: the preparation of both reagents is strictly based on the Instruction manual miRCURY® LNA® miRNA ISH Optimization

Kits (FFPE) Handbook (provided by Qiagen, page 20). Below is provided the corresponding procedure:

(a) To 15 mL PBS-T (0.1%), add 300 μL Sheep serum (2% final concentration) and label this tube "Blocking solution".
(b) Transfer 5 ml from the tube labeled "Blocking solution" to a new tube and label the second tube "Dilutant solution".
(c) To the tube labeled "Blocking solution", add 300 μl 30% BSA (final concentration 1%). The Blocking solution is ready to use.
(d) To the tube labeled "Dilutant solution", add 5 ml PBS (final concentration 0.05% Tween and 1% sheep serum) and 330 μl 30% BSA (final concentration 1%). The Dilutant solution is ready to use.

Ethanol 50%, EtOH 70%: dissolve EtOH 100% with dH2O to the appropriate final concentration.

30% BSA: Add 30 g of BSA to 100 ml PBS1x.

1% BSA: Add 1 g of BSA to 100 ml PBS1x.

DAB working solution: Add 1 drop of DAB Chromogen to 1 ml DAB Substrate and mix well. Avoid unnecessary exposure to light. The preparation of this solution is strictly based on the protocol provided in the Instructions for use of TripleStain IHC kit by Abcam (ab183286, page 9).

Antibody Blocker: dilute 1 part of Antibody Blocker with 39 parts of ddH2O. To dissolve it, employ water bath or hot plate to heat to 80 °C. The preparation of this solution is strictly based on the protocol provided in the Instructions for use of TripleStain IHC kit by Abcam (ab183286, page 10).

Blocker A, Blocker B (included in the IHC Kit provided by Abcam, Cat no. ab183286).

8.4 Methods

Formalin fixed and paraffin embedded (FFPE) tissues are placed on positively charged slides in order to reduce the detachment of processed tissues. The sections should be 5–7 μm thick.

Caution: the procedure is carried at room temperature (RT) unless otherwise specified.

1. **Tissue preparation**
   Place slides in staining/slide rack and proceed with **deparaffinization** by melting paraffin and dissolving it in xylene that is followed by **rehydration** in decreasing concentrations of ethanol solutions.

   – Incubation in lab oven at 60 °C for 30 min.

- Wash in Xylene for 10 min. Move sections to a new glass trough and incubate for another 10 min in fresh xylene.
- Incubate in 99.9% Ethanol for 10 min.
- Incubate in 95% Ethanol for 5 min.
- Incubate in 80% Ethanol for 5 min.
- Incubate in 70% Ethanol for 5 min.
- Incubate in 50% Ethanol for 5 min.
- Wash in TBS 1x for 5 min.

Caution: from this point on sections should not be dried out.

2. **In situ hybridization (ISH)** (Note 1)

2a. **Treatment with Proteinase K**: incubate slides with proteinase K buffer (final concentration: 20 mg/ml) for 10 min at 37 °C (Note 2).

2b. Wash in PBS 1x for 5 min.

2c. Pre-hybridization (optional): treat the slides with the hybridization buffer (i.e. 1X miRCURY LNA miRNA ISH Buffer Set (FFPE), without including the probe for 15 min at the hybridization temperature (T_m).

2d. **Preparation of the hybridization mixtures—Denaturation of the probes**: Place the probes onto separate RNase-free tubes and proceed to denaturation at 90 °C for 4 min employing a PCR cycler. Afterwards, place the denatured probes into a microcentrifuge and spin down. Immediately add the hybridization buffer [i.e. 1X miRCURY LNA miRNA ISH Buffer Set (FFPE)] to each tube and place the hybridization mixture on each section (an alternative procedure is provided in Note 3). It is necessary to prepare separately along with the hybridization mix including the probe(s) of interest, one hybridization mix including U6 snRNA probe and another one having the scramble-miR probe which are employed as positive and negative control respectively.

2e. **Hybridization**: Treat the sections with the hybridization mix for 1 h at the appropriate T_m in a hybridizer (Note 4).

After hybridization step, RNase-free context is no longer required.

2f. **Post-hybridization washes**: Wash sections in prewarmed SSC buffer series (Note 5). Quickly remove excess probe by rinsing slides in pre-warmed 5x SSC buffer. Then proceed as follows:

- Wash two times with pre-warmed 5x SSC for 5 min at the hybridization temperature (T_m),
- Wash two times with pre-warmed 1x SSC for 5 min at the hybridization temperature (T_m),
- Wash two times with pre-warmed 0.2x SSC for 5 min at the hybridization temperature (T_m),

2g. Wash two times with PBS-T; the first one rapidly and the second one for 5 min.

3. **Blocking**. Place the sections in a humidifying chamber.

   3a. Treatment with Hydrogen peroxide (H_2O_2): Block endogenous peroxidase activity with 3% H_2O_2 for 15 min at RT (incubation is performed in the dark). This step should be performed before incubating with anti-DIG-POD.

   3b. Wash two times with PBS-T; the first one rapidly and the second one for 5 min.

   3c. Incubate with the blocking solution for 15 min at RT.

4. **Incubation with sheep anti-DIG POD**

   4a. Treat sections with sheep anti-DIG-POD for 1 h at RT in a humidifying chamber: dilute sheep anti-DIG-POD at 1:800 in Dilutant buffer.

   4b. Wash with PBS-T, three times; the first one rapidly and the rest one for 3 min. each at RT.

5. **Visualization with DAB Chromogen**

   5a. Incubate sections with DAB working solution for 1 min at RT (Note 6). To block the reaction, transfer the sections to a Coplin staining jar filled with tap water, and rinse gently with running tap water. Treatment with DAB working solution generates brown color. To visualize the staining employ light microscope.

   Caution: Use appropriate care handling since DAB is potential carcinogenic. Avoid unnecessary light exposure.

6. **GL13 (SenTraGor™) staining**

   6a. **Permeabilization** with Triton X-100 0.5% for 3–5 min at RT.

   6b. Wash with PBS-T for 5 min.

   6c. **Dehydration with Ethanol**:

      Apply 50% Ethanol on the slides for 5 min at RT.
      Apply 70% Ethanol on the slides for 5 min at RT.

   6d. **Incubation with GL13 (SenTraGor™)**: Apply GL13 and incubate for 8 min at 37 °C in a humidifying chamber (Note 7).

   6e. **Post-GL13 (SenTraGor™) staining washes** (Note 8).

   - After incubation with GL13 reagent, remove the coverslip gently with a tip or a surgical blade (that is attached to a handle) or a thin edged forceps and wipe with the help of a soft paper extra GL13 stain from all sites. Then immediately rinse with 50% Ethanol for 1–3 min.
   - Wash with PBS 1x for 3 min.
   - Rinse with Triton X-100 0.5% for 1–2 min at RT.

7. **Blocking**

   7a. Treatment with Antibody Blocker (pre-heated at 80 °C, according to the setup of the solution) for 10 min in a humidifying chamber at 80 °C.

7b. Remove sections and cool for 5–10 s.
7c. Wash with PBS-T, three times; the first one rapidly and the rest for 3 min each at RT.
7d. Treatment with Blocker A for 30 min in a humidifying chamber at RT. Make sure to cover the sections completely.
7e. Wash with PBS-T, three times; the first one rapidly and the rest for 3 min each at RT.
7f. Treatment with Blocker B for 5 min in a humidifying chamber at RT. Make sure to cover the sections completely.
7g. Wash with PBS-T, three times; the first one rapidly and the rest for 3 min each at RT.

8. **Incubate with mouse anti-biotin** (Note 9)

8a. Treat sections with mouse anti-biotin (diluted 1:300 in PBS 1x) overnight at 4 °C. To avoid drying out, place the sections in a humidifying chamber.
8b. Wash with PBS-T, three times; the first one rapidly and the rest for 3 min each at RT.

9. **Incubate with mouse HRP polymer** (Note 10)

9a. Treat sections with mouse HRP polymer for 15 min at RT in a humidifying chamber.
9b. Wash with PBS-T, three times; the first one rapidly and the rest for 3 min each at RT.

10. **Hematoxylin counterstaining**

10a. Immerse biological samples in a glass trough with diluted filtered hematoxylin (in distilled water). 5–10 s of incubation with hematoxylin is enough to stain nuclei blue. To visualize the staining employ light microscope.
10b. Immediately rinse sections with running tap water for 1 min. In case the desired color contrast is not reached, repeat step 10a (Note 11).
10c. Wash with PBS-T, three times; the first one rapidly and the rest for 3 min each at RT.

11. **Visualization of GL13 (SenTraGor™) positivity with emerald chromogen** (Note 12)

11a. Treat samples with Emerald chromogen for 5–10 min in a humidifying chamber. For optimal staining monitor the reaction under light microscope. Emerald staining gives a bluish-green color.
11b. Wash with tap water for 1 min. Wipe off with the help of a soft paper the excessive water from all sites and air dry slides for 2 min.

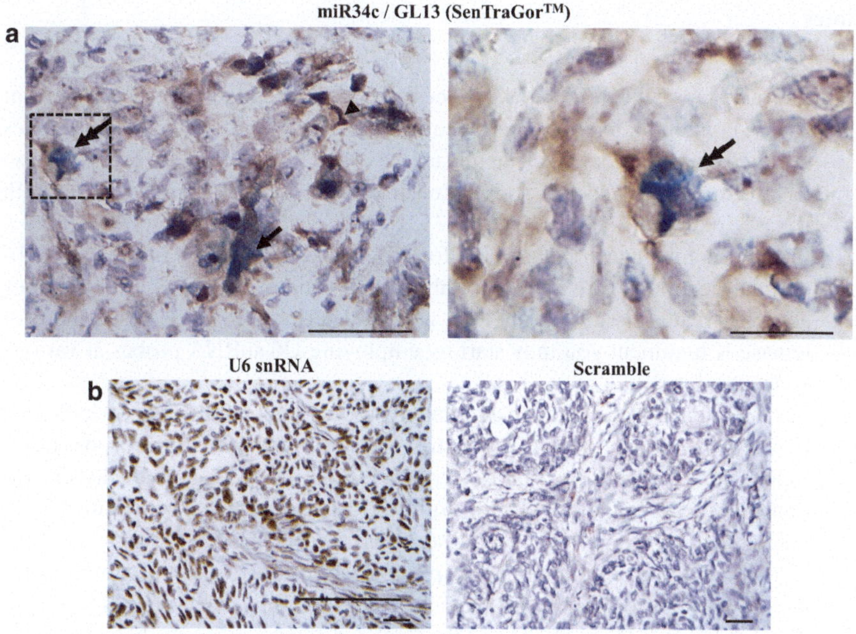

Fig. 8.2 **a**. In situ hybridization (ISH) targeting miR34c followed by GL13 (SenTraGor™) staining in FFPE human sample with head and neck squamous cell carcinoma after radiotherapy. Arrowhead denotes miR34c positive staining (with brown color); arrow demonstrates GL13 (SenTraGor™) positive staining (with bluish green color); double arrow depicts miR34c/GL13 double positive staining. Scale bar: 50 μm (25 μm). **b**. For the optimization of ISH U6 snRNA and scramble-miRNA probes serve as positive and negative control respectively. Scale bar: 50 μm

12. **Dehydration—Mounting**

12a. Place slides in staining/slide rack and proceed with dehydration in increasing concentrations of ethanol solutions followed by incubation in xylene.

- Incubate in 80% Ethanol for 20 s.
- Incubate in 95% Ethanol for 20 s.
- Incubate in 99.9% Ethanol for 1 min.
- Incubate in Xylene for 20 s.

12b. Mount sections by applying 1 drop of non-aqueous mounting medium. Then cover the section with coverslip, by avoiding air bubble traping. Prior visualizing leave the mounting medium at RT for 10–15 min. For optimal visualization of GL13 staining the light microscope should be equipped with high magnification lenses (magnification 400x; 630x) (Fig. 8.2).

A troubleshooting guide is provided in the end of the chapter (Table 8.1).

Notes

1. In every step of the miRNA ISH process specific handling is required to prevent RNase contamination both of the reagents and the tissue samples. Wear gloves during the entire process, clean all surfaces with RNase removal solutions, use RNase-free tubes and tips and make sure that all solutions are prepared with RNase-free water.
2. Incubation time and concentration for Proteinase K should be optimized. Sub-optimal treatment leads to faint or absence of staining, whereas if the sections are over-treated tissue morphology will be deteriorated. To optimize the proteinase K treatment you may start by employing U6 snRNA probes at various concentrations.
3. For the probe denaturation an alternative process can be followed. Specifically, first of all prepare the hybridization mixture by diluting the specific probes in 1X miRCURY LNA miRNA ISH Buffer Set (FFPE). Then denature the hybridization mixture at 90 °C for 4 min, followed by spin down. Then immediately apply the hybridization mix on the sections.
4. Hybridization temperature (T_m) is optimized according to each probe. A useful rule of thumb is to start 30 °C below the RNA Tm when detecting RNA targets. The Tm for miR34c is 56 °C and the final concentration ranges between 50 to 80 nM (dilution factor: 1:500–1:312.5).
5. The stringency of post-hybridization washing depends on the Tm, the concentration of the washing buffer and the duration per wash. Proper washing removes non-specific hybrids and unbound probes.
6. Duration of treatment with DAB solution should be optimized. Therefore, optimal staining should be verified by real-time observation under the light microscope.
7. Given that GL13 is dissolved in Ethanol, to avoid evaporation of Ethanol, cover the biological sample (i.e. sections) during the incubation with a coverslip as previously described (Evangelou and Gorgoulis 2017). Importantly, we receive the GL13 reagent by employing a syringe that carries a 2 mm filter (to avoid receiving aggregates). Incubation with GL13 is optimized; duration usually ranges from 4–14 min at 37 °C. For 14 min incubation at 37 °C, you may initially incubate for 7 min, remove the coverslip, add fresh GL13, place a new coverslip on the section and incubate additionally 7 min. Real-time observation under the light microscope is necessary in order to determine the optimal duration.
8. Optimize the washing step with Ethanol 50% and Triton X-100 0.5% by monitoring this step under the light microscope. Avoid increased incubation time, since this may lead to reduced staining intensity.
9. For tissues having high endogenous biotin (like liver), an additional step is required prior incubation with mouse anti-biotin in order to block endogenous biotin. For this purpose you may employ the Streptavidin/Biotin blocking kit (SP 2002, from Vector).

10. Prior incubation with HRP Polymer an additional step with hydrogen peroxide treatment for 10 min can be included. Additionally, when employing alkaline phosphatase TBS-T is recommended instead of PBS-T as the wash buffer, since phosphate in the PBS may interfere with the activity of alkaline phosphatase.
11. Avoid overstaining with hematoxylin since this will cover the brown staining.
12. Emerald is water soluble, and therefore hematoxylin counterstaining should be performed prior Emerald staining.

Acknowledgements Financial support was from the European Union's Horizon 2020 research and innovation programme under the Marie Sklodowska-Curie grants agreement No. 722729 (SYN-TRAIN); the Welfare Foundation for Social & Cultural Sciences (KIKPE), Greece; Pentagon Biotechnology Ltd, UK and NKUA-SARG grants No 70/3/8916, 70/3/12128.

Conflict of Interest The authors wish to declare no conflict of interest.

References

Abdelmohsen K, Gorospe M (2015) Noncoding RNA control of cellular senescence. Wiley Interdiscip Rev RNA 6(6):615–629

Allsopp RC, Chang E, Kashefi-Aazam M, Rogaev EI, Piatyszek MA, Shay JW et al (1995) Telomere shortening is associated with cell division in vitro and in vivo. Exp Cell Res 220:194–200

Bischof O, Martínez-Zamudio RI (2015) MicroRNAs and lncRNAs in senescence: a re-view. IUBMB Life 67(4):255–267

Campisi J (1997) The biology of replicative senescence. Eur J Cancer Part A 33(5):703–709

Campisi J, Kapahi P, Lithgow GJ, Melov S, Newman JC, Verdin E (2019) From discoveries in ageing research to therapeutics for healthy ageing. Nature 571(7764):183–192

Childs BG, Baker DJ, Kirkland JL, Campisi J, Deursen JM (2014) Senescence and apoptosis: dueling or complementary cell fates? EMBO Rep 15(11):1139–1153

Cooks T, Theodorou SD, Paparouna E, Rizou SV, Myrianthopoulos V, Gorgoulis VG et al (2019) Immunohisto(cyto)chemistry: an old time classic tool driving modern oncological therapies. Histol Histopathol 34(4):335–352

Debacq-Chainiaux F, Ben Ameur R, Bauwens E, Dumortier E, Toutfaire M (2016) Toussaint O Stress-Induced (Premature) senescence, 243–262

Disayabutr S, Kim EK, Cha SI, Green G, Naikawadi RP, Jones KD et al (2016) MIR-34 MIR-NAs regulate cellular senescence in type II alveolar epithelial cells of patients with idiopathic pulmonary fibrosis. PLoS ONE 11(6):1–15

Evangelou K, Gorgoulis VG (2017) Sudan black b, the specific histochemical stain for lipofuscin: a novel method to detect senescent cells. Methods Mol Biol 1534:111–119

Evangelou K, Lougiakis N, Rizou SV, Kotsinas A, Kletsas D, Muñoz-Espín D et al (2017) Robust, universal biomarker assay to detect senescent cells in biological specimens. Aging Cell 16(1):192–197

Gorgoulis VG, Halazonetis TD (2010) Oncogene-induced senescence: the bright and dark side of the response. Curr Opin Cell Biol 22(6):816–827

Gorgoulis V, Adams PD, Alimonti A, Bennett DC, Bischof O, Bishop C et al (2019) Cellular senescence: defining a path forward. Cell 179(4):813–827

Hayflick L, Moorhead PS (1961) The serial cultivation of human diploid cell strains. Exp Cell Res 25(7):585–621

He L, He X, Lowe SW, Hannon GJ (2007) MicroRNAs join the p53 network–another piece in the tumour-suppression puzzle. Nat Rev Cancer 7(11):819–822

Kirkland JL, Tchkonia T, Zhu Y, Niedernhofer LJ, Robbins PD (2017) The clinical potential of senolytic drugs. J Am Geriatr Soc 65(10):2297–2301

Komseli ES, Pateras IS, Krejsgaard T, Stawiski K, Rizou SV, Polyzos A et al (2018) A prototypical non-malignant epithelial model to study genome dynamics and concurrently monitor micro-RNAs and proteins in situ during oncogene-induced senescence. BMC Genom 19(1):1–22

Levy MZ, Allsopp RC, Futcher AB, Greider CW, Harley CB (1992) Telomere end-replication problem and cell aging. J Mol Biol 225(4):951–960

Munk R, Panda AC, Grammatikakis I, Gorospe M, Abdelmohsen K(2018) Senescence-associated microRNAs. In: International review of cell and molecular biology, vol 334, 1st edn. Elsevier Inc, pp 1–26

Muñoz-Espín D, Serrano M (2014) Cellular senescence: from physiology to pathology. Nat Rev Mol Cell Biol 15(7):482–496

Myrianthopoulos V, Evangelou K, Vasileiou PVS, Cooks T, Vassilakopoulos TP, Pangalis GA et al (2019) Senescence and senotherapeutics: a new field in cancer therapy. Pharmacol Ther 193:31–49

Vester B, Wengel J (2004) LNA (Locked Nucleic Acid): high-affinity targeting of complementary RNA and DNA. Biochemistry 43(42):13233–13241

Chapter 9
Novel Probes and Carriers to Target Senescent Cells

Beatriz Lozano-Torres, Juan F. Blandez, Félix Sancenón, and Ramón Martínez-Máñez

Abstract Detecting senescent cells has become a task of increasing importance in recent years. The implication of cellular senescence in a variety of physiological processes as well as in various diseases is the main reason why potential diagnostic tools for senescent cells are sought. This chapter clearly and concisely addresses the discussion of the published tools (molecular based probes and nanocarriers) to carry out the diagnosis of senescent cells both in vitro and in vivo and in tissues. The comprehensive reading of this chapter will allow the reader to overview the advances in the area of the detection of cellular senescence using probes and carriers as diagnostic methods.

Keywords Cellular senescence · Ageing · Detection · Probes · Nanocarriers

9.1 The Importance of Targeting Senescent Cells

Overcoming the gradual ageing of the world population remains a formidable challenge for health professionals. As life expectancy increases, the vulnerability to suffer

B. Lozano-Torres · J. F. Blandez · F. Sancenón · R. Martínez-Máñez (✉)
Instituto Interuniversitario de Investigación de Reconocimiento Molecular y Desarrollo Tecnológico (IDM), Universitat Politècnica de València, Universitat de València, Camino de Vera s/n, 46022 Valencia, Spain
e-mail: rmaez@qim.upv.es

Departamento de Química, Universitat Politècnica de València, Camino de Vera s/n, 46022 València, Spain

B. Lozano-Torres · F. Sancenón · R. Martínez-Máñez
CIBER de Bioingeniería, Biomateriales y Nanomedicina (CIBER-BBN), Madrid, Spain

B. Lozano-Torres · J. F. Blandez · F. Sancenón · R. Martínez-Máñez
Unidad Mixta UPV-CIPF de Investigación en Mecanismos de Enfermedades y Nanomedicina, Universitat Politècnica de València, Centro de Investigación Príncipe Felipe, Valencia, Spain

Unidad Mixta de Investigación en Nanomedicina y Sensores, Universitat Politècnica de València, IIS La Fe, Valencia, Spain

D. Muñoz-Espin and M. Demaria (eds.), *Senolytics in Disease, Ageing and Longevity*, Healthy Ageing and Longevity 11,
https://doi.org/10.1007/978-3-030-44903-2_9

from some type of chronic disease or pathology increases exponentially. Indeed, diseases such as atheroesclerosis, osteoarthritis, cataracts, cardiac hypertrophy, renal dysfunction, lipodystrophy, sarcopenia and neurodegenerative diseases are usually related with ageing (Niedernhofer and Robbins 2018). In recent years, the scientific community has made a great effort to understand the molecular mechanisms that lead to ageing. A pioneering work, in which the mice were genetically modified to produce high levels of p53 (a tumor suppressor protein), revealed that the mice had almost non-existent cancer rates, but their half-life was significantly shorter and had aging symptoms at an early age (Tyner et al. 2002). In contrast, "super p53" mice (carrying p53-tg alleles in addition to the two endogenous alleles) presented the same rate of cancer resistance but normal lifespan, probably because p53 is normally regulated in this model (García-Cao et al. 2002). It was concluded that there is a strong relationship between cellular senescence and ageing.

The main biological role of cellular senescence is to prevent the proliferation of damaged or stressed cells (Munoz-Espin and Serrano 2014). In addition, tissue damage and cellular senescence provide critical signals for cellular reprogramming and it has been reported to be beneficial in various tissue repair settings (Mosteiro et al. 2018; Ritschka et al. 2017). However, upon persistent damage or during ageing, the process of tissue repair becomes inefficient and senescent cells tend to accumulate in tissues promoting local inflammation, tissue ageing, dysfunction and destruction, and potentially, tumorigenesis and metastasis (Munoz-Espin and Serrano 2014; McHugh and Gil 2018). Besides, it has been demonstrated that the selective elimination of senescent cells in genetic models amends a variety of ageing-associated symptons, ameliorate long-term degenerative processes and extends both healthspan and lifespan (Baker et al. 2016; Baker et al. 2011). In fact, senotherapy (i.e. the selective elimination of senescent cells by using pharmacologically active compounds) has been shown to ameliorate and even revert certain diseases in mouse models (Soto-Gamez and Demaria 2017; Zhu et al. 2015; Paez-Ribes et al. 2019; Childs et al. 2017). A number of these pharmacologically active small compounds or senolytics have been reported to have remarkable therapeutic effects on multiple diseases in mice in association with the elimination of senescent cells (Paez-Ribes et al. 2019). In fact, the elimination of senescent cells is considered today as a promising strategy to treat ageing-related diseases and delay ageing (Paez-Ribes et al. 2019; Xu et al. 2018).

A related key issue in the senescence field is the design of methods to easily identify cellular senescence (Lozano-Torres et al. 2019). A first tool to detect senescent cells came from the discovery that senescent cells show high levels of lysosomal β-galactosidase (β-Gal) activity (Dimri et al. 1995; Biran et al. 2017). This is known as senescence-associated β-galactosidase (SA-β-Gal) and it has served as the basis for the design of a number of chromo-fluorogenic probes (vide infra) (Lozano-Torres et al. 2019). Some other markers for senescence, related to the expression of cell cycle inhibitors and/or tumour suppressors, have been reported. For instance, a key tool for the identification of cellular senescence in the context of ageing and age-related diseases was the discovery of $p16^{INK4A}$, a cyclin-dependent kinase inhibitor that acts as a regulator of cell cycle arrest in senescent cells (Serrano et al. 1993; Serrano et al.

1997). Another markers of senescent cells are p19^{ARF}, a positive regulator of the transcription factor p53, and intracellular lipofuscin accumulation (Sharpless and Sherr 2015; Hernandez-Segura et al. 2018). However, most of these markers cannot be used for the assessment of senescence in vivo (de Jesus and Blasco 2012). In this scenario, advances in the area of detection of cell senescence are necessary, especially as tools for assessment of treatment response of senotherapeuthics in senescence in vivo models and clinical trials (Lozano-Torres et al. 2019).

9.2 Chemical Probes for the Detection of Senescent Cells

9.2.1 SA-β-Gal-Dependent Chromo-Fluorogenic Probes

Proliferating cells express β-Gal encoded by the GLB1 gene that shows normal activity at pH 4. Since in 1995 Dimri and collaborators reported the overexpression of β-Gal enzyme activity at suboptimal pH (pH 6), due to the expansion of the lysosomal compartment of senescent cells (Lee et al. 2006), this enzymatic activity, known as senescence-associated β-galactosidase (SA-β-Gal), is one of the most commonly used markers for the detection of cellular senescence (Dimri et al. 1995). This marker is easy to detect and reliable both in vivo and in vitro (de Jesus and Blasco 2012; Lee et al. 2014; Zhang et al. 2017). SA-β-Gal activity is conveniently measured using chromogenic or fluorogenic probes taking advantage of the fact that the enzyme hydrolyses galactosidic bonds (Agostini et al. 2012). In this context, most reported molecular probes able to detect SA-β-Gal are composed by two subunits, i.e. (i) a β-galactose residue as reactive fragment and (ii) a chromophore or fluorophore as a signalling moiety. In most cases both subunits are linked through a *O*-glycosidic or *N*-glycosidic covalent bonds or through self-immolative fragments containing one glycosidic linkage (Lozano-Torres et al. 2017). In these probes the emission of the fluorophore at a certain wavelength is drastically reduced or shifted when galactose is linked to the selected signalling unit. However, in the presence of SA-β-Gal, the glycosidic bond is hydrolysed and the fluorophore or chromophore is released, restoring its emission wavelength or its colour (Lozano-Torres et al. 2019).

The first chromogenic probe for β-Gal, an *ortho*-nitrophenyl-β-galactoside (**ONPG**, Fig. 9.1), was reported by Aizawa (1939). This compound is colourless; however, in presence of β-Gal, the *O*-glycosidic bond is hydrolysed yielding galactose and *ortho*-nitrophenol, which is yellow. A further colorimetric probe for β-Gal detection was synthesized by Horwitz and collaborators in 1964, named **X-Gal**. It consisted of a galactose linked to a substituted indole and it is a well-known colorimetric probe for the detection of β-Gal (Horwitz et al. 1964). In presence of the enzyme, the galactosyl moiety in **X-Gal** is removed giving a 5-bromo-4-chloro-3-indolyl residue, which dimerizes spontaneously and is oxidized into the insoluble intensely blue product 5,5′-dibromo-4,4′-dichloro-indigo following the mechanism showed in Fig. 9.1. After **X-Gal** was reported, many variants have been synthesized

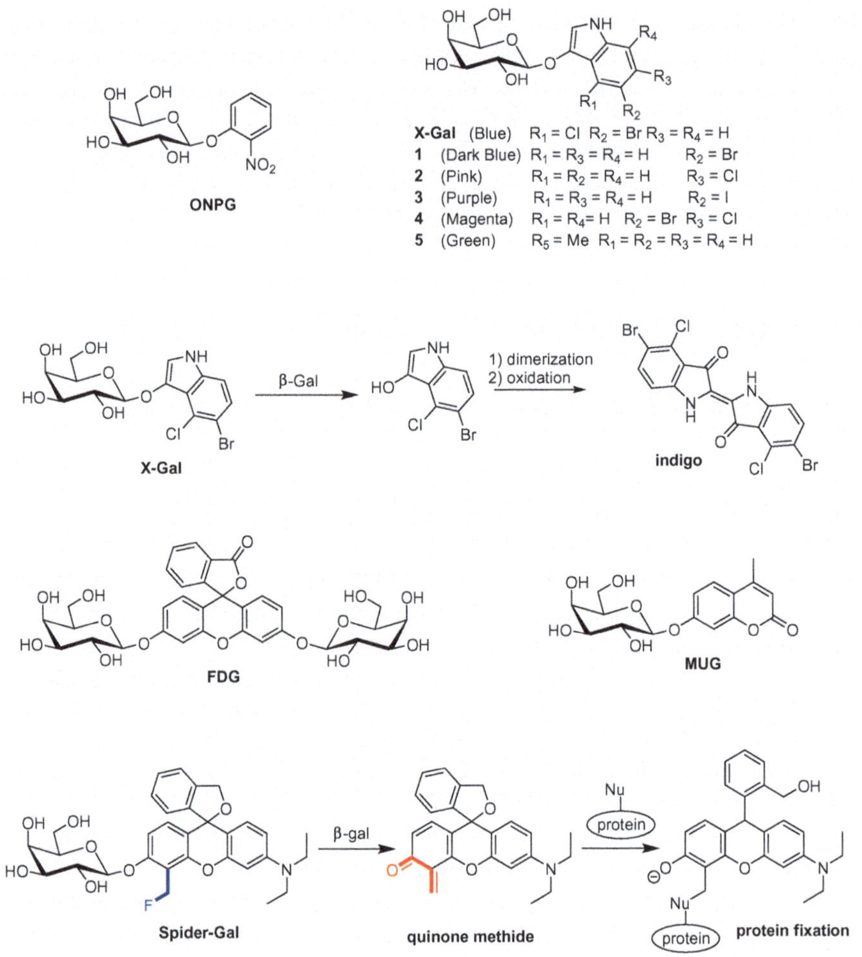

Fig. 9.1 Most commonly and commercially available probes for cellular senescence detection

with slight modifications giving final compounds that display different colours in the presence of β-Gal (see Fig. 9.1). All these derivatives are usually commercialized as β-Gal activity and senescence detection kits. It is important to remark that **ONPG** and **X-Gal** are not permeable to the cell membrane and they can only be used in lysed cells.

In a pioneering work, Rotman et al. described in 1961 the first fluorescent probe for measuring β-Gal activity, (fluorescein-di-(β-D-galactopyranoside)) (Rotman 1961; Rotman et al. 1963), **FDG** in Fig. 9.1. In this study the authors reported that PBS solutions (pH 7.2) of **FDG** were non-emissive (λ_{exc} = 400 nm) but a remarkable emission band at 490 nm was developed in the presence of β-Gal, due to the hydrolysis of glycosidic bonds and the subsequent release of fluorescein. These optical changes were also used to detect β-Gal activity in droplets of silicone oil containing **FDG** and the enzyme. Following a similar approach Hirchsmann and co-workers prepared in 1962 a compound named **MUG** (Woods and Sapp 1962) for the detection of β-Gal activity (see Fig. 9.1). Glycine buffer solutions of **MUG** (pH 7) did not show any emission ($\lambda_{exc} = 365$ nm) whereas in presence of the enzyme, the galactoside residue was removed yielding the highly fluorescent 4-methylumbelliferone fluorophore ($\lambda_{em} = 455$ nm), resulting in an overall OFF-ON response. Unfortunately, **MUG** is also impermeable to the cell membrane. A great number of other coumarin-based fluorophores for β-Gal activity detection were later described (Gee et al. 1999; Chilvers et al. 2001).

A recent commercially available probe was developed by Urano and co-workers, who described a fluorogenic hemicyanin probe for labelling living senescent cells in culture and in living tissues. The prepared probe, (E)-2-[2-(6-hydroxy-2,3-dihydro-1*H*-xanthen-4-yl)vinyl]-3,3-dimethyl-1-propyl-3*H*-indole-1-ium, is commonly known as **Spider-Gal** (Fig. 9.1) (Doura et al. 2016). PBS (pH 7.4) solutions of **Spider-Gal** present an absorption band at 525 nm and are almost non-emissive. However, addition of β-Gal enzyme induces the appearance of a highly fluorescent band centred at 560 nm (excitation at 525 nm). The observed emission enhancement was ascribed to the β-Gal-induced hydrolysis of **Spider-Gal** that yielded the **quinone methide** derivative shown in Fig. 9.1. In the original research, the response of **Spider-Gal** was tested in HEK cells transfected or not with *LacZ* gene and the probe was able to distinguish between different levels of β-Gal by confocal microscopy, flow cytometry and ex vivo fixed tissues. In the last years, **Spider-Gal** has become in a widely used tool by researchers in the field of cellular senescence due to its high sensitivity and its high retentivity inside cells.

Although only a few probes based on the hydrolysis of *O*-glycosidic, *N*-glycosidic covalent bonds by SA-β-Gal enzyme activity are marketed; many probes based on this same idea of binding a monosaccharide to a fluorophore have been described (Lozano-Torres et al. 2019). Some of these probes for SA-β-Gal detection have been validated in *lacZ* transfection models which are not exactly realistic senescence models. In fact, in *lacZ* transfection models the transfected cells highly overexpress *lacZ* (with excessively high concomitant activity of β-Gal), and the bacterial enzyme is cytosolic and not localized in lysosomes (the usual subcellular location of SA-β-Gal activity) which could lead to failures in the detection of senescence (Lozano-Torres et al. 2017, 2019). Moreover, many probes have been validated in vivo-*lacZ* models according to two approaches: (i) transfecting tumour cells with the pCMV-lacZ plasmid and injected them in the animal, or (ii) labelling tumours overexpressing with avidin-β-Gal (Kamiya et al. 2007; Gu et al. 2016). A more realistic approach to validate SA-β-Gal probes used cell lines with high lysosomal β-Gal activity per se, such as cells from congenital dyskeratosis patients, SHIN3, SKOV3, OVK18, OVCAR3, OVCAR4, OVCAR5, OVCAR8 or HUVEC (Agostini et al. 2012; Imai et al. 1990; Hung et al. 1992; Hamilton et al. 1983; Yuan et al. 1997).

Certain probes have also been tested to detect replicative senescence; senescence induced by reactive oxygen species or chemically induced senescence. Kim et al.

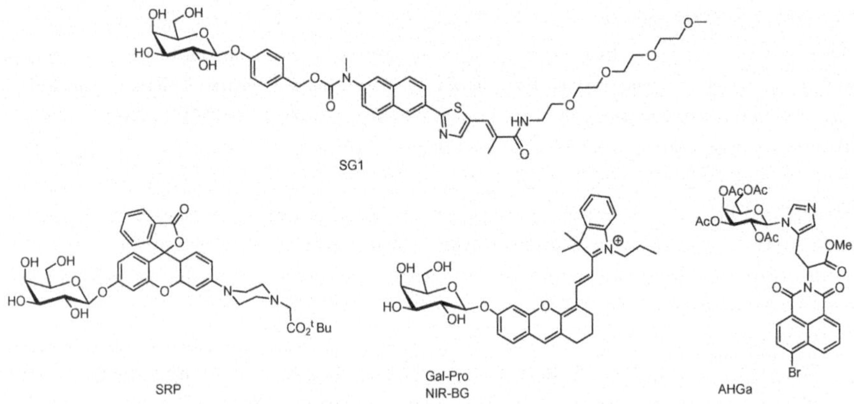

Fig. 9.2 Some other fluorogenic probes for senescence detection

reported the ratiometric two-photon fluorescent probe (**SG1**) composed of the fluorophore 6-(benzo[d]thiazol-2′-yl)-2-(methylamino)-naphthalene, a self immolative carbamate as linker, a polyether as solubilizing group and galactose as hydrolytic moiety (Fig. 9.2) (Lee et al. 2014). The authors found that the maximum of the emission spectra of **SG1** changed from 461 to 540 nm in the presence of β-Gal. β-Gal enzyme induced the hydrolysis of galactose in the anomeric carbon with the subsequent rupture of the self immolative linker that released the polyether-functionalized 6-(benzo[d]thiazol-2′-yl)-2-(methylamino)-naphthalene fluorophore. Probe **SG1** was tested in human diploid fibroblasts (HDF) in a replicative model of senescence. By treating the cells with **SG1**, the authors observed that the ratio F_{yellow}/F_{blue} increased from 0.30 to 1.13 when cells accumulated passages and were excited at a wavelength of 750 nm. Finally, the authors used the probe for staining 7-month-old and 26-month-old Sprague-Dawley rat skin tissues, showing an enhancement of the F_{yellow}/F_{blue} ratio.

A new senescence-specific fluorescent probe based in a rhodol core linked to galactose (**SRP**) was studied in vascular endothelial HUVEC cells, where senescence was induced with H_2O_2 (Fig. 9.2). The **SRP** probe showed a high chemoselectivity in the presence of β-Gal versus other enzymes. Senescent cultured HUVEC cells exhibited high fluorescent signal after treatment with the probe, in sharp contrast with non-senescent HUVEC cells (Kim et al. 2018). Recently, Zhang and co-workers, prepared a near-infrared fluorescent probe based on hemicyanine skeleton ((E)-2-(2-(6-hydroxy-2,3-dihydro-1*H*-xanthen-4-yl)vinyl)-3,3-dimethyl-1-propyl-3*H*-indol-1-ium) conjugated with a D-galactose moiety (**Gal-Pro** in Fig. 9.2). The absorption spectra of **Gal-Pro** in PBS presented an intense band at 596 nm and two weaker absorptions at 560 and 643 nm. Moreover, the emission spectra of the probe showed a weak band at 665 nm. In the presence of β-Gal, a new emission band at 703 nm appeared (12.8-fold enhancement). The authors tested the probe response in a premature senescence model of HDF cells induced by H_2O_2. Control and senescent cells were treated with the **Gal-Pro** (Zhang et al. 2017). No signal was observed for control cells, whereas in senescent cells clear turn-on NIR fluorescent images were captured. This molecule, also named **NIR-BG** (Fig. 9.2) was tested in cells with knocked-in *LacZ* and HeLa and MCF7 senescent cells where senescence was induced with camptothecin (CTP) or radiation therapy. The probe was validated in vivo in mice bearing *LacZ*(+) tumour and CTP drug treated HeLa xenograft models in order to corroborate the applicability of **NIR-BG** in genetically or drug-induced expressed β-Gal (Wang et al. 2019).

As it is apparent from the literature, only very few reports exist in which senescence probes have been validated in accurate in vivo models of senescence. An example is the probe **AHGa** (anhydride–histidine–galactose), which contains an *N*-glyosidic bond (Fig. 9.2) (Lozano-Torres et al. 2017). **AHGa** probe was tested in SK-MEL-103 xenografted mice in which senescence was induced by oral administration of palbociclib. The tumours in palbociclib-treated mice intravenously injected with the probe showed strong fluorescence relative to non-senescent tumours. Moreover, confocal microscopy images of fresh sections obtained from different organs (such as the liver, spleen, kidneys, lungs and heart) showed non-significant changes in fluorescence.

9.2.2 Other Hydrolases Probes

Apart from β-Gal, the overexpression of other lysosomal hydrolases (Knaś et al. 2012) such as α-L-fucosidase (α-Fuc), has been also employed for the detection of cellular senescence (Hildebrand et al. 2013). The overexpression of α-Fuc was deduced from measurements of the activity of hydrolases in various cell lines using different senescence models. From these studies it was apparent that α-Fuc activity was higher or, at least, comparable with SA-β-Gal in senescent cells. Based on the above, molecular probes containing selected chromophores or fluorophores linked with α-fucose have been developed (**X-Fuc** (Esterly et al. 1967) and **MUF** (Rushton and Dawson 1975) in Fig. 9.3) and used to measure α-fucosidase activity. Other enzymes, such as β-glucuronidase (Grabowska et al. 2019), acid phosphatase, β-hexosaminidase (Urbanell et al. 2014), α-mannosidase (Knaś et al. 2012) and *N*-acetyl-β-glucosaminidase (Agirbasli et al. 1996), have also been reported to be overexpressed in senescent cells, but in lower levels than those found for β-Gal and α-Fuc. In this context, *p*-nitrophenyl-α-D-mannopyranoside, 4-nitrophenyl *N*-acetyl-ß-D-glucosaminide and 4-nitrophenyl-ß-D-glucuronide (Fig. 9.3) have been employed for activity measurement of α-mannosidase, *N*-acetyl-β-glucosaminidase and β-glucuronidase, respectively, in lysed cells (Knaś et al. 2012). Detection of these enzymes using chromo-fluorogenic probes could be an alternative for effective senescence detection.

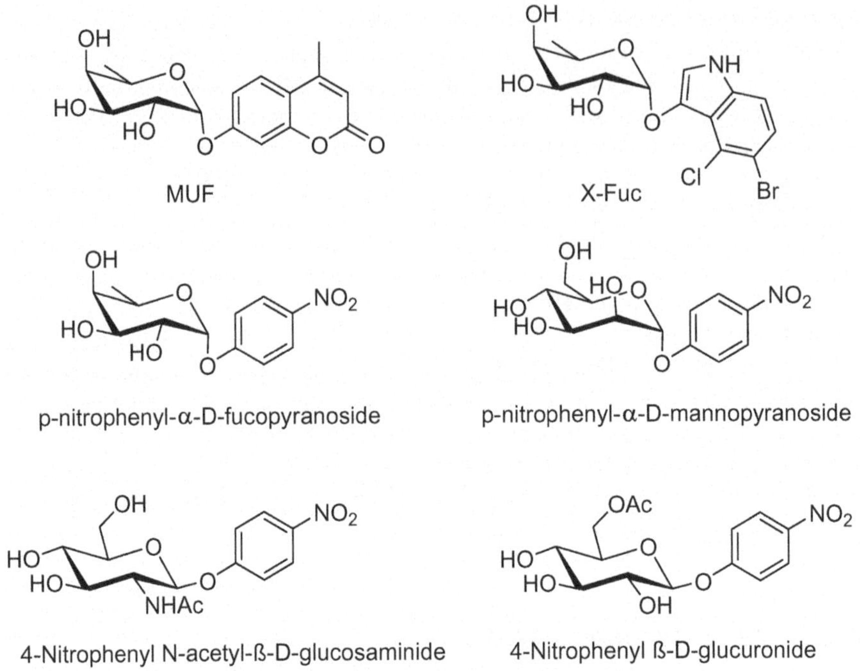

Fig. 9.3 Chromo-fluorogenic probes for the detection of other lysosomal hydrolases

9.2.3 Lipofuscin Detection

Another proposed marker of senescence is lipofuscin. Lipofuscin is a non-degradable aggregate of oxidized proteins, lipids and oligosaccharides that accumulates progressively mostly in aged post mitotic cells (Jung et al. 2007). Intracellular lipofuscin accumulation appears to be a universal correlate of animal senescence (Katz et al. 1984). Lipofuscin is an emissive aggregate that can be visualized using fluorescent microscopy. Moreover, histochemical methods, such as staining with Sudan Black B (**SBB**), are commonly used for lipofuscin detection. Lipofuscin stained with **SBB** is applicable for in vitro measurements (Gatenby and Moussa 1949; Rasmussen 1961; Jung et al. 2010). Studies demonstrated that **SBB**-stained lipofuscin is present in cells that expressed SA-β-Gal activity and is absent in SA-β-Gal-negative cells (Georgakopoulou et al. 2012). In a modification of this basic procedure, Gorgoulis, Bartek and co-workers recently developed a two-step staining procedure based on a biotin-antibiotin antibody-peroxidase-conjugated reaction, in order to improve the in vitro and ex vivo lipofuscin detection achieved with commercially available **SBB** staining (Fig. 9.4). Therefore, the authors used **SBB-biotin** conjugate to target lipofuscin (Evangelou et al. 2017). After binding of lipofuscin with **SBB-biotin** the staining of senescent cells is carried out upon addition of an antibiotin antibody functionalized with a peroxidase-conjugated polymeric backbone. Finally, addition

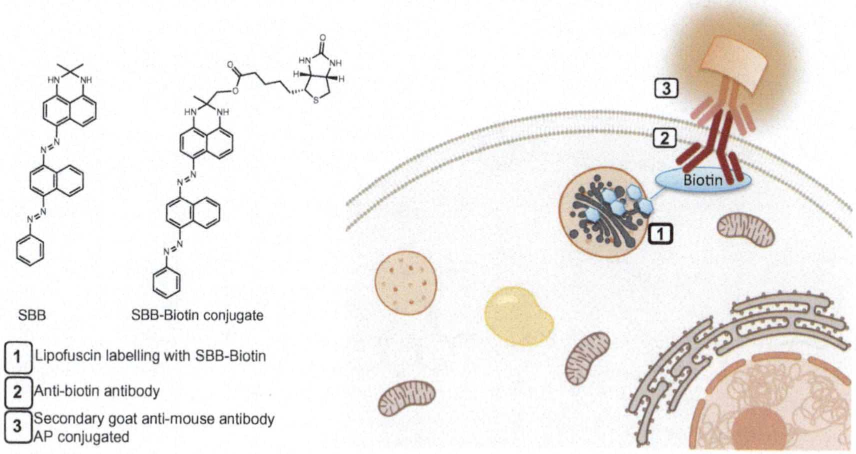

Fig. 9.4 Chemical structures of **SBB** and **SBB-biotin** conjugate developed by Gorgoulis, Bartek and co-workers to detect lipofuscin

of 3,3′-diaminobenzidine (DAB) induced the appearance of a dark blue-black colour (due to the peroxidase-catalysed oxidation of DAB) which indicated the presence of lipofuscin (Georgakopoulou et al. 2012).

9.2.4 Other Approaches for the Detection of Senescent Cells Using Chemical Probes

Based on the facts cited above, it seems clear that most of the described methods for the detection of cellular senescence are based on chromo-fluorogenic probes that evaluate SA-β-Gal activity. In spite of this fact, in recent years, other innovative methods for the detection of senescent cells have been proposed which do not rely on the presence of lysosomal β-Gal or do not display a final chromo-fluorogenic signal.

CyBC9 probe has been recently developed based on a Cy7 skeleton (Fig. 9.5) that was used for the detection of senescence using proliferative (young) and senescent (old) cells (Ang et al. 2019). The probe is a nontoxic membrane-permeable fluorescent molecule. The probe was tested in human early mesenchymal stromal cells (heMSCs) (young and old) from different tissues, such as bone marrow (BM) Wharton's Jelly (UC) and adipose tissue (AD). The authors demonstrated that **CyBC9** probe is accumulated in the mitochondria of senescent cells. The selectivity of the probe is based in the membrane potential of mitochondria. In senescence cells, there is an increase in ROS species that induced the depolarization of the mitochondria membrane allowing the accumulation of the probe **CyBC9**.

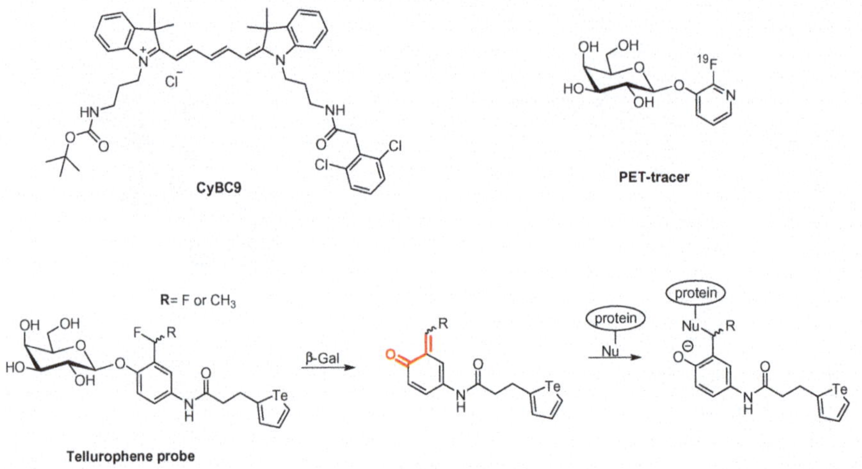

Fig. 9.5 Probes for detection of cellular senescence that are not chromo-fluorophores for SA-β-Gal activity

Lumba et al. reported a probe for lysosomal β-Gal activity capable of detecting cellular senescence in cultured cells by mass cytometry (Lumba et al. 2017). Mass cytometry is a similar technique to flow cytometry but equipped with an inductively coupled plasma mass spectrometry (ICP-MS) detector. The **tellurophene probe** contains a tellurene moiety as a reporter group which generates a quinone alkide when hydrolysed (Fig. 9.5). The quinone alkide is a reactive alkylating agent that forms covalent tellurophene-bearing conjugates with protein nucleophiles, allowing the quantification of lysosomal β-Gal activity in individual cells by mass cytometry. The authors used a retinal pigment epithelial (Rpe) cell line known to undergo senescence upon H_2O_2 treatment. Upon addition of **tellurophene probe**, they observed a higher labelling (ca. 1.75–fold) in senescent Rpe cells compared to control cells by mass cytometry analysis.

Another technique recently used in the detection of senescent cells is positron-emission tomography (PET), for which a radioactive molecule linked to β-galactoside is used (**PET-tracer** in Fig. 9.5) (Schwenck et al. 2019). **PET-tracer** was tested in two different in vitro models: (i) HCT-116 cells with doxorubicin-induced senescence and (ii) HRas driven liver progenitor cells with a doxycycline regulatable p53-specific shRNA. For this purpose both control and senescent cells of each cell line were treated with the **PET-tracer**, washed and measured in a gamma counter. Senescent cells from both lines incorporated more **PET-tracer** than control ones (ca. 3-fold higher accumulation). Besides, in vivo experiments showed higher uptake for senescent HCT-116 tumours (1.54-fold) compared to control ones whereas the observed value for HRas driven liver progenitor cells was 1.66-fold higher.

9.3 Nanodevices Targeting Senescent Cells

Additionally to molecular-based sensors, the use of smart nanodevices is an active field of research and several interesting examples has been published in the last years applied to the detection of certain cells (Aznar et al. 2016). These nanodevices are mainly constituted by inorganic or organic nanoparticles carrying a certain cargo and sometimes functionalized with (bio)molecules acting as targeting ligands (Lozano-Torres et al. 2017). These complex nanoarchitectures are finding promising biomedical applications. Taking this into account, it is not surprising that several functional nanoparticles targeting senescent cells have been described recently (Muñoz-Espín 2019). These nanodevices are based on mesoporous silica nanoparticles, carbon quantum dots (CQDs), porous $CaCO_3$ nanoparticles and molecularly imprinted polymer nanoparticles (nanoMIPs).

Martínez-Máñez et al. developed a nanodevice able to selectively release its cargo in senescent cells. The nanodevice consisted of mesoporous silica nanoparticles, loaded with rhodamine B (Rh B), functionalised with 3-aminopropyltrialcoxisilane (APTES) and capped with a galacto-oligosaccharide (Agostini et al. 2012). A similar system capped with lactose was previously reported by the same authors in 2009 (Bernardos et al. 2009). The release profile of the gated nanoparticles revealed a poor cargo delivery, whereas a remarkable amount of rhodamine B was released after 24 h in presence of β-Gal enzyme (Fig. 9.6a). Cargo release

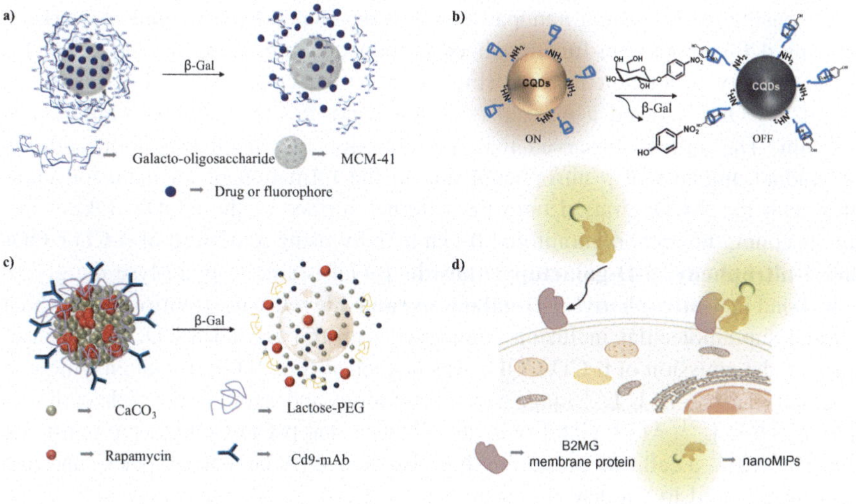

Fig. 9.6 **a** Mesoporous silica nanoparticles capped with a galacto-oligosaccharide. In the presence of lysosomal β-Gal the galacto-oligosaccharide cap is hydrolysed and cargo is released. **b** Mechanism of detection of β-Gal using probe **4-nitrophenyl-β-D-galactopyranoside** and β-CD-CQDs. **c** $CaCO_3$ nanocarriers that selectively deliver the cargo in senescent cells. **d** Molecularly imprinted nanoparticles (nanoMIPs) targeting B2MG membrane protein associated to senescence

in the presence of β-Gal was ascribed to the enzymatic hydrolysis of the capping galacto-oligosaccharide. Besides, the nanoparticles were tested in the cells lines DC1787, X-DC 1774 and X-DC4646 from patients suffering congenital dyskeratosis, in which lysosomal β-Gal is overexpressed. High emission signals were found in DC1787, X-DC 1774 and X-DC4646 cells lines, whereas poor cargo delivery was found in H460 control cells where endogenous β-Gal is not overexpressed. In a similar work, Martínez-Máñez and Serrano developed mesoporous silica nanoparticles loaded with Rh B or Indocyanine Green (ICG) and capped with a β-1,4-galactose hexamer (GalNP) (Muñoz-Espín et al. 2018). In the presence of the β-Gal enzyme, the capping oligosaccharide was hydrolysed and cargo was released. These nanoparticles were tested in a mouse model of pulmonary fibrosis and in a cancer mouse model of senescence induced by chemotherapy. The authors observed that the combination of palbociclib treatment with GalNP loaded with Rh B or ICG presented a more efficiently release of the fluorophore in senescent tumours, observing an enhancement of fluorescence. Moreover, the induction of pulmonary fibrosis by treating mice with bleomycin followed by intravenous injection with the fluorophore-loaded GalNP led to an increased fluorescent signal with respect to mice not treated with bleomycin, which did not present cellular senescence in lungs. Finally, drug-loaded nanoparticles with doxorubicin (GalNP(Dox)) or navitoclax (GalNP(Nav)) facilitated tumour regression in xenografted mice concomitantly treated with palbociclib and also the recovery in the pulmonary capacity in mice subjected to bleomycin-induced lung fibrosis. Besides, the authors found that the encapsulation of doxorubicin and navitoclax reduced their toxicity.

A sensing method using nanoparticles was developed by Feng and co-workers, who used β-cyclodextrin-functionalized carbon quantum dots (β-CD-CQDs) for β-Gal activity evaluation in vitro (Fig. 9.6b) (Tang et al. 2017). PBS solutions of β-CD-CQDs showed a marked emission band at 445 nm upon excitation at 365 nm. The authors observed that this emission was quenched upon addition of gradual amounts of *p*-nitrophenol due to the formation of an inclusion complex with the β-CDs grafted onto the external surface of the CQDs. Taking this into account, the authors monitored β-Gal activity using a mixture of β-CD-CQDs and **4-nitrophenyl-β-D-galactopyranoside**. β-Gal is able to hydrolyse the glycosidic bond in **4-nitrophenyl-β-D-galactopyranoside** releasing *p*-nitrophenol which formed supramolecular inclusion complexes with the grafted β-CD. As a consequence, the emission of β-CD-CQDs was quenched (Fig. 9.6b). β-Gal enzyme was detected in a 1.9–70 U L^{-1} range of concentrations and with a limit of detection as low as 0.6 U L^{-1}. MTT viability assays showed that β-CD-CQDs were non-toxic for OVCAR-3, a cell line presenting high lysosomal β-Gal activity per se and that is typically used to validate β-Gal probes. OVCAR-3 cells incubated with β-CD-CQDs showed a marked fluorescence but when exposed to both β-CD-CQDs and the galactose derivative **4-nitrophenyl-β-D-galactopyranoside** a marked emission quenching was observed.

Detection and elimination of senescent cells has also been achieved with porous calcium carbonate nanoparticles loaded with a fluorophore (coumarin-6) or a drug (rapamycin) (Thapa et al. 2017). The authors reacted lactose with a poly(ethylene glycol) derivative containing amine and carboxylate moieties in opposite sides and the poly(ethylene glycol) derivative was adsorbed onto the external surface of the loaded nanoparticles. Finally, an anti-CD-9 antibody (CD-9 receptor is overexpressed in senescent cells) was grafted onto the external surface of the nanoparticles using an EDC/NHS mediated esterification reaction (Fig. 9.6c). Aqueous suspensions of the coumarin-6-loaded nanoparticles at pH 7.4 showed a poor cargo release after 48 h. However, in the presence of β-Gal, a marked cargo release was observed due to lactose hydrolysis and subsequent detachment of the PEG chain from the surface of the nanoparticles. Confocal microscopy studies carried out with the coumarin-loaded-material showed that nanoparticles were preferentially internalized by old/senescent human dermal fibroblasts (HDF) when compared to the same young cells due to the overexpression of CD-9 membrane receptors, resulting in the subsequent release of coumarin-6 by β-Gal activity. The same nanoparticles loaded with rapamycin, an mTOR inhibitor that prevents senescence by affecting the p53/p21 pathway, displayed a senomorphic effect due to the rapamycin release, inducing marked reductions in β-Gal levels, p53/p21, CD-9 expression and also decreased senescence-associated secretory phenotypes (IL-6, IL-1 β) together with reduced senescent cells population numbers and prevention of cell cycle arrest.

More recently, another method for the detection of senescent cells based on molecularly imprinted nanoparticles (nanoMIPs) capable of recognizing a membrane protein associated with senescence was proposed (Ekpenyong-Akiba et al. 2019). It has been reported that senescent EJ bladder cancer cells (EJ), fibrosarcoma cells (HT1080) and human diploid fibroblasts (HDF) overexpressed the B2MG membrane protein (Althubiti et al. 2014). Bearing this idea in mind, the authors developed nanoMIPs recognising an extracellular epitope of B2MG. For nanoMIPs synthesis, in a first step, the epitope (amino acids 101-115) of β2 microglobulin (used as template) was covalently grafted onto the external surface of thiol-functionalised glass beads as a solid support by using succinimidyl-iodo acetate as linker. Then, in a second step, this hybrid material was suspended in a mixture of fluorescein-based acrylic monomers solution in order to obtain a polymer around the template (Fig. 9.6d). Finally, nanoMIPs were separated from the solid support washing with water at 65 °C. The affinity of nanoMIPs for B2MG protein was tested in EJ bladder cancer cells after genetically-induced senescence by overexpressing p16 with a tetracycline(tet)-regulatable system. Flow cytometry assays and fluorescence microscopy images revealed the preferential accumulation of nanoMIPs in the external membrane of senescent cells. Moreover, the senolytic activity of dasatinib-loaded nanoMIPs was demonstrated in EJp16 cells. Finally, in vivo studies revealed that nanoMIPs tagged with the dye DyLight 800 NHS Ester were able to distinguish between young (2 months) and old (11 months) C57/BL6J mice observing a preferential signal in the intestine, specifically in the jejunum.

9.4 Conclusions and Future Perspectives

The development of strategies to detect and eliminate senescent cells is an emerging research field that has received increasing attention in the last years. In this chapter, selected chromo-fluorogenic molecular probes and nanodevices targeting senescent cells are described. Most of the chromo-fluorogenic probes described for the detection of cellular senescence are based in the SA-β-Gal enzymatic activity. Normally these probes are composed by a fluorophore or a chromophore covalently linked to a galactopyranoside residue. In the presence of the enzyme, the glycosidic bond is hydrolysed inducing an easily chromogenic or fluorogenic observable and measurable change. Commercially available probes such as **X-Gal**, **C12FDG** or **Spider-Gal** are well known in the field of cellular senescence, but some of these probes still have a number of drawbacks. For this reason, researchers are still striving to find probes capable detecting senescent cells in different conditions and environments. Moreover, some recent probes based on the detection of the activity of other hydrolase enzymes, also overexpressed in senescent cells, such as β-glucuronidase, acid phosphatase, β-hexosaminidase, α-mannosidase and *N*-acetyl-β-glucosaminidase have been described. Also, several recently described probes detect cellular senescence through techniques such as mass cytometry or PET. On the other hand, the use of nanodevices for the detection of senescent cells has also been described. This is in fact a promising strategy that allows preparing nanoparticles not only for signalling but also for delivering senolytic drugs in senescent cells paving the way for the design theranostics systems to target senescent cells. In this field, examples of nanoparticles based in the use of mesoporous silica nanoparticles, CQDs, $CaCO_3$ and nanoMIPs have been described. Probes described above or others that will be synthesized in the next years are expected to be a fundamental instrument to detect senescent cells in aged or damaged tissues, and to become suitable tools to monitor the action of senolytics in multiple age-related disorders and longitudinally assess treatment response.

Acknowledgements B.L.-T. is grateful to the Spanish Ministry of Economy for her Ph.D. grant (FPU15/02707). J. F.-B. thanks to their postdoctoral fellowship (PAID-10-17). The authors are grateful to the Spanish Government (project RTI2018-100910-B-C41 (MCUI/AEI/FEDER, UE) and the Generalitat Valencia (Project PROMETEO/2018/024) for support.

References

Agirbasli M, Radhakrishnamurthy B, Jiang X, Bao W, Berenson GS (1996) Urinary n-acetyl-β-D-Glucosaminidase changes in relation to age, sex, race, and diastolic and systolic blood pressure in a young adult biracial population: The Bogalusa heart study. Am J Hypertens 9(2):157–161

Agostini A, Mondragón L, Bernardos A, Martínez-Máñez R, Marcos MD, Sancenón F et al (2012) Targeted cargo delivery in senescent cells using capped mesoporous silica nanoparticles. Angew Chem Int Ed 51(42):10556–10560

Aizawa K (1939) Studien über carbohydrasen II. die fermentative hydrolyse des *p*-Nitrophenol-β-Galactoside. Enzymologia 6:321–324

Althubiti M, Lezina L, Carrera S, Jukes-Jones R, Giblett SM, Antonov A et al (2014) Characterization of novel markers of senescence and their prognostic potential in cancer. Cell Death Dis 5:e1528

Ang J, Lee Y, Raghothaman D, Jayaraman P, Teo KL, Khan FJ et al (2019) Rapid detection of senescent mesenchymal stromal cells by a fluorescent probe. Biotechnol J 14(10):e1800691

Aznar E, Oroval M, Pascual L, Murguía JR, Martínez-Máñez R, Sancenón F (2016) Gated materials for on-command release of guest molecules. Chem Rev 116(2):561–718

Baker DJ, Wijshake T, Tchkonia T, LeBrasseur NK, Childs BG, van de Sluis B et al (2011) Clearance of $p16^{Ink4a}$-positive senescent cells delays ageing-associated disorders. Nature 479(7372):232–236

Baker DJ, Childs BG, Durik M, Wijers ME, Sieben CJ, Zhong J et al (2016) Naturally occurring $p16^{Ink4a}$-positive cells shorten healthy lifespan. Nature 530(9205):184–189

de Jesus BB, Blasco MA (2012) Assessing cell and organ senescence biomarkers. Circ Res 111(1):97–109

Bernardos A, Aznar E, Marcos MD, Martínez-Máñez R, Sancenón F, Soto J et al (2009) Enzyme-responsive controlled release using mesoporous silica supports capped with lactose. Angew Chem Int Ed 48(32):5884–5887

Biran A, Zada L, Abou Karam P, Vadai E, Roitman L, Ovadya Y et al (2017) Quantitative identification of senescent cells in aging and disease. Aging Cell 16(4):661–671

Childs BG, Gluscevic M, Baker DJ, Laberge RM, Marquess D, Dananberg J et al (2017) Senescent cells: an emerging target for diseases of ageing. Nat Rev Drug Discov. 16(10):718–735

Chilvers KF, Perry JD, James AL, Reed RH (2001) Synthesis and evaluation of novel fluorogenic substrates for the detection of bacterial β-Galactosidase. J Appl Microbiol 91(6):1118–1130

Dimri GP, Lee X, Basile G, Acosta M, Scott G, Roskelley C et al (1995a) Biomarker that identifies senescent human cells in culture and in aging skin in vivo. Proc Natl Acad Sci USA 92(20):9363–9367

Dimri GP, Lee X, Basile G, Acosta M, Scott G, Roskelley C et al (1995b) A biomarker that identifies senescent human cells in culture and in aging skin in vivo. PNAS USA 92(20):9363–9367

Doura T, Kamiya M, Obata F, Yamaguchi Y, Hiyama TY, Matsuda T et al (2016) Detection of *LacZ*-positive cells in living tissue with single-cell resolution. Angew Chem Int Ed 55(33):9620–9624

Ekpenyong-Akiba AE, Canfarotta F, Abd HB, Poblocka M, Casulleras M, Castilla-Vallmanya L et al (2019) Detecting and targeting senescent cells using molecularly imprinted nanoparticles. Nanoscale Horiz 4(3):757–768

Esterly JR, Standen AC, Pearson B (1967) The histochemical demonstration of intestinal β-D-fucosidase with 5-bromo-4-chloroindole-3-yl-β-d-fucopyranoside. J Histochem Cytochem 15(8):470–474

Evangelou K, Lougiakis N, Rizou SV, Kotsinas A, Kletsas D, Muñoz-Espín D et al (2017) Robust, universal biomarker assay to detect senescent cells in biological specimens. Aging Cell 16(1):192–197

García-Cao I, García-Cao M, Martín-Caballero J, Criado LM, Klatt P, Flores JM, Weill JC et al (2002) 'Super p53' mice exhibit enhanced DNA damage response, are tumor resistant and age normally. EMBO J 21(22):6225–6235

Gatenby JB, Moussa TAA (1949) The Sudan Black B technique in cytology: J Microsc. 69(2):72–75

Gee KR, Sun WC, Bhalgat MK, Upson RH, Klaubert DH, Latham KA et al (1999) Fluorogenic substrates based on fluorinated umbelliferones for continuous assays of phosphatases and β-Galactosidases. Anal Biochem 273(1):41–48

Georgakopoulou E, Tsimaratou K, Evangelou K, Fernandez MP, Zoumpourlis V, Trougakos I et al (2012) Specific lipofuscin staining as a novel biomarker to detect replicative and stress-induced senescence. A method applicable in cryo-preserved and archival tissues. Aging 5(1):37–50

Grabowska W, Mosieniak G, Achtabowska N, Czochara R, Litwinienko G, Bojko A et al (2019) Curcumin induces multiple signaling pathways leading to vascular smooth muscle cell senescence. Biogerontology 20(6):783–798

Gu K, Xu Y, Li H, Guo Z, Zhu SS, Zhu SS et al (2016) Real-time tracking and in vivo visualization of β-Galactosidase activity in colorectal tumor with a ratiometric near-infrared fluorescent probe. J Am Chem Soc 138(16):5334–5340

Hamilton TC, Young RC, McKoy WM, Grotzinger KR, Green JA, Chu EW et al (1983) Characterization of a human ovarian carcinoma cell line (NIH:OVCAR-3) with androgen and estrogen receptors. Cancer Res 43(11):5379–5389

Hernandez-Segura A, Nehme J, Demaria M (2018) Hallmarks of cellular senescence. Trends Cell Biol 28(6):436–453

Hildebrand D, Lehle S, Borst A, Haferkamp S, Essmann F, Schulze-Osthoff K (2013) α-Fucosidase as a novel convenient biomarker for cellular senescence. Cell Cycle 12(12):1922–1927

Horwitz JP, Chua J, Curby RJ, Tomson AJ, Da Rooge MA, Fisher BE et al (1964) Substrates for cytochemical demonstration of enzyme activity. I. Some substituted 3-Indolyl-β-D-Glycopyranosides. J Med Chem 7(4):574–75

Hung MC, Zhang X, Yan DH, Zhang HZ, He GP, Zhang TQ et al (1992) Aberrant expression of the C-ErbB-2/Neu protooncogene in ovarian cancer. Cancer Lett 61(2):95–103

Imai S, Kiyozuka Y, Maeda H, Noda T, Hosick HL (1990) Establishment and characterization of a human ovarian serous cystadenocarcinoma cell line that produces the tumor markers CA-125 and tissue polypeptide antigen. Oncology 47(2):177–184

Jung T, Bader N, Grune T (2007) Lipofuscin: formation, distribution, and metabolic consequences. Ann N Y Acad Sci 1119(1):97–111

Jung T, Höhn A, Grune T (2010) Lipofuscin: detection and quantification by microscopic techniques. Methods Mol Biol 594:173–193

Kamiya M, Kobayashi H, Hama Y, Koyama Y, Bernardo M, Nagano T et al (2007) An enzymatically activated fluorescence probe for targeted tumor imaging. J Am Chem Soc 129(13):3918–3929

Katz ML, Robinson WG, Herrmann RK, Groome AB, Bieri JG (1984) Lipofuscin accumulation resulting from senescence and Vitamin E deficiency: spectral properties and tissue distribution. Mech Ageing Dev 25(1–2):149–159

Kim EJ, Podder A, Maiti M, Lee JM, Chung BG, Bhuniya S (2018) Selective monitoring of vascular cell senescence via β-Galactosidase detection with a fluorescent chemosensor. Sens Actuator B-Chem 274(18):194–200

Knaś M, Zalewska A, Krętowski R, Niczyporuk M, Waszkiewicz N, Cechowska-Pasko M et al (2012) The profile of lysosomal exoglycosidases in replicative and stress-induced senescence in early passage human fibroblasts. Folia Histochem Cytobiol 50(2):220–227

Lee BY, Han JA, Im JS, Morrone A, Johung K, Goodwin EC et al (2006) Senescence-Associated β-Galactosidase is lysosomal β-Galactosidase. Aging Cell 5(2):187–195

Lee HW, Heo CH, Sen D, Byun HO, Kwak IH, Yoon G et al (2014) Ratiometric two-photon fluorescent probe for quantitative detection of β-Galactosidase activity in senescent cells. Anal Chem 86(20):10001–10005

Lozano-Torres B, Galiana I, Rovira M, Garrido E, Chaib S, Bernardos A et al (2017a) An OFF–ON two-photon fluorescent probe for tracking cell senescence in vivo. J Am Chem Soc 139(26):8808–8811

Lozano-Torres B, Pascual L, Bernardos A, Marcos MD, Jeppesen JO, Salinas Y et al (2017b) Pseudorotaxane capped mesoporous silica nanoparticles for 3,4-methylenedioxymethamphetamine (MDMA) detection in water. Chem Commun 53(25):3559–3562

Lozano-Torres B, Estepa-Fernández A, Rovira M, Orzáez M, Serrano M, Martínez-Máñez R et al (2019) The chemistry of senescence. Nat Rev Chem 3(7):426–441

Lumba MA, Willis LM, Santra S, Rana R, Schito L, Rey S et al (2017) A β-Galactosidase probe for the detection of cellular senescence by mass cytometry. Org Biomol Chem 15(30):6388–6392

McHugh D, Gil J (2018) Senescence and aging: causes, consequences, and therapeutic avenues. J Cell Biol 217(1):65–77

Mosteiro L, Pantoja C, de Martino A, Serrano M (2018) Senescence promotes in vivo reprogramming through $P16^{INK\ 4a}$ and IL-6. Aging Cell 17(2):e12711

Muñoz-Espín D (2019) Nanocarriers targeting senescent cells. Trans Med Aging 3:1–5

Munoz-Espin D, Serrano M (2014) Cellular senescence: from physiology to pathology. Nat Rev Mol Cell Biol 15(7):482–496

Muñoz-Espín D, Rovira M, Galiana I, Giménez C, Lozano-Torres B, Paez-Ribes M et al (2018) A versatile drug delivery system targeting senescent cells. EMBO Mol Med 10(9):e9355

Niedernhofer LJ, Robbins PD (2018) Senotherapeutics for healthy ageing. Nat Rev Drug Discov. 17(5):377–379

Paez-Ribes M, González-Gualda E, Doherty GJ, Muñoz-Espín D (2019) Targeting senescent cells in translational medicine. EMBO Mol Med. 11(12):e10234

Rasmussen GL (1961) A method of staining the statoacoustic nerve in bulk with Sudan Black B. Anat Rec 139:465–469

Ritschka B, Storer M, Mas A, Heinzmann F, Ortells MC, Morton JP et al (2017) The senescence-associated secretory phenotype induces cellular plasticity and tissueregeneration. Genes Dev 31(2):172–183

Rotman B (1961) Measurement of activity of single molecules of β-D-Galactosidase. Proc Natl Acad Sci USA 47(12):1981–1991

Rotman B, Zderic JA, Edelstein M (1963) Fluorogenic substrates for β-D-Galactosidases and phosphatases derived from fluorescein (3,6-dihydroxyfluoran) and its monomethylether. PNAS USA 50(3):1–6

Rushton AR, Dawson G (1975) Glycosphingolipid β-Galactosidases of cultured mammalian cells: characterization of the enzymes from mouse cell line LMTK and human Lesch-Nyhan fibroblasts. Biochim Biophys Acta Lipids Lipid Metab 388(1):92–105

Schwenck J, Cotton J, Zhou B, Wolter K, Kuehn A, Fuchs K et al (2019) In vivo imaging of tumor senescence with a novel beta-Galactosidase specific PET tracer. Nuklearmedizin 58(02):106

Serrano M, Hannon GJ, Beach DA (1993) A new regulatory motif in cell-cycle control causing specific inhibition of cyclin D/CDK4. Nature 366(6456):704–707

Serrano M, Lin AW, McCurrach ME, Beach D, Lowe SW (1997) Oncogenic Ras provokes premature cell senescence associated with accumulation of p53 and $p16^{INK4a}$. Cell 88(5):593–602

Sharpless NE, Sherr CJ (2015) Forging a signature of in vivo senescence. Nat Rev Cancer 15(7):397–408

Soto-Gamez A, Demaria M (2017) Therapeutic interventions for aging: the case of cellular senescence. Drug Discov Today 22(5):786–795

Tang C, Zhou J, Qian Z, Ma Y, Huang Y, Feng HA (2017) Universal fluorometric assay strategy for glycosidases based on functional carbon quantum dots: β-Galactosidase activity detection in vitro and in living cells. J Mater Chem B 5(10):1971–1979

Thapa RK, Nguyen HT, Jeong JH, Kim JR, Choi HG, Yong CS et al (2017) Progressive slowdown/prevention of cellular senescence by CD9-targeted delivery of rapamycin using lactose-wrapped calcium carbonate nanoparticles. Sci Rep 7(1):43299

Tyner SD, Venkatachalam S, Choi J, Jones S, Ghebraniousk N, Igelmann H, Lu X et al (2002) p53 mutant mice that display early ageing-associated phenotypes. Nature 415(3):45–53

Urbanell L, Magini A, Ercolani L, Sagini K, Polchi A, Tancini B et al (2014) Oncogenic H-Ras up-regulates acid β-hexosaminidase by a mechanism dependent on the autophagy regulator TFEB. PLoS One 9(2):e89485

Wang Y, Liu J, Ma X, Cui C, Deenik PR, Henderson PKP et al (2019) Real-time imaging of senescence in tumors with DNA damage. Sci Rep 9(1):2102–2113

Woods LL, Sapp J (1962) A New one-step synthesis of substituted coumarins. J Org Chem 27(10):3703–3705

Xu M, Pirtskhalava T, Farr JN, Weigand BM, Palmer AK, Weivoda MM et al (2018) Senolytics improve physical function and increase lifespan in old age. Nat Med 24(8):1246–1256

Yuan Y, Kim WH, Han HS, Lee JH, Park HS, Chung JK, Kang SB, Park JG (1997) Establishment and characterization of human ovarian carcinoma cell lines. Gynecol Oncol 66(3):378–387

Zhang J, Li C, Dutta C, Fang M, Zhang S, Tiwari A et al (2017) A novel near-infrared fluorescent probe for sensitive detection of β-Galactosidase in living cells. Anal Chim Acta 968:97–104
Zhu Y, Tchkonia T, Pirtskhalava T, Gower AC, Ding H, Giorgadze N et al (2015) The achilles' heel of senescent cells: from transcriptome to senolytic drugs. Aging Cell 14(4):644–658

Chapter 10
Potential Applications of Aptamers for Targeting Senescent Cells

Antonio Leonardi, Francesco Pacifico, and Elvira Crescenzi

Abstract Cellular senescence is a stress response characterized by a permanent loss of proliferative ability. Different types of senescence reflect different forms of stress. Although cellular senescence plays a key role in maintaining tissue homeostasis and in suppressing tumorigenesis, it is widely appreciated that accumulation of senescent cells in tissue contributes to development of age-related dysfunctions and limits the efficacy of chemo- and radiotherapy. Consequently, novel therapeutic strategies targeting senescent cells, or their pro-inflammatory secretome, have been recently developed. The application of these novel therapeutic approaches requires biomarkers for in vivo detection of senescent cells for ageing and age-associated diseases, as well as for therapy-induced senescence in cancer patients. Aptamers are short oligonucleotide or peptide sequences able to bind with high affinity and specificity a wide range of cellular targets, including cell surface epitopes, thanks to their unique three-dimensional folding. Oligonucleotide aptamers have been widely applied for the discovery of biomarkers, since the early 1990s, and Affimers have recently emerged as valuable tools for biomarker discovery. In this chapter we briefly introduce applications of aptamer technologies for recognition of cell surface antigens and suggest that aptamers provide the perfect technology to identify novel senescence-specific biomarkers.

Keyword Senescence SASP · Senotherapy · Biomarkers · Aptamers · Affimers

---

A. Leonardi
Dipartimento di Medicina Molecolare e Biotecnologie Mediche, "Federico II" University of Naples, 80131 Naples, Italy

F. Pacifico · E. Crescenzi (✉)
Istituto per l'Endocrinologia e l'Oncologia Sperimentale, CNR, 80131 Naples, Italy
e-mail: e.crescenzi@ieos.cnr.it

D. Muñoz-Espin and M. Demaria (eds.), *Senolytics in Disease, Ageing and Longevity*, Healthy Ageing and Longevity 11,
https://doi.org/10.1007/978-3-030-44903-2_10

10.1 Cellular Senescence and the Growing Field of Senotherapy

Cellular senescence is a stress response characterized by a permanent loss of proliferative ability that can be activated by different damaging stimuli. As such, cellular senescence represents an evolutionarily conserved mechanism that prevents the proliferation of damaged and potentially tumorigenic cells. Senescence also participates in mammalian embryonic development and critically contributes to embryonic remodeling and patterning (Muñoz-Espín et al. 2013; Storer et al. 2013). The activation of $p16^{INK4a}$-pRb and p53-$p21^{CIP1}$ signaling pathways sustain the senescence cell cycle arrest. Cellular senescence is frequently associated with irreparable DNA damage and a persistent DNA damage response (DDR) (Hewitt et al. 2012; Fumagalli et al. 2012; Sedelnikova et al. 2004), which is triggered as a result of telomere shortening during replicative senescence (d'Adda di Fagagna et al. 2003), oncogene activation in oncogene-induced senescence (OIS) (Bartkova et al. 2006; Di Micco et al. 2006; Hills and Diffley 2014), and radiotherapy or chemotherapy in therapy-induced senescence (TIS) (te Poele et al. 2002; Mirzayans et al. 2005). However, senescence may also be induced in the absence of DNA damage. For instance, PTEN inactivation induces senescence in the absence of DDR activation: PTEN-loss activates mTOR which enhances p53 translation and p21 expression, thus leading to permanent cell cycle arrest (Alimonti et al. 2010). Induction of cellular senescence with drugs that modify chromatin structure, such as histone deacetylase inhibitors, occurs in the absence of DNA damage (Pazolli et al. 2012). Sodium butyrate induces p21 expression and causes cellular senescence. Interestingly, sodium butyrate induced-senescent cells display activation of DDR proteins, such as gamma-H2AX foci, without detectable DNA damage (Pospelova et al. 2009). Curcumin treatment induces senescence in vascular smooth muscle cells in a DNA damage-independent manner, although activation of ATM is observed (Grabowska et al. 2015). Finally, developmental senescence is not associated with DNA damage (Muñoz-Espín et al. 2013; Storer et al. 2013).

Cellular organelles undergo senescence-specific alterations and critically contribute to the development of the senescent phenotype. Enlargement of the lysosomal compartment likely explain the increased activity of a well-established marker of senescence, the so-called Senescence-Associated beta-galactosidase (SA-β-gal) (Kurz et al. 2000). The DDR promotes PGC-1b-dependent mitochondrial biogenesis, thereby increasing mitochondrial content in senescent cells. Mitochondria critically contribute to cellular senescent-associated changes (Correia-Melo et al. 2016).

Senescent cells are transcriptionally and metabolically active, and develop a distinctive phenotype characterized by a flat and enlarged morphology. In addition, senescent cells acquire an enhanced secretory ability, called senescence-associated secretory phenotype (SASP) or senescence messaging secretome (SMS) (Coppé et al. 2008; Kuilman and Peeper 2009). The release of proinflammatory SASP factors is critical for reinforcing the growth arrest, for triggering immune-mediated removal of senescent cells and for activating wound healing (Salama et al. 2014; Serrano 2014).

Main transcriptional regulators of the SASP include NF-κB and C/EBPβ that cooperatively induce SASP expression (Acosta et al. 2008; Kuilman et al. 2008; Chien et al. 2011; Crescenzi et al. 2011). For instance, Acosta and colleagues demonstrated that NF-κB and C/EBPβ cooperatively induce the expression of CXCR2 ligands IL-8 and GROα in fibroblasts undergoing OIS (Acosta et al. 2008). Similarly, IL6 expression is induced by the transcription factor C/EBPβ upon exposure of cells to oncogenic stress (Kuilman et al. 2008). Both IL-8 and IL-6 play an important role in OIS reinforcing the senescence growth arrest. A critical role of RelA/p65 as a major transcription factor that accumulates on chromatin of senescent cells was demonstrated by Chien et al. (2011). Interestingly, suppression of RelA/p65 function not only abrogates the induction of proinflammatory cytokines IL-6, IL-8, CXCL1, but also reduces expression of intercellular adhesion molecules, such as ICAM-1, and several matrix metalloproteinases (Chien et al. 2011). RelA/p65 induces the expression of TNF-α and IFN-γ in drug-induced tumor senescent cells thereby sensitizing cells to Fas-mediated apoptosis (Crescenzi et al. 2011).

Activation of NF-κB during senescence depends on the cGAS-STING cytosolic DNA sensor pathway, in which cGAS detects endogenous senescence-specific cytoplasmic chromatin fragments (reviewed in Glück and Ablasser 2019). More recent data demonstrate that during OIS cGAS-STING controls the induction of TLR2 which acts as critical upstream mediator for the initiation of the SASP. TLR2 is activated by senescence-associated Damage-Associated Molecular Patterns (identified as acute-phase serum amyloids A1 and A2) and signals, downstream of STING, through NF-κB and $p38^{MAPK}$ to regulate the SASP and to reinforce the senescence growth arrest (Hari et al. 2019). Finally, a role for the JAK/STAT pathway in modulation of proinflammatory SASP components has been demonstrated (Novakova et al. 2010; Xu et al. 2015).

The SASP develops over several days in culture. Hence, expression of several SASP factors, notably the proinflammatory cytokines IL-6 and IL-8, requires up to 20d after DDR initiation before reaching maximal level (Coppé et al. 2008; Crescenzi et al. 2011). Furthermore, secretome composition is temporally modulated, changing with time from a TGF-β-driven to a C/EBPβ-driven profile (Hoare et al. 2016).

Interestingly, emerging evidences show that senescent cells manage to communicate with their neighbors not only through the SASP, but also through the release of small extracellular vesicles (EVs), namely exosomes and microvesicles (Takasugi 2018). Accordingly, various stresses known to induce cellular senescence increase the secretion of EVs (Takasugi 2018). EVs produced by senescent cells play a double role by removing harmful senescence-related cytoplasmic DNA fragments (Takahashi et al. 2017) and by delivering cargos of cell type-specific bioactive molecules (Takasugi et al. 2017; Sagini et al. 2018). Both normal and neoplastic senescent cells have been shown to modulate local tissue microenvironment through EVs. For instance, EVs secreted from senescent chondrocytes, isolated from osteoarthritis patients, induce paracrine senescence in nearby cells, thereby inhibiting cartilage extracellular matrix deposition. These effects are mediated by osteoarthritis-specific microRNA (miRNA) packaged in the EVs (Jeon et al. 2019). Similarly, primary human foreskin fibroblasts undergoing Ras-mediated senescence secrete exosomes which mediate paracrine senescence. The interferon-induced transmembrane protein 3 (IFITM3), specifically accumulated within EVs, is responsible for transmitting senescence to normal neighboring cells (Borghesan et al. 2019). Interestingly, while EVs secreted from senescent cells elicit paracrine senescence, EVs isolated from human induced pluripotent stem cells reduce reactive oxygen species levels in senescent mesenchymal stem cells in culture, thereby reversing the senescent phenotype. Stem cells derived EVs also alleviate senescence in a progerin-induced premature aging cell model. High levels of EVs antioxidant enzymes are responsible for the anti-aging effects (Liu et al. 2019). In line with this, are the results of Zhang and colleagues that demonstrate that EVs isolated from hypothalamic neural stem cells ameliorate age-related cognitive declines in mice. Exosomal miRNA mediate these anti-aging effects (Zhang et al. 2017). Notably, senescent EVs can also promote protumorigenic effects: EVs enriched in EphA2 receptor bind to ephrin-A1 ligand expressed on cancer cells and promote their proliferation through reverse signaling (Takasugi et al. 2017). Notably, increased expression of EphA2 was detected in the stroma of human breast and ovarian cancer (Takasugi et al. 2017). Finally, there is also evidence that EVs deliver membrane-derived bioactive lipids that impact on immune system functions (Sagini et al. 2018).

An additional form of cell communication described in senescent cells is a direct intercellular protein transfer through cytoplasmic bridges. Although the functional consequences of direct intercellular transfer are not completely understood, is interesting to note that this kind of communication facilitates the clearance of senescent cells by NK cells (Biran et al. 2015).

It is now widely accepted that a prolonged permanence of senescent cells in tissue and their sustained secretory activity, result in chronic inflammation and critically contributes to both age-related dysfunctions and to a protumoral microenvironment (Sun et al. 2018; Wei and Ji 2018; Jackson et al. 2012). In the context of TIS, concerns have been raised regarding the ability of cancer cells to escape the senescence arrest, to resume proliferation and to drive tumor recurrence after radio- and chemotherapy (Roberson et al. 2005; Karimi-Busheri et al. 2010). Cancer cells that evade senescent growth arrest are endowed with stem cell properties, are more malignant, and resistant to drug therapies (Yang et al. 2017), representing a likely risk for tumor relapse. Notably, the SASP promotes cell plasticity and stemness also through paracrine actions (Mosteiro et al. 2016; Ritschka et al. 2017). In line with these observations, in vivo studies support the idea that accumulation of senescent cells results in detrimental tissue remodelling, both in normal and in neoplastic tissues. For instance, selective elimination of senescent cells by genetic approaches delays the onset and attenuates progression of age-related disorders, and extends lifespan (Jeon et al. 2017; Bussian et al. 2018; Baker et al. 2016; Baker et al. 2011). Similarly, specific ablation of TIS cells in the tumor microenvironment reduces both tumor relapse and the adverse effects of chemotherapy (Demaria et al. 2017). Notably, TIS induction was shown to be associated with adverse treatment outcome and decreased overall survival in cancer patients (Wu et al. 2012; Supiot et al. 2008; Sidi et al. 2011; Kim et al. 2016). Hence, available data suggest that senescent cells and the SASP contribute to organismal ageing and limit the efficacy of chemo- and radiotherapy (Muñoz-Espín and Serrano 2014). Consequently, removal of senescent cells or the neutralization of their secretome, have become attractive therapeutic strategies to prevent senescence-associated dysfunctions, thus opening the new field of senotherapy research. Drugs that suppress or modulate the SASP without interfering with senescent growth arrest and senolytic agents that selectively induce apoptosis in senescent cells have been studied and are under development.

The possibility to pharmacologically reprogram the SASP towards an anti-tumorigenic rather than pro-tumorigenic profile was first highlighted by Toso and colleagues in the context of docetaxel-induced senescence in PTEN-null tumors, where pharmacological inhibition of the Jak2/Stat3 pathway resulted in activation of an antitumor immune response and enhanced chemotherapy efficacy (Toso et al. 2014). The neutralization of detrimental SASP components in the tumor microenvironment (TME), where senescent stromal cells promote proliferation and survival of nearby neoplastic cells, has also been studied focusing on senescent fibroblasts. In this setting, inhibition of $p38^{MAPK}$ was found to inhibit SASP-mediated tumor growth driven by senescent fibroblasts (Alspach et al. 2014). Pharmacological inhibition of the chromatin regulator MLL1 in fibroblasts also abrogated the SASP preventing its pro-carcinogenic effects. Furthermore, similar results were obtained in etoposide-induced senescent MCF7 breast cancer cells, mimicking the clinical scenario of TIS in tumor cells (Capell et al. 2016). Simvastatin, an HMG-coA reductase inhibitor, decreased the SASP of senescent human fibroblasts, thereby reducing endocrine resistance in breast cancer (Liu et al. 2015). Similarly, the flavonoid apigenin suppressed the SASP and reduced detrimental paracrine effects of fibroblasts on breast cancer cells (Perrott et al. 2017). Rapamycin selectively suppressed the mTOR and NF-κB-dependent secretion of proinflammatory cytokines in senescent fibroblasts, thereby suppressing prostate tumor growth in mice (Laberge et al. 2015). Trabectedin, a DNA minor-groove binding anticancer drug, which causes a gene- and promoter-dependent modulation of transcription (Larsen et al. 2016), markedly decreased the expression of proinflammatory SASP cytokines in tumor cells and limited their pro-tumoral activities in vitro (Camorani et al. 2018). Importantly, Georgilis and colleagues identified 50 druggable, potential targets for therapeutic intervention, whose knockdown suppressed pro-tumorigenic effects of the senescent fibroblasts secretome on tumor growth (Georgilis et al. 2018).

However, expression profiles of different senescent cells are not uniform and the SASP appears to be stimulus-, time- and cell type-dependent (Hernandez-Segura et al. 2017; Wiley et al. 2017; Maciel-Barón et al. 2016) which makes the modulation of the SASP a complex task. Furthermore, in the context of TIS, the SASP has both tumor-suppressive and tumor-promoting effects: SASP cytokines are known to induce immune-mediated clearance of senescent tumor cells (Iannello et al. 2013; Tasdemir et al. 2016; Eggert et al. 2016) and interfering with these immunostimulatory signals might result in detrimental effects. For instance, in a murine model of lymphoma, the reduction of SASP proteins through NF-κB inhibition hampered immunosurveillance following TIS and led to drug resistance, early relapse and reduced survival (Chien et al. 2011). Together, these observations suggest that activation of cell-, time- and stimulus-specific senescence signaling pathways in tumor cells results in different and even functionally-opposite SASPs (Burgess 2011; Llanos and Serrano 2016). Hence, therapeutic interference with key SASP regulators might require individual molecular profiles and monitoring of SASP signatures in cancer patients post-treatment.

As an alternative therapeutic strategy to prevent senescence-associated dysfunctions, senolysis, i.e. the selective removal of senescent cells, has been approached. Senolytic agents have been tested in pre-clinical models and have proved to be effective in alleviating age-related diseases and symptoms of progeria (Xu et al. 2018; Yousefzadeh et al. 2018; Farr et al. 2017; Roos et al. 2016). These studies support the clinical potential of senolysis for promoting healthy ageing. Comprehensive reviews have recently been published on this subject (Kirkland and Tchkonia 2017; Myrianthopoulos et al. 2019). In contrast, the effects of senolytic agents in TIS appear to be more multifaceted. For instance, the senolytic agent panobinostat effectively removed chemotherapy-induced senescent non-small cell lung cancer and head and neck squamous cell carcinoma cells, and proved more efficacious than multiple cycles of chemotherapy (Samaraweera et al. 2017). In contrast, the senolytic cocktail dasatinib plus quercetin, which effectively eliminated senescent fibroblasts (Schafer et al. 2017), not only showed no efficacy on TIS liver cancer cells but also displayed pro-tumorigenic effects in the absence of chemotherapy (Kovacovicova et al. 2018). These data again underline the complexity of senescent cancer cells and emphasize the need for further in vivo investigation of their features and their role(s) in the tumor microenvironment.

Future steps in senotherapy research would include the study of combination therapy with radiotherapy-chemotherapy plus senotherapy for cancer treatment, and advancing senotherapeutics from preclinical studies into the clinic for age-related diseases. Several clinical trials are ongoing for age-related diseases, aimed at evaluate the feasibility of senotherapy. For instance, a two-center, open-label study (ClinicalTrials.gov Identifier: NCT02874989) of senolytic treatment with Dasatinib and Quercetin was conducted in patients with idiopathic pulmonary fibrosis, supporting safety of the treatment and providing initial evidence of senolytic efficacy (Justice et al. 2019). In an open label Phase 1 pilot study (ClinicalTrials.gov Identifier: NCT02848131), the combination of Dasatinib plus Quercetin, administered to subjects with diabetic kidney disease, significantly decreased senescent cell burden (Hickson et al. 2019). A phase 1 clinical trial testing the safety and tolerability of a senolytic small molecule (UBX0101) for osteoarthritis treatment is ongoing (ClinicalTrials.gov Identifier: NCT03513016).

Application of senotherapies to cancer and age-related diseases will require specific and sensitive biomarkers for the detection of normal and neoplastic senescent cells in patients. In particular, assessment of senescence biomarkers in biological samples of cancer patients following neoadjuvant chemo-radiotherapy could help predict treatment responses and contribute to designing personalized therapies.

10.2 Markers for Senescence

The lack of specific markers for the in vivo detection of senescent cells hampers both the molecular characterization of cellular senescence in ageing, in ageing-related diseases and in cancer patients, and the development of senescence-specific molecular tools and probes for diagnostic, prognostic and therapeutic purposes. Assessment of SA-β-gal activity (Kurz et al. 2000) still represents the most commonly used method for identification of senescent cells, although this marker lacks appropriate specificity, being constantly expressed by senescent cells but also in several other conditions such as in confluent cells in culture (Severino et al. 2000; Yang and Hu 2005). Also, the use of fresh tissue, required for the detection of enzymatic activity of SA-β-gal, is incompatible with routine formalin-fixation and paraffin-embedding of clinical specimens. However, the high lysosomal β-galactosidase activity of senescent cells was successfully exploited for the delivery of galacto-oligosaccharide-encapsulated cytotoxic drugs to live senescent cells in vitro and in vivo, given that release of encapsulated material depended on digestion of the sugar-coat by the senescent-associated lysosomal β-galactosidase (Muñoz-Espín et al. 2018).

Combinations of markers indicative of cell cycle arrest, active DDR and distinctive morphological changes have been sometimes employed to detect senescent cells. For instance, co-staining of either SA-β-gal activity or histone H3K9 tri-methylation (H3K9me3, a repressive heterochromatin mark denoting cell cycle arrest) together with low Ki67 expression has allowed for detection of senescence in therapy-exposed cancer cells (Reimann et al. 2010). An irregular morphology and enlargement of nuclei is frequently observed in senescent cells, which has been related to alterations in components of the nuclear envelope and in specific downregulation of lamin B1 (Freund et al. 2012). Accordingly, the combination of SA-β-gal positivity and double Ki67-Lamin B1 negativity was used to detect senescent cells within untreated human invasive breast carcinomas (Cotarelo et al. 2016). However, although panels of biomarkers might improve our ability to identify senescent cells, there are technical challenges for the practical use of such panels in the clinic and in research.

While senescent cells differ from proliferating and quiescent cells in several aspects (Rodier and Campisi 2011), alterations in cell surface proteins are of greatest interest since surface proteins can be used both as senescence-specific biomarkers and as potential druggable targets. In addition, cell surface markers are easily detected using conventional methods on fresh samples and on fixed specimens. A screening for plasma membrane-associated proteins preferentially expressed in senescent cells led to the identification of DEP1, STX4 and B2MG proteins, whose co-expression marked senescent cells in culture and in tissue samples (Althubiti et al. 2014), although it was noted that STX4 was a better marker of $p21^{CIP1}$-induced senescence, whereas DEP1 was more specific for $p16^{INK4a}$-induced senescence. The combination of DEP1 and B2MG also allowed for flow cytometric identification of senescent cells. Mass spectrometry analysis of membrane- and cell surface-associated proteins from replicative senescent human diploid fibroblasts identified DPP4 (CD26) as a senescence surface marker. Interestingly, anti-DPP4 antibodies induced efficient antibody-dependent cellular cytotoxicity against senescent fibroblasts, thus enabling their preferential elimination by natural killer cells (Kim et al. 2017). More recently, the plasma membrane-localized fatty acid translocase CD36 was found upregulated in various cell types during replicative senescence, OIS, and TIS (Chong et al. 2018).

Proteomic approaches unveiled not only specific quantitative changes but also qualitative modification of surface plasma membrane-associated proteins in senescent cells, such as protein glycation (or glycoxidation), carbonylation, and carbamylation (Itakura et al. 2016; Vanhooren et al. 2015). Indeed, specific non-enzymatic post-translational modifications of proteins and lipids have been proposed as hallmarks of ageing (Friguet 2002). For instance, the products of non-enzymatic glycation and oxidation (glycoxidation) of proteins and lipids, called advanced glycation end-products (AGEs), are known to accumulate in organisms in an age-dependent manner. It is also well known that reactive oxygen species induce progressive accumulation of oxidized and carbonylated proteins during aging (Baraibar et al. 2013). This age-related accumulation of oxidized proteins likely depends on a decreased efficacy of repair systems and/or failure of removal mechanisms (Chondrogianni et al. 2014). Among oxidized products, a non-degradable material referred to as lipofuscin, accumulates in the lysosomal compartment in aged tissues, and has been used as a senescence marker (von Zglinicki et al. 1995; Georgakopoulou et al. 2013). Recently, a novel sensitive technique for the detection of lipofuscin has been developed and applied for the detection of senescent cells in fresh, frozen and formalin-fixed biological specimens (Evangelou et al. 2017). Carbamylation, the spontaneous binding of isocyanic acid to proteins, also occurs continuously throughout life (Gorisse et al. 2016), and is easily detected on long-lifespan proteins such as extracellular matrix structural proteins. In order to characterize novel aging biomarkers, the key next step is to identify specific cell surface proteins targeted by these modifications in senescent cells. In line with this are the results of Frescas and colleagues that recently described the identification of a senescence-specific oxidative modification of vimentin (malondialdehyde (MDA)-modified vimentin), which is exposed on the surface of senescent fibroblasts and is secreted into the blood of senescence-prone SAMP8 mice (Frescas et al. 2017).

It is important to note that the acquisition of a senescent phenotype is associated to cell type- and stimulus-dependent variations in gene expression. For instance, by using differential display and DNA arrays, Toussaint and coworkers (Pascal et al. 2005) showed that, despite displaying a wide variety of common biomarkers of senescence at the morphological level, WI-38 foetal lung human diploid fibroblasts acquired different expression profiles when induced to undergo senescence by replicative exhaustion or by treatment with tert-butylhydroperoxide or ethanol. Similarly, using RNA-seq, Purcell compared expression profiles of replicative senescence with that of various types of stress-induced senescence in fibroblasts, and highlighted both similarities and differences in expression between different pro-senescence stimuli (Purcell et al. 2014). Interestingly, comparative bioinformatic analyses of high-throughput microarray data from nerve and muscle biopsies from young and aged healthy donors showed a limited number of common differentially expressed genes among the skeletal muscle and nervous tissues, thus confirming the great heterogeneity that characterizes cellular senescence in vitro and in vivo (Voutetakis et al. 2015). Using numerous whole-transcriptome profiles of different fibroblast strains, Hernandez-Segura not only confirmed the stress- and cell-specific transcriptional response to senescence-inducing stimuli, but also showed that the gene expression profiles are temporally modulated (Hernandez-Segura et al. 2017). Recently, transcriptional heterogeneity of replicative senescent or stress-induced premature senescent fibroblasts was confirmed at single cell level using droplet-based single-cell mRNA sequencing (Tang et al. 2019). In this context, it is worth noting that gene expression analyses showed dramatically different profiles between senescent fibroblasts and senescent epithelial cells (Zhang et al. 2003). This observation, together with the stimulus-specific regulation of cellular senescence, suggests that the search for biomarkers of senescence for ageing and age-associated diseases, as well as for TIS in cancer patients, might require a large sampling effort, or that different biomarkers would eventually be used to identify different senescent phenotypes.

10.3 Potential Applications of Aptamers for Senescent Cells Biomarker Discovery and Targeting

The term aptamer (derived from the Latin aptus = to fit) was coined by Ellington and Szostak (http://ellingtonlab.org/blog/2014/12/1/on-aptamers) to describe molecules (oligonucleotides and, later, peptides) that bind to specific targets with high affinity and specificity, through structural recognition, thanks to their particular three-dimensional conformation, in the same way an antibody binds to an antigen (Ellington and Szostak 1990; Tuerk and Gold 1990; Ellington and Szostak 1992). Non-covalent interactions such as hydrophobic interactions, electrostatic interactions and van der Waals bonds mediate the binding of aptamers to their target molecules. Nucleic acid aptamers, usually ranging from 20 to 100 nt, are isolated from pools of random oligonucleotide sequences through an iterative process of selection and amplification, called SELEX (systemic evolution of ligands by exponential enrichment) (Tuerk and Gold 1990). Both RNA and single-strand DNA aptamers have been successfully selected using SELEX that can bind similar targets with similar affinity (McKeague and Derosa 2012; Lin and Patel 1997). However DNA aptamers are more stable than RNA aptamers (White et al. 2000), and the latter are usually chemically modified in order to enhance their stability (Kratschmer and Levy 2017; Germer et al. 2013).

Compared to antibodies, nucleic acid aptamers have a number of advantages. First, aptamers are chemically synthesized in the lab so their production does not require animals, is relatively easy and cheap, with low variability between batches. Aptamers can also be raised against non-immunogenic targets, such as small inorganic and organic molecules (https://www.aptagen.com/aptamer-index). Due to their small size, aptamers are minimally immunogenic (for instance, Zheng et al. 2017). Their small size also results in enhanced tissue penetration and reduced circulation time (Kaur et al. 2018; Keefe et al. 2010). Aptamers are amenable to chemical modification in the sugar, base or phosphate backbone, which can improve both their resistance to nuclease degradation and their affinity for targets (Ni et al. 2017). Finally, aptamers can be easily coupled to different probes, nanoparticles or other biomolecules.

Focusing on the potential application of nucleic acid aptamer technology for the discovery of senescent cells biomarkers, it is important to note that SELEX can be carried out using as targets both purified proteins or complex targets, such as whole cells (cell-SELEX) and plasma (Zhuo et al. 2017; Shamah et al. 2008; Fitter and James 2005). Hence, SELEX could be employed to generate aptamers that specifically recognize (and possibly modulate the function(s)) of already known senescence biomarkers, such as DEP1 (Althubiti et al. 2014), DPP4 (Kim et al. 2017) and CD36 (Chong et al. 2018). On the other hand, cell-SELEX is a powerful strategy for discovery of unknown biomarkers and the cell-SELEX method could be used to identify aptamers that show high affinity for still unidentified membrane molecules on the surface of senescent cells (Berezovski et al. 2008; Sefah et al. 2010). The inclusion of a counter-selection step, with proliferating and/or quiescent cells as negative controls, would allow for enrichment of senescence-specific aptamers. It is important to note that aptamers generated through cell-SELEX recognize various kinds of biological molecules, for instance proteins, lipids and carbohydrates, in their native conformation (Li 2007). Aptamers are able to distinguish between closely related molecules. For instance, Chen and colleagues isolated an RNA aptamer able to distinguish the single amino acid mutant protein $p53^{R175H}$ from the wild type p53 protein (Chen et al. 2015). Furthermore, aptamers are able to distinguish post-translational modifications (PTM) of target proteins (Díaz-Fernández et al. 2018; Gilbert et al. 1997). For instance, aptamers were produced that could recognize the hydrophobic farnesyl moiety in K-Ras (Gilbert et al. 1997). Highly specific DNA aptamers that specifically bound to histone H4 acetylated at lysine 16 (H4-K16Ac) were identified by Williams et al. (2009). DNA aptamers able to distinguish between peptides differing by a single glycosylation were developed, using N-glycosylated vascular endothelial growth factor (VEGF) (Rose et al. 2010). Notably, these aptamers were not specific for the sugar moiety itself, but rather the combination of the peptide sequence and the glycosylation. Since many types of PTM present in mammalian proteins are absent or different in bacteria (for instance ubiquitination and acetylation), it is important that recombinant proteins used as targets for SELEX are purified from mammalian cells if the aptamers are supposed to be used in mammalian cell systems, as demonstrated with the anti-cyclophilin B aptamer M9-5 (Ray et al. 2013). By binding to their target proteins, aptamers can modulate their functions, and aptamers working as both antagonists or agonists have been isolated. For example, an antagonistic DNA aptamer targeting PD-L1 efficiently blocked the binding between human PD-1 and PD-L1, thus restoring T cell functions and enhancing anti-tumor immune responses (Lai et al. 2016). An agonistic DNA aptamer, which bound to and stimulated the immune inhibitory receptor CD200R1, was shown to suppress inflammatory responses in vivo (Prodeus et al. 2018). Interestingly, this aptamer was capable of agonizing both murine and human CD200R1 (Prodeus et al. 2018).

Since their development in early 1990s nucleic acid aptamers have become important tools in a wide range of both clinical and research applications, such as ELISA and pull-down assays, flow cytometry, immunohistochemistry, drug delivery, functionalization of nanoparticles, and bioimaging (Zhu et al. 2015; Vandghanooni et al. 2018; He et al. 2018; Zhou and Rossi 2017). These observations strongly support the use of aptamers in the growing field of senotherapy. Although only one aptamer, Pegaptanib sodium, (Macugen; Eyetech Pharmaceuticals/Pfizer), an RNA aptamer directed against vascular endothelial growth factor (VEGF)-165, has been approved for therapeutic use in humans (Ng et al. 2006), other nucleic acid aptamers are at advanced stages of clinical trials (Kaur et al. 2018; Maier and Levy 2016).

Peptide aptamers were first described by Brent and coworkers (Colas et al. 1996), who based their approach on the concept that short peptides anchored at both amino and carboxy termini (i.e. conformationally constrained) are capable of high-affinity and specific interactions with other molecules. By using *Escherichia coli* Thioredoxin A protein as a rigid scaffold to display 20-residue random peptides, Brent and coworkers were able to select peptide aptamers that specifically bound human Cdk2, and, notably, inhibited its kinase activity. Since then, many other non-antibody protein scaffolds have been designed, which constrain variable peptide sequences for display, reviewed in Yu et al. 2017 and Simeon and Chen 2018. Here, we will focus on a recently developed, promising peptide aptamer technology known as Adhiron or Affimer system (Tiede et al. 2014, Tiede et al. 2017), developed by the BioScreening Technology Group (BSTG), from University of Leeds (UK) and their industrial partner Avacta (https://www.avacta.com/). Two different scaffolds were developed in this system: the first derived from human protease inhibitor Stefin A, the other derived from Cystatin A, a plant cysteine protease inhibitor. Each scaffold is capable of displaying one or two peptide variable regions of varying lengths (Tiede et al. 2014). Peptide aptamers derived from either of these two scaffolds are collectively called Affimers. Compared to antibodies, Affimers have a number of advantages. First, Affimers are easily expressed in *Escherichia coli* as well as in mammalian cells, and their production requires no animals (Dos Remedios and Peckham 2017). Both scaffolds are monomeric, highly stable (to both pH and temperature, with melting temperature between 70 and 100 °C) and small (~100 amino acids, 11 kDa, 3 nm in diameter). Their small size allows for better penetration into cellular and subcellular structures and tissue. Finally, the Affimer scaffolds show low immunogenicity (https://www.avacta.com/blogs/no-immunogenicity-issues-Affimer-technology), and can be subjected to chemical modification and conjugation.

Affimers recognizing specific targets are selected using a phage display strategy: a library of phage display Affimer reagents is screened against either purified target molecules or complex targets, such as whole cells, and a step of counter-screening may be used to increase specificity of isolated Affimers (Tang et al. 2017). Affimer libraries have been successfully screened against more than 500 target proteins and used as reagents in diagnostics, imaging and biomedical applications (recently reviewed in Kyle 2018).

Similar to nucleic acid aptamers, Affimers specifically recognize proteins with different types of post-translation modifications and are able to inhibit protein function and/or protein-protein interactions by binding to their specific targets. For instance, Affimers directed against VEGFR2 inhibited VEGF-stimulated receptor phosphorylation and downstream signaling in HUVEC cells, eventually reducing VEGF-induced HUVEC tubulogenesis (Tiede et al. 2017). Affimers that specifically bound SUMO-1 or SUMO-2, or both, inhibited SUMO-mediated protein-protein interactions (Hughes et al. 2017). Inhibition of protein-protein interaction by Affimers was also described by Robinson and colleagues, who generated Affimers against the immunoglobulin receptor FcγRIIIa. These Affimers were able to selectively block IgG immune complex binding to FcγRIIIa (but not to FcγRIIa or FcγRIIIb), and abrogated FcγRIIIa-dependent effector functions in macrophage (Robinson et al. 2018). Affimers were also developed for in vivo imaging of tumor markers. For instance, high affinity Affimers recognizing Tenascin C, an extracellular matrix glycoprotein that is highly expressed in the tumor stroma of most epithelial malignancies (Yoshida et al. 2015), were isolated by Tiede et al. (2017). This Affimer, conjugated with biotin, allowed for the detection of Tenascin C in human colorectal cancer and glioblastoma xenograft tissue sections, with a specificity similar to antibodies. In addition, Rhodamine Red-labelled Affimer accumulated in Tenascin C expressing tumour in vivo. Interestingly, Affimers were quickly cleared from the circulatory system, compared to antibodies, allowing for rapid imaging of tumor (Tiede et al. 2017).

On the whole, these studies highlight the potentiality of Affimers as tools for the identification of novel senescence-specific cell surface biomarkers.

10.4 Conclusions

In this chapter we have highlighted the need for novel senescence biomarkers and the development of diagnostic, prognostic and therapeutic interventions for aging and cancer-related TIS. A wealth of data from high-throughput transcriptomics and proteomics studies indicate that the acquisition of a senescent phenotype, in normal as well as neoplastic cells, is associated with alterations of cell surface proteins and/or their modifications. These studies support the possibility of utilizing these quantitative and qualitative differences for developing senescence-specific markers.

Since the first reports in the 1990s, nucleic acid aptamers and the SELEX technology have been widely applied for the discovery of biomarkers, useful in research and in the clinic. More recently, the peptide aptamer (Affimers) technology has emerged as valuable tools for biomarker discovery. We propose that aptamers represent a promising strategy to succeed in identifying new senescence-specific markers.

Acknowledgements We wish to thank Dr Darren Tomlinson and Dr Thomas A. Hughes (University of Leeds, UK) for introducing us to Affimers technology.

References

Acosta JC, O'Loghlen A, Banito A, Guijarro MV, Augert A, Raguz S et al (2008) Chemokine signaling via the CXCR2 receptor reinforces senescence. Cell 133(6):1006–1018

Alimonti A, Nardella C, Chen Z, Clohessy JG, Carracedo A, Trotman LC et al (2010) A novel type of cellular senescence that can be enhanced in mouse models and human tumor xenografts to suppress prostate tumorigenesis. J Clin Invest. 120(3):681–693

Alspach E, Flanagan KC, Luo X, Ruhland MK, Huang H, Pazolli E et al (2014) $p38^{MAPK}$ plays a crucial role in stromal-mediated tumorigenesis. Cancer Discov 4(6):716–729

Althubiti M, Lezina L, Carrera S, Jukes-Jones R, Giblett SM, Antonov A et al (2014) Characterization of novel markers of senescence and their prognostic potential in cancer. Cell Death Dis 5:e1528

Baker DJ, Wijshake T, Tchkonia T, LeBrasseur NK, Childs BG, van de Sluis B et al (2011) Clearance of $p16^{Ink4a}$-positive senescent cells delays ageing-associated disorders. Nature 479(7372):232–236

Baker DJ, Childs BG, Durik M, Wijers ME, Sieben CJ, Zhong J et al (2016) Naturally occurring p16(Ink4a)-positive cells shorten healthy lifespan. Nature 530(7589):184–189

Baraibar MA, Ladouce R, Friguet B (2013) Proteomic quantification and identification of carbonylated proteins upon oxidative stress and during cellular aging. J Proteomics 92:63–70

Bartkova J, Rezaei N, Liontos M, Karakaidos P, Kletsas D, Issaeva N et al (2006) Oncogene-induced senescence is part of the tumorigenesis barrier imposed by DNA damage checkpoints. Nature 444(7119):633–637

Berezovski MV, Lechmann M, Musheev MU, Mak TW, Krylov SN (2008) Aptamer-facilitated biomarker discovery (AptaBiD). J Am Chem Soc 130(28):9137–9143

Biran A, Perelmutter M, Gal H, Burton DG, Ovadya Y, Vadai E et al (2015) Senescent cells communicate via intercellular protein transfer. Genes Dev 29(8):791–802

Borghesan M, Fafián-Labora J, Eleftheriadou O, Carpintero-Fernández P, Paez-Ribes M, Vizcay-Barrena G et al (2019) Small extracellular vesicles are key regulators of non-cell autonomous intercellular communication in senescence via the interferon protein IFITM3. Cell Rep. 27(13):3956–3971.e6

Burgess DJ (2011) Senescence. NF-κB shows its beneficial side. Nat Rev Cancer 11(12):832–833

Bussian TJ, Aziz A, Meyer CF, Swenson BL, van Deursen JM, Baker DJ (2018) Clearance of senescent glial cells prevents tau-dependent pathology and cognitive decline. Nature 562(7728):578–582

Camorani S, Cerchia L, Fedele M, Erba E, D'Incalci M, Crescenzi E (2018) Trabectedin modulates the senescence-associated secretory phenotype and promotes cell death in senescent tumor cells by targeting NF-κB. Oncotarget. 9(28):19929–19944

Capell BC, Drake AM, Zhu J, Shah PP, Dou Z, Dorsey J et al (2016) MLL1 is essential for the senescence-associated secretory phenotype. Genes Dev 30(3):321–336

Chen L, Rashid F, Shah A, Awan HM, Wu M, Liu A et al (2015) The isolation of an RNA aptamer targeting to p53 protein with single amino acid mutation. Proc Natl Acad Sci USA 112(32):10002–10007

Chien Y, Scuoppo C, Wang X, Fang X, Balgley B, Bolden JE et al (2011) Control of the senescence-associated secretory phenotype by NF-κB promotes senescence and enhances chemosensitivity. Genes Dev 25(20):2125–2136

Chondrogianni N, Petropoulos I, Grimm S, Georgila K, Catalgol B, Friguet B et al (2014) Protein damage, repair and proteolysis. Mol Aspect Med 35:1–71

Chong M, Yin T, Chen R, Xiang H, Yuan L, Ding Y et al (2018) CD36 initiates the secretory phenotype during the establishment of cellular senescence. EMBO Rep 19(6)

Colas P, Cohen B, Jessen T, Grishina I, McCoy J, Brent R (1996) Genetic selection of peptide aptamers that recognize and inhibit cyclin-dependent kinase 2. Nature 380(6574):548–550

Coppé JP, Patil CK, Rodier F, Sun Y, Muñoz DP, Goldstein J et al (2008) Senescence-associated secretory phenotypes reveal cell-nonautonomous functions of oncogenic RAS and the p53 tumor suppressor. PLoS Biol 6(12):2853–2868

Correia-Melo C, Marques FD, Anderson R, Hewitt G, Hewitt R, Cole J et al (2016) Mitochondria are required for pro-ageing features of the senescent phenotype. EMBO J 35(7):724–742

Cotarelo CL, Schad A, Kirkpatrick CJ, Sleeman JP, Springer E, Schmidt M et al (2016) Detection of cellular senescence within human invasive breast carcinomas distinguishes different breast tumor subtypes. Oncotarget 7(46):74846–74859

Crescenzi E, Pacifico F, Lavorgna A, De Palma R, D'Aiuto E, Palumbo G et al (2011) NF-κB-dependent cytokine secretion controls Fas expression on chemotherapy-induced premature senescent tumor cells. Oncogene 30(24):2707–2717

d'Adda di Fagagna F, Reaper PM, Clay-Farrace L, Fiegler H, Carr P, Von Zglinicki T et al (2003) A DNA damage checkpoint response in telomere-initiated senescence. Nature 426(6963):194–198

Demaria M, O'Leary MN, Chang J, Shao L, Liu S, Alimirah F et al (2017) Cellular senescence promotes adverse effects of chemotherapy and cancer relapse. Cancer Discov 7(2):165–176

Di Micco R, Fumagalli M, Cicalese A, Piccinin S, Gasparini P, Luise C et al (2006) Oncogene-induced senescence is a DNA damage response triggered by DNA hyper-replication. Nature 444(7119):638–642

Díaz-Fernández A, Miranda-Castro R, de-Los-Santos-Álvarez N, Lobo-Castañón MJ (2018) Post-translational modifications in tumor biomarkers: the next challenge for aptamers? Anal Bioanal Chem 410(8):2059–2065

Dos Remedios C, Peckham M (2017) 3Rs and biophysics. Biophys Rev 9(4):277–278

Eggert T, Wolter K, Ji J, Ma C, Yevsa T, Klotz S et al (2016) Distinct functions of senescence-associated immune responses in liver tumor surveillance and tumor progression. Cancer Cell 30(4):533–547

Ellington AD, Szostak JW (1990) In vitro selection of RNA molecules that bind specific ligands. Nature 346(6287):818–822

Ellington AD, Szostak JW (1992) Selection in vitro of single-stranded DNA molecules that fold into specific ligand-binding structures. Nature 355(6363):850–852

Evangelou K, Lougiakis N, Rizou SV, Kotsinas A, Kletsas D, Muñoz-Espín D et al (2017) Robust, universal biomarker assay to detect senescent cells in biological specimens. Aging Cell 16(1):192–197

Farr JN, Xu M, Weivoda MM, Monroe DG, Fraser DG, Onken JL et al (2017) Targeting cellular senescence prevents age-related bone loss in mice. Nat Med 23(9):1072–1079

Fitter S, James R (2005) Deconvolution of a complex target using DNA aptamers. J Biol Chem 280(40):34193–34201

Frescas D, Roux CM, Aygun-Sunar S, Gleiberman AS, Krasnov P, Kurnasov OV et al (2017) Senescent cells expose and secrete an oxidized form of membrane-bound vimentin as revealed by a natural polyreactive antibody. Proc Natl Acad Sci USA 114(9):E1668–E1677

Freund A, Laberge RM, Demaria M, Campisi J (2012) Lamin B1 loss is a senescence-associated biomarker. Mol Biol Cell 23(11):2066–2075

Friguet B (2002) Aging of proteins and the proteasome. Prog Mol Subcell Biol 29:17–33

Fumagalli M, Rossiello F, Clerici M, Barozzi S, Cittaro D, Kaplunov JM et al (2012) Telomeric DNA damage is irreparable and causes persistent DNA-damage-response activation. Nat Cell Biol 14(4):355–365

Georgakopoulou EA, Tsimaratou K, Evangelou K, Fernandez Marcos PJ, Zoumpourlis V, Trougakos IP et al (2013) Specific lipofuscin staining as a novel biomarker to detect replicative and stress-induced senescence. A method applicable in cryo-preserved and archival tissues. Aging (Albany NY) 5(1):37–50

Georgilis A, Klotz S, Hanley CJ, Herranz N, Weirich B, Morancho B et al (2018) PTBP1-mediated alternative splicing regulates the inflammatory secretome and the pro-tumorigenic effects of senescent cells. Cancer Cell 34(1):85–102.e9

Germer K, Leonard M, Zhang X (2013) RNA aptamers and their therapeutic and diagnostic applications. Int J Biochem Mol Biol 4(1):27–40
Gilbert BA, Sha M, Wathen ST, Rando RR (1997) RNA aptamers that specifically bind to a K Ras-derived farnesylated peptide. Bioorg Med Chem 5(6):1115–1122
Glück S, Ablasser A (2019) Innate immunosensing of DNA in cellular senescence. Curr Opin Immunol 56:31–36
Gorisse L, Pietrement C, Vuiblet V, Schmelzer CE, Köhler M, Duca L et al (2016) Protein carbamylation is a hallmark of aging. Proc Natl Acad Sci USA 113(5):1191–1196
Grabowska W, Kucharewicz K, Wnuk M, Lewinsk A, Suszek M, Przybylska D et al (2015) Curcumin induces senescence of primary human cells building the vasculature in a DNA damage and ATM-independent manner. Age (Dordr) 37(1):9744
Hari P, Millar FR, Tarrats N, Birch J, Quintanilla A, Rink CJ et al (2019) The innate immune sensor Toll-like receptor 2 controls the senescence-associated secretory phenotype. Sci Adv 5(6):eaaw0254
He F, Wen N, Xiao D, Yan J, Xiong H, Cai S et al (2018) Aptamer based targeted drug delivery systems: current potential and challenges. Curr Med Chem [Epub ahead of print]
Hernandez-Segura A, de Jong TV, Melov S, Guryev V, Campisi J, Demaria M (2017) Unmasking Transcriptional Heterogeneity in Senescent Cells. Curr Biol 27(17):2652–2660.e4
Hewitt G, Jurk D, Marques FD, Correia-Melo C, Hardy T, Gackowska A et al (2012) Telomeres are favoured targets of a persistent DNA damage response in ageing and stress-induced senescence. Nat Commun 3:708
Hickson LJ, Langhi Prata LGP, Bobart SA, Evans TK, Giorgadze N, Hashmi SK et al (2019) Senolytics decrease senescent cells in humans: Preliminary report from a clinical trial of Dasatinib plus Quercetin in individuals with diabetic kidney disease. EBioMedicine 47:446–456
Hills SA, Diffley JF (2014) DNA replication and oncogene-induced replicative stress. Curr Biol 24(10):R435–R444
Hoare M, Ito Y, Kang TW, Weekes MP, Matheson NJ, Patten DA et al (2016) NOTCH1 mediates a switch between two distinct secretomes during senescence. Nat Cell Biol 18(9):979–992
Hughes DJ, Tiede C, Penswick N, Tang AA, Trinh CH, Mandal U et al (2017) Generation of specific inhibitors of SUMO-1- and SUMO-2/3-mediated protein-protein interactions using Affimer (Adhiron) technology. Sci Signal 10(505):pii:eaaj2005
Iannello A, Thompson TW, Ardolino M, Lowe SW, Raulet DH (2013) p53-dependent chemokine production by senescent tumor cells supports NKG2D-dependent tumor elimination by natural killer cells. J Exp Med 210(10):2057–2069
Itakura Y, Sasaki N, Kami D, Gojo S, Umezawa A, Toyoda M (2016) N- and O-glycan cell surface protein modifications associated with cellular senescence and human aging. Cell Biosci 6:14
Jackson JG, Pant V, Li Q, Chang LL, Quintás-Cardama A, Garza D et al (2012) p53-mediated senescence impairs the apoptotic response to chemotherapy and clinical outcome in breast cancer. Cancer Cell 21(6):793–806
Jeon OH, Kim C, Laberge RM, Demaria M, Rathod S, Vasserot AP et al (2017) Local clearance of senescent cells attenuates the development of post-traumatic osteoarthritis and creates a pro-regenerative environment. Nat Med 23(6):775–778
Jeon OH, Wilson DR, Clement CC, Rathod S, Cherry C, Powell B et al (2019) Senescence cell-associated extracellular vesicles serve as osteoarthritis disease and therapeutic markers. JCI Insight 4(7):pii:125019
Justice JN, Nambiar AM, Tchkonia T, LeBrasseur NK, Pascual R, Hashmi SK et al (2019) Senolytics in idiopathic pulmonary fibrosis: results from a first-in-human, open-label, pilot study. EBioMedicine 40:554–563
Karimi-Busheri F, Rasouli-Nia A, Mackey JR, Weinfeld M (2010) Senescence evasion by MCF-7 human breast tumor-initiating cells. Breast Cancer Res 12(3):R31
Kaur H, Bruno JG, Kumar A, Sharma TK (2018) Aptamers in the therapeutics and diagnostics pipelines. Theranostics 8(15):4016–4032
Keefe AD, Pai S, Ellington A (2010) Aptamers as therapeutics. Nat Rev Drug Discov 9(7):537–550

Kim SB, Bozeman RG, Kaisani A, Kim W, Zhang L, Richardson JA et al (2016) Radiation promotes colorectal cancer initiation and progression by inducing senescence-associated inflammatory responses. Oncogene 35(26):3365–3375

Kim KM, Noh JH, Bodogai M, Martindale JL, Yang X, Indig FE et al (2017) Identification of senescent cell surface targetable protein DPP4. Genes Dev 31(15):1529–1534

Kirkland JL, Tchkonia T (2017) Cellular senescence: a translational perspective. EBioMedicine 21:21–28

Kovacovicova K, Skolnaja M, Heinmaa M, Mistrik M, Pata P, Pata I et al (2018) Senolytic cocktail Dasatinib+Quercetin (D+Q) does not enhance the efficacy of senescence-inducing chemotherapy in liver cancer. Front Oncol 8:459

Kratschmer C, Levy M (2017) Effect of chemical modifications on aptamer stability in serum. Nucleic Acid Ther 27(6):335–344

Kuilman T, Peeper DS (2009) Senescence-messaging secretome: SMS-ing cellular stress. Nat Rev Cancer 9(2):81–94

Kuilman T, Michaloglou C, Vredeveld LC, Douma S, van Doorn R, Desmet CJ et al (2008) Oncogene-induced senescence relayed by an interleukin-dependent inflammatory network. Cell 133(6):1019–1031

Kurz DJ, Decary S, Hong Y, Erusalimsky JD (2000) Senescence-associated (beta)-galactosidase reflects an increase in lysosomal mass during replicative ageing of human endothelial cells. J Cell Sci 113(Pt 20):3613–3622

Kyle S (2018) Affimer proteins: theranostics of the future? Trends Biochem Sci 43(4):230–232

Laberge RM, Sun Y, Orjalo AV, Patil CK, Freund A, Zhou L et al (2015) MTOR regulates the pro-tumorigenic senescence-associated secretory phenotype by promoting IL1A translation. Nat Cell Biol 17(8):1049–1061

Lai WY, Huang BT, Wang JW, Lin PY, Yang PC (2016) A novel PD-L1-targeting antagonistic DNA aptamer with antitumor effects. Mol Ther Nucleic Acids 5(12):e3977

Larsen AK, Galmarini CM, D'Incalci M (2016) Unique features of trabectedin mechanism of action. Cancer Chemother Pharmacol 77(4):663–671

Li L (2007) Detection of protein biomarkers using RNA aptamer microarrays and enzymatically amplified surface plasmon resonance imaging. Anal Chem 79(3):1082–1088

Lin CH, Patel DJ (1997) Structural basis of DNA folding and recognition in an AMP-DNA aptamer complex: distinct architectures but common recognition motifs for DNA and RNA aptamers complexed to AMP. Chem Biol 4(11):817–832

Liu S, Uppal H, Demaria M, Desprez PY, Campisi J, Kapahi P (2015) Simvastatin suppresses breast cancer cell proliferation induced by senescent cells. Sci Rep 5:17895

Liu S, Mahairaki V, Bai H, Ding Z, Li J, Witwer KW et al (2019) Highly purified human extracellular vesicles produced by stem cells alleviate aging cellular phenotypes of senescent human cells. Stem Cells 37(6):779–790

Llanos S, Serrano M (2016) Senescence and cancer: in the name of immunosuppression. Cancer Cell 30(4):507–508

Maciel-Barón LA, Morales-Rosales SL, Aquino-Cruz AA, Triana-Martínez F, Galván-Arzate S, Luna-López A et al (2016) Senescence associated secretory phenotype profile from primary lung mice fibroblasts depends on the senescence induction stimuli. Age (Dordr) 38(1):26

Maier KE, Levy M (2016) From selection hits to clinical leads: progress in aptamer discovery. Mol Ther Methods Clin Dev 5:16014

McKeague M, Derosa MC (2012) Challenges and opportunities for small molecule aptamer development. J Nucleic Acids 2012:748913

Mirzayans R, Scott A, Cameron M, Murray D (2005) Induction of accelerated senescence by gamma radiation in human solid tumor-derived cell lines expressing wild-type TP53. Radiat Res 163(1):53–62

Mosteiro L, Pantoja C, Alcazar N, Marión RM, Chondronasiou D, Rovira M et al (2016) Tissue damage and senescence provide critical signals for cellular reprogramming in vivo. Science 354(6315):pii:aaf4445

Muñoz-Espín D, Serrano M (2014) Cellular senescence: from physiology to pathology. Nat Rev Mol Cell Biol 15(7):482–496

Muñoz-Espín D, Cañamero M, Maraver A, Gómez-López G, Contreras J, Murillo-Cuesta S et al (2013) Programmed cell senescence during mammalian embryonic development. Cell 155(5):1104–1118

Muñoz-Espín D, Rovira M, Galiana I, Giménez C, Lozano-Torres B, Paez-Ribes M et al (2018) A versatile drug delivery system targeting senescent cells. EMBO Mol Med 10(9):pii:e9355

Myrianthopoulos V, Evangelou K, Vasileiou PVS, Cooks T, Vassilakopoulos TP, Pangalis GA et al (2019) Senescence and senotherapeutics: a new field in cancer therapy. Pharmacol Ther 193:31–49

Ng EW, Shima DT, Calias P, Cunningham ET Jr, Guyer DR, Adamis AP (2006) Pegaptanib, a targeted anti-VEGF aptamer for ocular vascular disease. Nat Rev Drug Discov 5(2):123–132

Ni S, Yao H, Wang L, Lu J, Jiang F, Lu A et al (2017) Chemical modifications of nucleic acid aptamers for therapeutic purposes. Int J Mol Sci 18(8):pii:E1683

Novakova Z, Hubackova S, Kosar M, Janderova-Rossmeislova L, Dobrovolna J, Vasicova P et al (2010) Cytokine expression and signaling in drug-induced cellular senescence. Oncogene 29(2):273–284

Pascal T, Debacq-Chainiaux F, Chrétien A, Bastin C, Dabée AF, Bertholet V et al (2005) Comparison of replicative senescence and stress-induced premature senescence combining differential display and low-density DNA arrays. FEBS Lett 579(17):3651–3659

Pazolli E, Alspach E, Milczarek A, Prior J, Piwnica-Worms D, Stewart SA (2012) Chromatin remodeling underlies the senescence-associated secretory phenotype of tumor stromal fibroblasts that supports cancer progression. Cancer Res 72(9):2251–2261

Perrott KM, Wiley CD, Desprez PY, Campisi J (2017) Apigenin suppresses the senescence-associated secretory phenotype and paracrine effects on breast cancer cells. Geroscience 39(2):161–173

Pospelova TV, Demidenko ZN, Bukreeva EI, Pospelov VA, Gudkov AV, Blagosklonny MV (2009) Pseudo-DNA damage response in senescent cells. Cell Cycle 8(24):4112–4118

Prodeus A, Sparkes A, Fischer NW, Cydzik M, Huang E, Khatri I et al (2018) A synthetic cross-species CD200R1 agonist suppresses inflammatory immune responses in vivo. Mol Ther Nucleic Acids 12:350–358

Purcell M, Kruger A, Tainsky MA (2014) Gene expression profiling of replicative and induced senescence. Cell Cycle 13(24):3927–3937

Ray P, Sullenger BA, White RR (2013) Further characterization of the target of a potential aptamer biomarker for pancreatic cancer: cyclophilin B and its posttranslational modifications. Nucleic Acid Ther 23(6):435–442

Reimann M, Lee S, Loddenkemper C, Dörr JR, Tabor V, Aichele P et al (2010) Tumor stroma-derived TGF-beta limits myc-driven lymphomagenesis via Suv39h1-dependent senescence. Cancer Cell 17(3):262–272

Ritschka B, Storer M, Mas A, Heinzmann F, Ortells MC, Morton JP et al (2017) The senescence-associated secretory phenotype induces cellular plasticity and tissue regeneration. Genes Dev 31(2):172–183

Roberson RS, Kussick SJ, Vallieres E, Chen SY, Wu DY (2005) Escape from therapy-induced accelerated cellular senescence in p53-null lung cancer cells and in human lung cancers. Cancer Res 65(7):2795–2803

Robinson JI, Baxter EW, Owen RL, Thomsen M, Tomlinson DC, Waterhouse MP et al (2018) Affimer proteins inhibit immune complex binding to FcγRIIIa with high specificity through competitive and allosteric modes of action. Proc Natl Acad Sci USA 115(1):E72–E81

Rodier F, Campisi J (2011) Four faces of cellular senescence. J Cell Biol 192(4):547–556

Roos CM, Zhang B, Palmer AK, Ogrodnik MB, Pirtskhalava T, Thalji NM et al (2016) Chronic senolytic treatment alleviates established vasomotor dysfunction in aged or atherosclerotic mice. Aging Cell 15(5):973–977

Rose CM, Hayes MJ, Stettler GR, Hickey SF, Axelrod TM, Giustini NP et al (2010) Capillary electrophoretic development of aptamers for a glycosylated VEGF peptide fragment. Analyst 35(11):2945–2951

Sagini K, Costanzi E, Emiliani C, Buratta S, Urbanelli L (2018) Extracellular vesicles as conveyors of membrane-derived bioactive lipids in immune system. Int J Mol Sci 19(4).pii:E1227

Salama R, Sadaie M, Hoare M, Narita M (2014) Cellular senescence and its effector programs. Genes Dev 28(2):99–114

Samaraweera L, Adomako A, Rodriguez-Gabin A, McDaid HM (2017) A novel indication for panobinostat as a senolytic drug in NSCLC and HNSCC. Sci Rep 7(1):1900

Schafer MJ, White TA, Iijima K, Haak AJ, Ligresti G, Atkinson EJ et al (2017) Cellular senescence mediates fibrotic pulmonary disease. Nat Commun 8:14532

Sedelnikova OA, Horikawa I, Zimonjic DB, Popescu NC, Bonner WM, Barrett JC (2004) Senescing human cells and ageing mice accumulate DNA lesions with unrepairable double-strand breaks. Nat Cell Biol 6(2):168–170

Sefah K, Shangguan D, Xiong X, O'Donoghue MB, Tan W (2010) Development of DNA aptamers using Cell-SELEX. Nat Protoc 5(6):1169–1185

Serrano M (2014) Senescence helps regeneration. Dev Cell 31(6):671–672

Severino J, Allen RG, Balin S, Balin A, Cristofalo VJ (2000) Is beta-galactosidase staining a marker of senescence in vitro and in vivo? Exp Cell Res 257(1):162–171

Shamah SM, Healy JM, Cload ST (2008) Complex target SELEX. Acc Chem Res 41(1):130–138

Sidi R, Pasello G, Opitz I, Soltermann A, Tutic M, Rehrauer H et al (2011) Induction of senescence markers after neo-adjuvant chemotherapy of malignant pleural mesothelioma and association with clinical outcome: an exploratory analysis. Eur J Cancer 47(2):326–332

Simeon R, Chen Z (2018) In vitro-engineered non-antibody protein therapeutics. Protein Cell 9(1):3–14

Storer M, Mas A, Robert-Moreno A, Pecoraro M, Ortells MC, Di Giacomo V et al (2013) Senescence is a developmental mechanism that contributes to embryonic growth and patterning. Cell 155(5):1119–1130

Sun Y, Coppé JP, Lam EW (2018) Cellular senescence: the sought or the unwanted? Trends Mol Med 24(10):871–885

Supiot S, Shubbar S, Fleshner N, Warde P, Hersey K, Wallace K et al (2008) A phase I trial of pre-operative radiotherapy for prostate cancer: clinical and translational studies. Radiother Oncol 88(1):53–60

Takahashi A, Okada R, Nagao K, Kawamata Y, Hanyu A, Yoshimoto S et al (2017) Exosomes maintain cellular homeostasis by excreting harmful DNA from cells. Nat Commun 8:15287

Takasugi M (2018) Emerging roles of extracellular vesicles in cellular senescence and aging. Aging Cell 17(2)

Takasugi M, Okada R, Takahashi A, Virya Chen D, Watanabe S, Hara E (2017) Small extracellular vesicles secreted from senescent cells promote cancer cell proliferation through EphA2. Nat Commun 8:15729

Tang AA, Tiede C, Hughes DJ, McPherson MJ, Tomlinson DC (2017) Isolation of isoform-specific binding proteins (Affimers) by phage display using negative selection. Sci Signal 10(505):pii:eaan0868

Tang H, Geng A, Zhang T, Wang C, Jiang Y, Mao Z (2019) Single senescent cell sequencing reveals heterogeneity in senescent cells induced by telomere erosion. Protein Cell 10(5):370–375

Tasdemir N, Banito A, Roe JS, Alonso-Curbelo D, Camiolo M, Tschaharganeh DF et al (2016) BRD4 connects enhancer remodeling to senescence immune surveillance. Cancer Discov 6(6):612–629

te Poele RH, Okorokov AL, Jardine L, Cummings J, Joel SP (2002) DNA damage is able to induce senescence in tumor cells in vitro and in vivo. Cancer Res 62(6):1876–1883

Tiede C, Tang AA, Deacon SE, Mandal U, Nettleship JE, Owen RL et al (2014) Adhiron: a stable and versatile peptide display scaffold for molecular recognition applications. Protein Eng Des Sel 27(5):145–155

Tiede C, Bedford R, Heseltine SJ, Smith G, Wijetunga I, Ross R et al (2017) Affimer proteins are versatile and renewable affinity reagents. Elife 6:pii:e24903

Toso A, Revandkar A, Di Mitri D, Guccini I, Proietti M, Sarti M et al (2014) Enhancing chemotherapy efficacy in Pten-deficient prostate tumors by activating the senescence-associated antitumor immunity. Cell Rep 9(1):75–89

Tuerk C, Gold L (1990) Systematic evolution of ligands by exponential enrichment: RNA ligands to bacteriophage T4 DNA polymerase. Science 249(4968):505–510

Vandghanooni S, Eskandani M, Barar J, Omidi Y (2018) Recent advances in aptamer-armed multimodal theranostic nanosystems for imaging and targeted therapy of cancer. Eur J Pharm Sci 117:301–312

Vanhooren V, Navarrete Santos A, Voutetakis K, Petropoulos I, Libert C, Simm A et al (2015) Protein modification and maintenance systems as biomarkers of ageing. Mech Ageing Dev 51:71–84

von Zglinicki T, Nilsson E, Döcke WD, Brunk UT (1995) Lipofuscin accumulation and ageing of fibroblasts. Gerontology 41(Suppl 2):95–108

Voutetakis K, Chatziioannou A, Gonos ES, Trougakos IP (2015) Comparative meta-analysis of transcriptomics data during cellular senescence and in vivo tissue ageing. Oxid Med Cell Longev 2015:732914

Wei W, Ji S (2018) Cellular senescence: molecular mechanisms and pathogenicity. J Cell Physiol 233(12):9121–9135

White RR, Sullenger BA, Rusconi CP (2000) Developing aptamers into therapeutics. J Clin Invest 106(8):929–934

Wiley CD, Flynn JM, Morrissey C, Lebofsky R, Shuga J, Dong X et al (2017) Analysis of individual cells identifies cell-to-cell variability following induction of cellular senescence. Aging Cell 16(5):1043–1105

Williams BA, Lin L, Lindsay SM, Chaput JC (2009) Evolution of a histone H4-K16 acetyl-specific DNA aptamer. J Am Chem Soc 131(18):6330–6331

Wu PC, Wang Q, Grobman L, Chu E, Wu DY (2012) Accelerated cellular senescence in solid tumor therapy. Exp Oncol 34(3):298–305

Xu M, Tchkonia T, Ding H, Ogrodnik M, Lubbers ER, Pirtskhalava T et al (2015) JAK inhibition alleviates the cellular senescence-associated secretory phenotype and frailty in old age. Proc Natl Acad Sci USA 112(46):E6301–E6310

Xu M, Pirtskhalava T, Farr JN, Weigand BM, Palmer AK, Weivoda MM et al (2018) Senolytics improve physical function and increase lifespan in old age. Nat Med 24(8):1246–1256

Yang NC, Hu ML (2005) The limitations and validities of senescence associated-beta-galactosidase activity as an aging marker for human foreskin fibroblast Hs68 cells. Exp Gerontol 40(10):813–819

Yang L, Fang J, Chen J (2017) Tumor cell senescence response produces aggressive variants. Cell Death Discov 3:17049

Yoshida T, Akatsuka T, Imanaka-Yoshida K (2015) Tenascin-C and integrins in cancer. Cell Adh Migr 9(1–2):96–104

Yousefzadeh MJ, Zhu Y, McGowan SJ, Angelini L, Fuhrmann-Stroissnigg H, Xu M et al (2018) Fisetin is a senotherapeutic that extends health and lifespan. EBioMedicine 36:18–28

Yu X, Yang YP, Dikici E, Deo SK, Daunert S (2017) Beyond antibodies as binding partners: the role of antibody mimetics in bioanalysis. Annu Rev Anal Chem (Palo Alto Calif) 10(1):293–320

Zhang H, Pan KH, Cohen SN (2003) Senescence-specific gene expression fingerprints reveal cell-type-dependent physical clustering of up-regulated chromosomal loci. Proc Natl Acad Sci USA 100(6):3251–3256

Zhang Y, Kim MS, Jia B, Ya J, Zuniga-Hertz J P, Han C et al (2017). Hypothalamic stem cells control ageing speed partly through exosomal miRNAs. Nature 548(7665):52–57

Zheng J, Zhao S, Yu X, Huang S, Liu HY (2017) Simultaneous targeting of CD44 and EpCAM with a bispecific aptamer effectively inhibits intraperitoneal ovarian cancer growth. Theranostics 7(5):1373–1388

Zhou J, Rossi J (2017) Aptamers as targeted therapeutics: current potential and challenges. Nat Rev Drug Discov. 16(3):181–202

Zhu Q, Liu G, Kai M (2015) DNA aptamers in the diagnosis and treatment of human diseases. Molecules 20(12):20979–20997

Zhuo Z, Yu Y, Wang M, Li J, Zhang Z, Liu J et al (2017) Recent Advances in SELEX Technology and Aptamer Applications in Biomedicine. Int J Mol Sci 18(10):pii:E2142

Chapter 11
Mitochondria: Potential Targets for Interventions to Counteract Senescence

Anthony Lagnado, Stella Victorelli, and João F. Passos

Abstract Cellular senescence, an irreversible cell-cycle arrest, has been shown to occur in a variety of physiological contexts and play multiple functions. Importantly, recent data indicates that senescence is not only associated with aging and age-related diseases, but also plays a causal role in these processes. Senescence is thought to contribute to tissue dysfunction mostly through the generation of a senescence associated secretory phenotype (SASP) which can induce chronic inflammation, changes in tissue architecture, stem cell depletion and spread senescence in otherwise young healthy cells. An important feature of cellular senescence is mitochondrial dysfunction, which has been shown not only to contribute to the cell-cycle arrest but also to tightly regulate the SASP. In this book chapter, we review some of the most up-to-date literature on the role of mitochondria during cellular senescence. Finally, we propose that by targeting mitochondria we may be able to counteract the detrimental effects of cellular senescence during aging and associated diseases.

Keywords Cellular senescence · Aging · Mitochondria · Telomeres · SASP

11.1 Introduction

11.1.1 What Is Cellular Senescence?

Cellular senescence was discovered in the 1960s by Hayflick and Moorhead. They observed that cells did not grow indefinitely and ceased proliferating after undergoing a certain number of cell doublings in vitro, a process they called cellular senescence (Hayflick and Moorhead 1961; Hayflick 1965; Lagouge and Larsson 2013). Although

---

A. Lagnado · S. Victorelli · J. F. Passos (✉)
Department of Physiology and Biochemical Engineering, Mayo Clinic, Rochester, USA
e-mail: Passos.Joao@mayo.edu

J. F. Passos
Institute for Cell and Molecular Biosciences, Newcastle University Institute for Ageing, Newcastle University, Newcastle upon Tyne, UK

D. Muñoz-Espin and M. Demaria (eds.), *Senolytics in Disease, Ageing and Longevity*, Healthy Ageing and Longevity 11,
https://doi.org/10.1007/978-3-030-44903-2_11

senescent cells can no longer divide, they remain viable and metabolically active for a long period of time (Wang et al. 1994).

This breakthrough discovery opened a new field of research which aimed to investigate the mechanisms underlying this process as well as its role in vivo. Scientists have since deciphered many of the complex pathways that drive senescence. It has been shown that senescence can be trigged not only by replicative exhaustion, a process thought to be mostly driven by telomere shortening (Levy et al. 1992) but also in response to various stresses (Passos et al. 2007; Kirkwood and Cremer 1982). Early on, it was suggested that one of the main functions of cellular senescence is to act as an anti-tumor mechanism. Consistently, senescence can be triggered in response to aberrant activation of mutated oncogenes, which halts uncontrolled cell division (Serrano et al. 1997). Two main pathways are thought to be involved in senescence, which depend on the activation of cyclin-dependent kinase inhibitors (CDKi). The p53-p21 pathway relies on the activation of the "master gene" TP53 that will induce the expression p21, which prevents phosphorylation and inactivation of the anti-tumor protein pRb. Hypophosphorylated pRb remains bound to E2F transcription factor, preventing it from regulating expression of genes required for the cell cycle, consequently leading to a G1 arrest (Ohtani et al. 1995; Geng et al. 1996; Brugarolas et al. 1995; el-Deiry et al. 1994). Additionally, the p16-pRb pathway is believed to act as a second barrier to stop the cell-cycle. The CDKi p16 inhibits Cdk4/6/cyclin D, causing pRb to remain in its hypophosphorylated state (Serrano et al. 1993). It is thought that p16 acts as a point of no return ensuring the irreversibility of the arrest (Beauséjour et al. 2003; Mirzayans et al. 2010).

Apart from the cell-cycle arrest, senescent cells undergo multiple phenotypic changes in all cellular compartments; these include an increase in cytoplasmic and nuclear size, changes in lipid metabolism and deposition, increased lysosomal and mitochondrial content, amongst many others (Cristofalo et al. 1989; Cristofalo and Pignolo 1993). Importantly, senescent cells have been shown to produce and secrete a number of pro-inflammatory cytokines, chemokines, growth factors, as well as proteins responsible for degrading the extracellular matrix, collectively known as the Senescence-associated secretory phenotype (Coppé et al. 2008). Although the SASP has been shown to have beneficial effects in some biological processes such as wound healing (Demaria et al. 2014) and mammalian development (Munoz-Espin et al. 2013), chronic exposure to these factors can induce senescence in neighboring cells, contribute to tissue dysfunction, and even promote tumorigenesis (Demaria et al. 2014; Munoz-Espin et al. 2013; Storer et al. 2013; Nelson et al. 2012; Acosta et al. 2013; Krtolica et al. 2001; Gasser et al. 2005; Xue et al. 2007; Krizhanovsky et al. 2008; Watanabe et al. 2017).

Senescent cells have also been shown to be resistant to apoptosis in conditions of genotoxic stress and serum starvation. This characteristic has been shown to be connected with the specific expression of pro-survival pathways, which protect them against apoptosis (Wang et al. 2005; Seluanov et al. 2001).

11.1.2 Why Is Senescence Important?

As mentioned above, senescence has been implicated in several processes and has been shown to have beneficial as well as detrimental effects depending on the physiological context. Senescence has been shown to play beneficial roles during development, tumor suppression, wound healing and tissue repair (Demaria et al. 2014; Munoz-Espin et al. 2013). In contrast, studies have consistently shown that senescent cells accumulate in various tissues with age, as well as in several age-related diseases (Birch et al. 2015; Birch et al. 2016; Ogrodnik et al. 2017; Schafer et al. 2017). More recently, mouse models were developed which allowed the genetic clearance of $p16^{ink4a}$ positive senescent cells (Baker et al. 2011). Since then, senescence has been demonstrated to play a causal role in a variety of age-related diseases such as osteoarthritis (Jeon et al. 2017; Xu et al. 2017), osteoporosis (Farr et al. 2017), neurodegeneration (Bussian et al. 2018; Chinta et al. 2018; Musi et al. 2018), atherosclerosis (Childs et al. 2016; Roos et al. 2016), fatty liver disease (Ogrodnik et al. 2017), myocardial infarction (Walaszczyk et al. 2019) heart failure with preserved ejection fraction (HFpEF) (Anderson et al. 2019); diabetes (Palmer et al. 2019), pulmonary fibrosis (Schafer et al. 2017), neuropsychiatric disorders (Ogrodnik et al. 2019) amongst others.

Senescent cells are thought to contribute to tissue dysfunction by inducing senescence in neighboring cells (Nelson et al. 2012; Acosta et al. 2013), by disrupting the extracellular matrix and inducing fibrosis (Schafer et al. 2017; Anderson et al. 2019), by recruiting and activating immune cells exacerbating inflammation (Jurk et al. 2014) and by interfering with stem cell and progenitor cells (Ogrodnik et al. 2019; Yosef et al. 2016). Transplanting relatively small numbers of senescent cells into young mice has been shown to accelerate physical dysfunction potentially by spreading senescence to host tissues (Xu et al. 2018).

Collectively, these studies ultimately led to the development of "senolytic" drugs, which are drugs capable of inducing apoptosis specifically in senescent but not young cells. Currently, several drugs with senolytic properties have been discovered (Zhu et al. 2015; Yousefzadeh et al. 2018) and shown to have beneficial effects in the context of aging and age-related diseases in mice (Chapman et al. 2019). The first results from a small human clinical trial in patients with idiopathic pulmonary fibrosis (IPF) using senolytic drug cocktail dasatinib and quercetin were published this year (Justice et al. 2019).

Another way to target senescence is by alleviating the SASP without impacting on the cell-cycle machinery, which is essential for the tumor suppressive capabilities of senescent cells (de Magalhaes and Passos 2018). So far, a number of drugs, which seem to have these properties, have been identified. These are termed "senostatic" and will be discussed in more detail later on.

11.2 Mitochondrial Dysfunction Is a Hallmark of Cellular Senescence

11.2.1 Concerning Mitochondria

Mitochondria are organelles of prokaryotic origin and it is believed that the eukaryotic cell originated from endosymbiosis between a host cell and ingested bacteria. As a result, mitochondria share common features with their ancestors; they contain a distinct circular genome organized in nucleoids in the matrix and their own mitochondrial DNA (mtDNA) maintenance machinery. Mammalian mitochondria contain over 1500 proteins, but mtDNA encodes only 11 mRNAs for 13 subunits of the oxidative phosphorylation system, 22 transfer RNAs (tRNAs) and two ribosomal RNAs (rRNAs) which are essential for mitochondrial translation. Therefore, mitochondria have to rely on nuclear DNA for the production of the rest of their proteins. Consequently, mitochondria and the nucleus are tightly connected regarding production and import of mitochondrial proteins (Gaston et al. 2009; Mootha et al. 2003; Pagliarini et al. 2008; Sickmann et al. 2003; Neupert and Herrmann 2007; Schmidt et al. 2010).

Mitochondria are also structurally similar to bacteria in that they contain two membranes (the inner and the outer membranes) which separate the matrix and the inter-membrane compartment. The outer membrane is responsible for molecular exchanges with the cytosol and contains various integrated proteins at the surface such as the GTPases mitofusin (1 and 2) involved in mitochondrial fusion (Chen et al. 2003).

Generally, mitochondria are well known for their role in the production of energy (ATP) via oxidative phosphorylation (OXPHOS). The electron transport chain (ETC) and ATP synthase are located on the inner membrane which is highly folded into cristae (McBride et al. 2006). The OXPHOS system is composed of five multi-subunit enzymatic complexes: NADH-ubiquinone oxidoreductase (complex I), succinate-ubiquinone oxidoreductase (complex II), ubiquinone-cytochrome c oxidoreductase (complex III), cytochrome c oxidase (complex IV) and the ATP synthase (complex V) (Signes and Fernandez-Vizarra 2018).

Electrons from reduced substrates pass from complex I up to complex IV and ultimately react with oxygen, forming water as a byproduct and causing protons to be pumped across the mitochondrial inner membrane. The proton pumping force drives protons to complex V in the inner membrane forming ATP from ADP and phosphate. The total driving force driving protons (Δp) to mitochondria is a combination of the mitochondrial membrane potential ($\Delta\psi m$) and the mitochondria pH gradient (ΔpHm) which typically ranges between 180 and 220 mV (Mitchell and Moyle 1969; Nicholls 1974; Zorova et al. 2018). However, during this process, electrons may inadvertently leak and react with oxygen to form superoxide anion, which can lead to damage in several macromolecules including DNA, proteins and lipids. Superoxide anions can additionally react with other free radicals species such as nitric oxide or with transition metals to form highly reactive secondary ROS such as hydroxyl radical (Murphy 2009). Nevertheless, cells possess mechanisms capable of counteracting ROS. Superoxide anions can be converted to hydrogen peroxide by antioxidant enzymes superoxide dismutase (Mn-SOD, present in the mitochondrial matrix and Cu/Zn-SOD in the cytosol) and then further broken down by enzymes catalase or glutathione peroxidase to oxygen and water. It has been proposed that ROS generated during this process damages mtDNA and other mitochondrial components, which in turn leads to respiratory chain dysfunction.

Denham Harman was the first to hypothesize that mitochondrial dysfunction and consequent generation of ROS could be a driver of the aging process (Harman 1972). It has been argued that mtDNA may be more sensitive to oxidative modifications than the nuclear genome, due to its close proximity to the ETC and to ROS. mtDNA has been shown to have a mutation rate 2–3 fold greater than nuclear DNA (Yakes and Van Houten 1997; Brandon et al. 2006; Cline 2012). Products from mtDNA oxidation include 8-oxo-2′-deoxyguanosine (8-oxo-dG), and thimine glycol (TG), the former is a highly mutagenic lesion leading to misrepair and tranversion of G:C → A:T. Increased base modifications have been reported in human cells and this increases with age (Grollman and Moriya 1993; Michikawa et al. 1999; Ames 2009).

Apart from energy production, mitochondria are also involved in other essential cellular processes such as several metabolic reactions, cell-death, autophagy, cell-cycle control and intracellular signaling (Ryan and Hoogenraad 2007; Daiber 2010; Nunnari and Suomalainen 2012). All of these processes play important roles in cellular senescence as we will discuss below.

11.2.2 Senescent Cells Contain Dysfunctional Mitochondria

During senescence, mitochondria have been shown to experience a variety of changes in terms of their structure, dynamics and function (Chapman et al. 2019).

Mitochondria from senescent cells exhibit an increased proton leak, decreased mitochondrial membrane potential and increased generation of ROS (Passos et al. 2010; Passos et al. 2007). Additionally, senescent cells have been shown to experience an increase in mitochondrial mass and in the abundance of tricarboxylic acid (TCA) cycle metabolites (Kaplon et al. 2013), however, this may be cell-type specific. For instance, while mitochondrial mass increase during senescence has been shown in fibroblasts (Passos et al. 2007, 2010; Moiseeva et al. 2009), hepatocytes (Correia-Melo et al. 2016) and pancreatic β-cells (Helman et al. 2016), the opposite seems to happen in a subset of $CD8^{+}$ T-cells, which show several features of cellular senescence (Henson et al. 2014). Furthermore, senescent cells show differences in the organization of their mitochondrial networks. Studies have shown that mitochondria in senescent cells exist in a hyper-fused state, which may be a consequence of decreased expression of proteins involved in mitochondrial fission such as Fis1 (mitochondrial fission protein1) and Drp1 (dynamin related protein 1) (Mai et al. 2010). Additionally, live-cell imaging of mitochondria from senescent cells revealed that they have significantly reduced fission and fusion rates (Dalle Pezze et al. 2014).

Senescent cells have also been shown to activate a retrograde response (Passos et al. 2007). The mitochondrial retrograde response is a way of communication between the mitochondria and the nucleus which allow the activation of adaptive processes to counteract mitochondrial dysfunction (Butow and Avadhani 2004). It is characterized by a major reprogramming of nuclear gene expression and has been shown to trigger Ca^{2+} dependent signaling, modulation of mitochondrial biogenesis and metabolic and anti-apoptotic changes (Butow and Avadhani 2004; Miceli and Jazwinski 2005).

The observation that mitochondria become dysfunctional during senescence raises several important questions: (i) do mitochondria contribute to the senescent phenotype and if so, in what way?; (ii) what factors contribute to the loss of mitochondrial function during senescence?; (iii) can we target mitochondrial dysfunction as an anti-senescence therapy? In the remainder of this book chapter we will endeavor to answer some of these questions.

11.2.3 Mitochondrial-Derived ROS Is a Driver of Telomere-Dependent Senescence

Mitochondrial ROS have been shown to drive telomere-dependent cellular senescence by different mechanisms. Mild oxidative stress has been shown to accelerate the rate of telomere shortening contributing to premature senescence (von Zglinicki 2002; von Zglinicki et al. 1995). Telomeric DNA, which is rich in guanine triplets, is thought to be particularly sensitive to oxidative modifications when compared to the rest of the genome, resulting in the accumulation of single-stranded breaks at telomeres (Petersen et al. 1998). These single-stranded breaks are thought to contribute to accelerated telomere shortening upon cell division (von Zglinicki 2002). Consistently, interventions which increase mitochondrial ROS have been shown to accelerate the rate of telomere shortening (von Zglinicki 2001). In contrast, interventions that decrease oxidative stress such as low oxygen, antioxidants, overexpression of antioxidant enzymes, nicotinamide and mild mitochondrial uncoupling have been shown to slow down the rate of telomere shortening and extend replicative lifespan in vitro (Passos et al. 2007; Forsyth et al. 2003; Richter and Zglinicki 2007; von Zglinicki et al. 2000; Serra et al. 2003; Kang et al. 2006).

Acute exposure to pro-oxidant agents can also result in premature senescence. This process was thought originally to occur independently of telomeres since no telomere shortening was observed (de Magalhães et al. 2002). More recently, studies have shown that exposure to hydrogen peroxide, high doses of X-ray irradiation or genotoxic agents result in an enrichment of DNA damage response (DDR) proteins at telomeric regions, independently from telomere length (Fumagalli et al. 2012; Hewitt et al. 2012). Contrary to non-telomeric DNA damage, which is rapidly repaired, telomeric DNA damage has been shown to elicit a persistent DDR which is likely to contribute to the stability of the senescence arrest (Anderson et al. 2019;

Victorelli and Passos 2017). Mechanistically, it has been shown that proteins that bind to telomeres (called collectively the shelterin complex), such as TRF2, can inhibit non-homologous end joining, thereby inhibiting telomere repair (Fumagalli et al. 2012). Oxidative stress can also affect telomeres in different ways. For instance, ROS has been shown to affect the binding of the shelterin proteins TRF1 and TRF2 to telomeres, leading to telomere uncapping and subsequent activation of a DDR (Opresko et al. 2005). Recently, it was shown that specific induction of 8-oxoguanine exclusively at telomere regions contributes to shorter telomeres and impaired cell growth (Fouquerel et al. 2019).

While most of the aforementioned studies have been performed in vitro, only recently studies have investigated the relationship between oxidative stress and telomere damage in vivo. Aged catalase$^{-/-}$, MnSOD$^{+/-}$ and Polg mutator mice (which contain a high frequency of mtDNA mutations) have been shown to experience an age-dependent increase in cells containing dysfunctional telomeres (Anderson et al. 2019). Furthermore, cardiac specific overexpression of pro-oxidant monoamine oxidase has been shown to accelerate telomere dysfunction, which can be prevented by administration of antioxidant N-acetyl-cysteine (NAC) (Anderson et al. 2019). Finally, mice with chronic inflammation due to enhanced NF-κB signaling have also been shown to experience age-dependent increase in telomere damage as well as increased ROS, which could be prevented by treatment with antioxidant BHA (Jurk et al. 2014).

Altogether, there is evidence to suggest that mitochondrial ROS is a key player in telomere-induced senescence, however, the relationship between ROS and senescence is considerably more complex as we will examine in the following section.

11.2.4 Mitochondrial ROS Act as Signaling Molecules During Senescence

While ROS can accelerate telomere damage inducing senescence, several studies have shown that ROS can also act as important signaling molecules during senescence, a process which occurs downstream of the activation of the cell-cycle arrest (Correia-Melo and Passos 2015). For example, it has been reported that induction of senescence via genotoxic stress or telomere uncapping resulted in an increased mitochondrial ROS generation, a process which occurs 2–3 days following the original insult (Passos et al. 2010). Other studies have shown that over-expression of p53, p21 and p16 resulted in increased ROS production, presumably of mitochondrial origin (Macip et al. 2003; Macip et al. 2002; Takahashi et al. 2006). Moreover, studies on oncogene-induced senescent cells have reported increased ROS generation in response to RAS and BRAFV600E over-expression, and this was dependent on p53 and pRb (Kaplon et al. 2013; Moiseeva et al. 2009; Lee et al. 1999). In the majority of the aforementioned studies, the cell cycle arrest was prevented by treatment with antioxidants, suggesting that ROS plays an important role in maintaining the senescence arrest.

It is still unclear how ROS impacts on the cell-cycle arrest. However, studies have shown that mitochondrial ROS can actively contribute to the replenishment of DDR foci via a positive feed-back loop (Passos et al. 2010; Correia-Melo et al. 2016). Live-cell imaging experiments using reporter systems which allow the visualization of the dynamics of DDR foci have shown that short-lived DDR foci present in senescent cells are ROS-dependent (Correia-Melo et al. 2016). Activation of senescence through the p16/Rb pathway has been shown to irreversibly block cytokinesis through a positive feedback loop involving ROS-dependent activation of protein kinase C delta (PKCdelta) (Takahashi et al. 2006).

Altogether, these studies indicate that the major signaling pathways driving senescence rely on ROS as signaling molecules to induce the senescence program.

11.2.5 The Role of Mitochondrial-Derived Metabolites in Senescence

The mitochondrial ETC produces various metabolites and essential cofactors which are necessary for normal cellular physiology. In this section, we will review some of the information regarding the impact of certain mitochondrial-derived metabolites in the control of cellular senescence, NAD^+ plays an important role in a variety of pathways such as the tricarboxylic acid (TCA) cycle, protein acetylation and DNA repair. It plays a role in OXPHOS by accepting electrons generated along the ETC, becoming reduced to NADH. Low NAD^+/NADH ratios have been shown to be a feature of cellular senescence and mitochondrial dysfunction (Lee et al. 2012; Wiley et al. 2016). It has been shown that the enzyme nicotinamide phosphoribosyltransferase (NAMPT) the rate-limiting enzyme for NAD^+ salvage from nicotinamide is decreased in senescent cells. Consistent with a causal role for NAMPT in senescence, decreased NAMPT activity led to premature senescence whereas its overexpression extended replicative lifespan (van der Veer et al. 2007). Additionally, decreased NAD^+/NADH ratio, induced by depletion of the cytosolic malate dehydrogenase (MDH1) a key component of the malate aspartate shuttle, has been shown to activate AMPK and lead to senescence (Lee et al. 2012). Low NAD^+/NADH levels have also been associated with the regulation of the SASP. It was shown that induction of mitochondrial dysfunction resulted in an atypical SASP, which lacked the interleukin 1 regulated arm (Wiley et al. 2016). NAD^+ also plays a role in other intracellular processes which have been implicated in senescence. NAD^+ is required for PARP (poly-ADP ribose polymerase) mediated DNA damage repair, which has been shown to be involved in cellular senescence (Efimova et al. 2010). NAD^+ also acts as a co-factor in the sirtuin family of proteins. Sirtuins are regulated by NAD^+/NADH ratios and have been suggested to serve as sensors of the metabolic state of cells, and implicated in aging and longevity (Lin et al. 2000; Longo and Kennedy 2006).

In the context of cellular senescence, sirtuins have been implicated in various processes such as DNA repair, metabolism and inflammation (Grabowska et al. 2017). For example, Sirtuin1 (Sirt1), a histone deacetylase that regulates p53 function and silences expression of NF-κB (a major regulator of the SASP), has been shown to decline during aging and senescence (Sasaki et al. 2006; Imai and Guarente 2014).

Recent evidence suggests that other pathways are involved in the relationship between senescence and NAD metabolism. It was recently shown that the SASP could induce increased expression of NADase CD38 (cluster of differentiation 38) in non-senescent cells, such as bone marrow derived macrophages or endothelial cells (Chini et al. 2019). This led to the hypothesis that during the aging process, accumulation of senescent cells and the SASP could contribute to a decline in NAD^+ levels via up-regulation of CD38 (Hogan et al. 2019). Consistently, pharmacological inhibition of CD38 was shown to prevent the decline in NAD^+ levels observed during murine aging and importantly reduce telomere-associated DNA damage, a major driver of senescence, in muscle and liver (Tarrago et al. 2018). Mechanistically, this process may involve NAD regulation of Sirtuins. Accordingly, a more recent study has demonstrated that the administration of NAD^+ precursor nicotinamide mononucleotide prevented telomere dysfunction in mice, a process which was partially dependent on Sirt1 (Amano et al. 2019).

Mitochondrial dysfunction can also reduce ATP synthesis. It has been suggested that inhibition of ATP synthesis induces senescence with upregulation of p21 and p16 pathways (Stockl et al. 2006). Other studies have shown that depletion of ATP levels, which results in the elevation of AMP to ATP ratio, or addition of extracellular AMP promotes growth arrest and senescence (Stockl et al. 2006; Wang et al. 2003; Zwerschke et al. 2003). Increase of ADP (or AMP) to ATP ratio activates AMP-activated protein kinase (AMPK), which is an energy sensor and a central mediator of cell metabolism (Mihaylova and Shaw 2011). Activation of AMPK has been shown to trigger a cell cycle arrest in a variety of cell types from different species including human, mouse and fly (Jones et al. 2005; Rattan et al. 2005; Owusu-Ansah et al. 2008; Humbert et al. 2010; Mandal et al. 2010; Li et al. 2011; Peyton et al. 2012). Constant activation of AMPK has been shown to increase (Peyton et al. 2012) and to induce senescence in a p53 dependent manner (Jones et al. 2005; Jiang et al. 2013). In addition, AMPK activation has been shown to down-regulate genes involved in proliferation such as cyclin A, cyclin B1 and cyclin E (Wang et al. 2002, 2003; Mandal et al. 2010; Peyton et al. 2012). Furthermore, AMPK has also been implicated in the increase of p16 through the inhibition of the RNA stabilizing factor human antigen R (HuR). By blocking its translocation to the cytoplasm from the nucleus, the expression of target mRNAs involved in the cell cycle arrest is enhanced (Wang et al. 2003). In addition, since AMPK can directly phosphorylate pRb at Ser^{804}, a decrease in AMPK activity is involved in the reduction of pRB, leading to a decrease in proliferation and senescence (Wang et al. 2003; Peyton et al. 2012; Wang et al. 2002; Dasgupta and Milbrandt 2009). Moreover, AMPK has been shown to be increased in oncogene-induced senescence, and its inactivation has been show to promote transformation of cells and cancer, supporting the role of AMPK in the establishment of a cell cycle arrest (Moiseeva et al. 2009; Bardeesy et al. 2002; Huang et al. 2008; Shackelford and Shaw 2009; Zhou et al. 2009).

11.2.6 Mitochondria Are Key Regulators of the SASP

As discussed in the previous sections, senescent cells are characterized by mitochondrial dysfunction. Additionally, mitochondrial-derived metabolites, including ROS, have been shown to play key roles in the induction of the senescence arrest. However, given the complexity of the pathways involved, the multitude of functions regulated by mitochondria and its close interaction with the nucleus it is challenging to investigate experimentally which specific aspects of mitochondrial biology contribute to senescence. For that reason, our group designed a proof-of-principle experiment aiming to investigate the role of mitochondria in the development of the senescent phenotype. We utilized the property of PINK1/Parkin to induce widespread mitophagy in order to generate human cells without mitochondria (Correia-Melo et al. 2017). Using this method, we demonstrated that senescent cells where mitochondria had been eliminated via widespread mitophagy, could not induce a SASP but still underwent an irreversible cell-cycle arrest (Correia-Melo et al. 2016). This work supported the concept that mitochondria are key regulators of the SASP. Consequently, a better understanding of the pathways involved in this process, could lead to potential pharmacological interventions to suppress the pro-inflammatory phenotype of senescent cells while keeping the cells in permanent cell cycle arrest (tumor suppressive capacity) (Birch and Passos 2017).

11.2.7 Possible Mechanisms by Which Mitochondria Regulate the SASP

Thus far, not much is known about the possible mechanisms by which mitochondria regulate the SASP (Fig. 11.1). ROS generated by mitochondria could be a potential contributor as some studies indicate. It has been suggested that the persistent DDR signaling observed during senescence, which is partially due to mitochondrial ROS, could contribute to the SASP (Passos et al. 2010). This could act directly, as studies indicate that ROS can activate NF-κB (Morgan and Liu 2011) or indirectly via ROS-dependent induction of DNA damage and a DDR (Rodier et al. 2009). Consistently, inhibition of a main initiator of the DDR, the serine/threonine protein kinase ATM has been shown to reduce the SASP by suppressing the activation of the pathway NEMO, IKK complex and nuclear translocation of NF-κB (a key transcription factor of a plethora of SASP factors) (Rodier et al. 2009; McCool and Miyamoto 2012). Additionally, increased mitochondrial biogenesis during senescence, a process regulated by mTORC1 has been implicated in the development

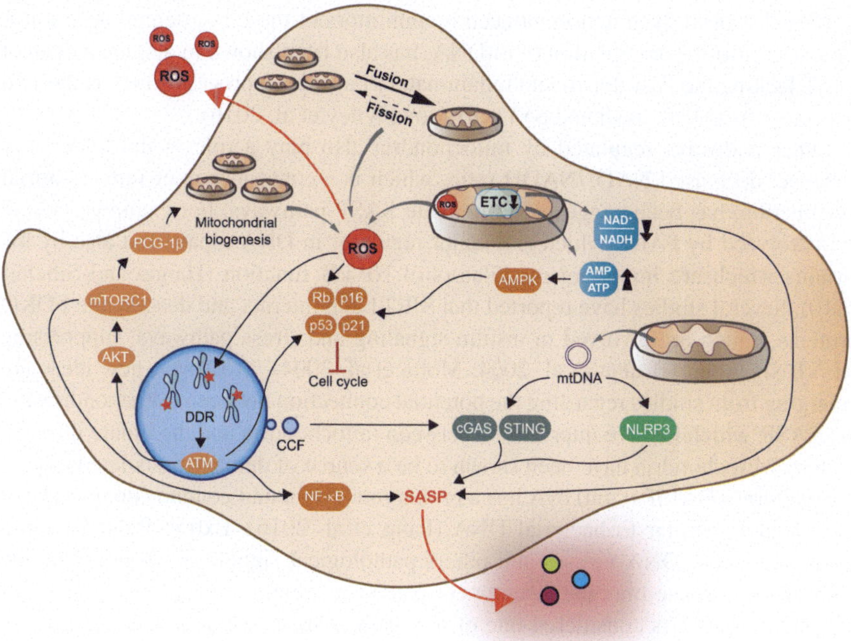

Fig. 11.1 Mitochondria regulate cellular senescence. Cumulative evidence suggests that mitochondria from senescent cells have an impaired electron transport chain (ETC) which results in increased Reactive Oxygen Species (ROS), decreased NAD^+/NADH ratios and increased AMP/ATP ratios. Furthermore, mitochondria in senescent cells have decreased fusion and fission rates and increased mitochondrial biogenesis which has been shown to contribute to mitochondrial dysfunction and ROS generation. Finally, evidence suggests that mitochondrial derived metabolites can contribute to both the cell-cycle arrest as well as the Senescence-associated secretory phenotype (SASP)

of the senescent phenotype (Correia-Melo et al. 2016). mTOR is a serine threonine kinase which has been suggested to be important in the aging process since its inhibition has been shown to extend lifespan and healthspan in several model organisms (Wilkinson et al. 2012; Harrison et al. 2009). During senescence, it has been shown that the DDR can directly activate mTOR via AKT, subsequently leading to PGC1-β-dependent mitochondrial biogenesis, ROS generation and the development of the SASP. Consistently, mTORC1 inhibitor rapamycin reduced mitochondrial biogenesis, ROS and the SASP both in vitro and in vivo (Correia-Melo et al. 2016, 2019). However, mTOR can also impact on the SASP via alternative mechanisms. It has been suggested that mTOR inhibition results in decreased translation of the membrane-bound cytokine IL-1A (Laberge et al. 2015) or MAPKAPK2 (Herranz et al. 2015), both of which control this SASP. Furthermore, mTOR inhibition has also been shown to enhance Parkin-mediated mitophagy, reducing mitochondrial dysfunction and senescence (Manzella et al. 2018).

Mitochondrial dysfunction induced by inhibitors of the ETC, depletion of mitochondrial Sirtuins or depletion of mtDNA, has also been shown to display a distinct SASP lacking the IL-1 dependent inflammatory pathway, a process which is thought to occur via AMPK-mediated p53 activation (Wiley et al. 2016).

Other pathways regulated by mitochondria also play a role in the SASP. For instance, decreased NAD^+/NADH ratio, which is a consequence of mitochondrial dysfunction, has been suggested to regulate SASP pathways. For example, NAD+ is hydrolyzed by PARP, which is a major regulator in DNA repair, and also by the sirtuins which are important modulators of NF-κB function (Haigis and Sinclair 2010). Several studies have reported that SIRT1 can interact and deacetylate FOXO proteins, which are involved in insulin signaling and stress pathways, suppressing NF-κB signaling (Brunet et al. 2004; Motta et al. 2004). Moreover, new ideas are emerging from studies regarding the potential connection between mitochondria and the SASP, which involve interactions between mitochondria and the innate immune system. Mitochondria have been shown to be a source of damage associated molecular patterns (DAMPs). mtDNA has a prokaryote origin and contains unmethylated CpG islands similar to bacterial DNA (Fang et al. 2016). Extracellular bacterial DNA and RNA, DNA viruses and others pathological signals are detected by the cells through innate immune receptors or pattern recognition receptors (PRRs). For instance, mtDNA is considered one of the main DAMPs when is detected outside the mitochondrial matrix (Fang et al. 2016). Therefore, cells have developed several signaling pathways which sense foreign DNA, including the NLRP3 inflammasome, which eventually activates NF-κB activity (Groslambert and Py 2018). The cGAS-STING pathway has also been described to trigger an antiviral immune response and expression of type I interferon genes (Xu et al. 2000; West et al. 2015). cGAS-STING pathway has been shown to identify fragments of nuclear DNA commonly referred to as cytoplasmic chromatin fragments (CCFs) (Ivanov et al. 2013). Recognition of CCFs by cGAS has been shown to drive the SASP (Dou et al. 2017; Gluck et al. 2017). Why CCFs arise in senescent cells is still not completely understood. One study suggests that disruption of the nuclear envelope which occurs during senescence may contribute to their formation (Dou et al. 2017). cGAS/STING pathway has also been shown to interact with cytosolic mtDNA and induce inflammation (Riley et al. 2018), however, it is still unknown if this contribute to senescence or the SASP.

Recent studies have also shown additional links between mitochondrial metabolites and the regulation of the SASP. Cardiolipin, which is a glycerophospholipid present in the inner mitochondrial membrane and acts as a DAMP, has been shown not only to increase in senescence but also to induce senescence (Arivazhagan et al. 2004; Wan et al. 2014). Furthermore, it was shown that mitochondrial derived peptides humanin and MOTS-c were increased in senescent cells, and that their administration to senescent cells resulted in higher mitochondrial respiration and an enhancement of certain components of the SASP (Kim et al. 2018).

Altogether, published data shows that mitochondrial may regulate the SASP via different mechanisms; however, additional studies are needed to further understand the complex pathways involved.

11.3 Targeting Mitochondria—A New Senotherapy?

As previously mentioned, mitochondria have been shown to be play a role in the cell cycle arrest and the gradual development of the SASP. Consequently, they are an essential regulator of the senescence response. Additionally, mitochondria are involved in the apoptosis resistance characteristic of senescent cells. For example, the B-cell lymphoma 2 (Bcl-2) Homology 3 (BH3) mimetics drugs, such as ABT-263 (navitoclax), which act as senolytics, target the anti-apoptotic BCL-2 protein family (Zhu et al. 2015, 2016; Chang et al. 2016; Thompson et al. 2019). BCL2 anti-apoptotic proteins are key regulators of apoptosis; they protect the mitochondrial outer membrane integrity by binding to the pro-apoptotic Bcl-2 family members, Bax (BCL-2-associated x protein) and Bak (BCL-2 homologous antagonist killer). Upon oligomerization, Bax and Bak form homodimers that permeabilize the outer membrane of mitochondria by forming large macro-pores allowing the release of cytochrome c, caspase activation and programmed cell death.

A recent study has demonstrated that a mitochondria-targeted tamoxifen (MitoTam) compound induced apoptosis in senescent cells. The mechanism of action was found to be associated with a low level of expression of adenine nucleotide translocase (ANT2) in senescent cells when compared to young cells. Moreover, MitoTam was shown to decrease the expression of senescent markers p16 and p19 in the lungs and kidneys of 19 month old mice (Hubackova et al. 2019).

Although targeting anti-apoptotic proteins mostly located at mitochondria appears to be an effective strategy to eliminate senescent cells, some studies indicate that some of these drugs may also have detrimental side effects. For instance, a recent study has shown that treatment with the BH3 mimetic drug, ABT737, induced minority mitochondrial outer membrane permeabilization (miMOMP) in young cells (Ichim et al. 2015). This response was not sufficient to trigger apoptosis but led to the activation of caspases (such as pro-caspase-3) which resulted in DNA damage and genomic instability (Ichim et al. 2015). Therefore, the long-term consequences of treatment with some of the senolytic drugs currently being tested on young, non-senescent cells should be further investigated.

Another senotherapy that is currently being investigated involves inhibiting the SASP, without inducing apoptosis or restoring proliferation of senescent cells. These drugs are commonly known as senostatic. As previously mentioned, the mitochondria are key regulators of the SASP during senescence, making them an ideal candidate to target the pro-inflammatory signature of senescent cells (Correia-Melo et al. 2016; Correia-Melo and Passos 2015). Indeed, a few pre-clinical studies have shown that drugs which have senostatic properties also impact on mitochondrial function. For example, rapamycin which is an mTOR inhibitor, has been shown to reduce mitochondrial ROS, and decrease the SASP in vitro and in vivo while maintaining the cell cycle arrest characteristic of senescent cells. Treatment with rapamycin has been shown to extend lifespan and improve the heathspan of aging mice (Correia-Melo et al. 2016, 2019; Harrison et al. 2009).

In addition, metformin which is an inhibitor of mitochondrial complex I and widely used in treatment of type 2 diabetes has been shown to have an effect on mitochondrial oxidative phosphorylation, attenuate ROS production and DNA damage in oncogene induced senescence (OIS), while maintaining the cell cycle arrest (Owen et al. 2000; Algire et al. 2012). Moreover, metformin has been shown to reduce the SASP by preventing the translocation of the transcription factor NF-κB and inhibiting the phosphorylation of IκB and IKKα/β, kinase (Deschênes-Simard et al. 2013).

In conclusion, mitochondria are key regulators of senescence and they are important targets for both senolytic and senostatic therapies. We predict that by further investigating the mechanisms by which mitochondria drive senescence will help us identify novel therapies to counteract senescence.

Acknowledgements We would like to thank the Mayo Clinic Robert and Arlene Kogod Center and the Ted Nash Long Life Foundation for support.

References

Acosta J, Banito A, Wuestefeld T, Georgilis A, Janich P, Morton J et al (2013) A complex secretory program orchestrated by the inflammasome controls paracrine senescence. Nat Cell Biol 15(8):978–990

Algire C, Moiseeva O, Deschenes-Simard X, Amrein L, Petruccelli L, Birman E et al (2012) Metformin reduces endogenous reactive oxygen species and associated DNA damage. Cancer Prev Res (Philadelphia, PA) 5(4):536–543

Amano H, Chaudhury A, Rodriguez-Aguayo C, Lu L, Akhanov V, Catic A et al (2019) Telomere dysfunction induces sirtuin repression that drives telomere-dependent disease. Cell Metab 29(6):1274–1290.e9

Ames BN (2009) Endogenous oxidative DNA damage, aging, and cancer. Free Radical Res Commun 7(3–6):121–128

Anderson R, Lagnado A, Maggiorani D, Walaszczyk A, Dookun E, Chapman J et al (2019) Length-independent telomere damage drives post-mitotic cardiomyocyte senescence. EMBO J 38(5):e100492

Arivazhagan P, Mizutani E, Fujii M, Ayusawa D (2004) Cardiolipin induces premature senescence in normal human fibroblasts. Biochem Biophys Res Commun 323:739–742

Baker D, Wijshake T, Tchkonia T, LeBrasseur N, Childs B, van de Sluis B et al (2011) Clearance of $p16^{Ink4a}$-positive senescent cells delays ageing-associated disorders. Nature 479(7372):232–236

Bardeesy N, Sinha M, Hezel AF, Signoretti S, Hathaway NA, Sharpless NE et al (2002) Loss of the Lkb1 tumour suppressor provokes intestinal polyposis but resistance to transformation. Nature 419(6903):162–167

Beauséjour C, Krtolica A, Galimi F, Narita M, Lowe S, Yaswen P et al (2003) Reversal of human cellular senescence: roles of the p53 and p16 pathways. EMBO J 22(16):4212–4222

Birch J, Passos JF (2017) Targeting the SASP to combat ageing: Mitochondria as possible intracellular allies? BioEssays News Rev Mol Cell Dev Biol 39(5):1600235

Birch J, Anderson RK, Correia-Melo C, Jurk D, Hewitt G, Marques FM et al (2015) DNA damage response at telomeres contributes to lung aging and chronic obstructive pulmonary disease. Am J Physiol Lung Cell Mol Physiol 309(10):L1124–L1137

Birch J, Victorelli S, Rahmatika D, Anderson RK, Jiwa K, Moisey E et al (2016) Telomere dysfunction and senescence-associated pathways in bronchiectasis. Am J Respir Crit Care Med 193(8):929–932

Brandon M, Baldi P, Wallace DC (2006) Mitochondrial mutations in cancer. Oncogene 25:4647

Brugarolas J, Chandrasekaran C, Gordon JI, Beach D, Jacks T, Hannon GJ (1995) Radiation-induced cell cycle arrest compromised by p21 deficiency. Nature 377(6549):552–557

Brunet A, Sweeney LB, Sturgill JF, Chua KF, Greer PL, Lin Y et al (2004) Stress-dependent regulation of FOXO transcription factors by the SIRT1 deacetylase. Science 303(5666):2011–2015

Bussian TJ, Aziz A, Meyer CF, Swenson BL, van Deursen JM, Baker DJ (2018) Clearance of senescent glial cells prevents tau-dependent pathology and cognitive decline. Nature 562(7728):578–582

Butow RA, Avadhani NG (2004) Mitochondrial signaling: the retrograde response. Mol Cell 14(1):1–15

Chang J, Wang Y, Shao L, Laberge R-M, Demaria M, Campisi J et al (2016) Clearance of senescent cells by ABT263 rejuvenates aged hematopoietic stem cells in mice. Nat Med 22(1):78–83

Chapman J, Fielder E, Passos JF (2019) Mitochondrial dysfunction and cell senescence: deciphering a complex relationship. FEBS Lett 593(13):1566–1579

Chen H, Detmer SA, Ewald AJ, Griffin EE, Fraser SE, Chan DC (2003) Mitofusins Mfn1 and Mfn2 coordinately regulate mitochondrial fusion and are essential for embryonic development. J Cell Biol 160(2):189–200

Childs BG, Baker DJ, Wijshake T, Conover CA, Campisi J, van Deursen JM (2016) Senescent intimal foam cells are deleterious at all stages of atherosclerosis. Science 354(6311):472–477

Chini C, Hogan KA, Warner GM, Tarragó MG, Peclat TR, Tchkonia T et al (2019) The NADase CD38 is induced by factors secreted from senescent cells providing a potential link between senescence and age-related cellular NAD+ decline. Biochem Biophys Res Commun 513(2):486–493

Chinta SJ, Woods G, Demaria M, Rane A, Zou Y, McQuade A et al (2018) Cellular senescence is induced by the environmental neurotoxin paraquat and contributes to neuropathology linked to Parkinson's disease. Cell Rep 22(4):930–940

Cline SD (2012) Mitochondrial DNA damage and its consequences for mitochondrial gene expression. Biochimica et Biophysica Acta (BBA)-Gene Regulatory Mechanisms 1819(9):979–991

Coppé J-P, Patil CK, Rodier F, Sun Y, Muñoz DP, Goldstein J et al (2008) Senescence-associated secretory phenotypes reveal cell-nonautonomous functions of oncogenic RAS and the p53 tumor suppressor. PLoS Biol 6(12):e301

Correia-Melo C, Passos JF (2015) Mitochondria: are they causal players in cellular senescence? Biochem Biophys Acta 1847(11):1373–1379

Correia-Melo C, Ichim G, Tait Stephen WG, Passos JF (2017) Depletion of mitochondria in mammalian cells through induction of widespread mitophagy. Nat Protoc 12(1):183–194

Correia-Melo C, Birch J, Fielder E, Rahmatika D, Taylor J, Chapman J et al (2019) Rapamycin improves healthspan but not inflammaging in nfkappab1$^{-/-}$ mice. Aging Cell 18(1):e12882

Correia-Melo C, Marques FDM, Anderson R, Hewitt G, Hewitt R, Cole J et al (2016) Mitochondria are required for pro-ageing features of the senescent phenotype. EMBO J

Cristofalo VJ, Pignolo RJ (1993) Replicative senescence of human fibroblast-like cells in culture. Physiol Rev 73(3):617–638

Cristofalo VJ, Doggett DL, Brooks-Frederich KM, Phillips PD (1989) Growth factors as probes of cell aging. Exp Gerontol 24(5–6):367–374

Daiber A (2010) Redox signaling (cross-talk) from and to mitochondria involves mitochondrial pores and reactive oxygen species. Biochimica et Biophysica Acta (BBA)-Bioenergetics 1797(6):897–906

Dalle Pezze P, Nelson G, Otten EG, Korolchuk VI, Kirkwood TBL, von Zglinicki T et al (2014) Dynamic modelling of pathways to cellular senescence reveals strategies for targeted interventions. PLoS Comput Biol 10(8):e1003728-e

Dasgupta B, Milbrandt J (2009) AMP-activated protein kinase phosphorylates retinoblastoma protein to control mammalian brain development. Dev Cell 16(2):256–270
de Magalhaes JP, Passos JF (2018) Stress, cell senescence and organismal ageing. Mech Ageing Dev 170:2–9
de Magalhães JP, Chainiaux F, Remacle J, Toussaint O (2002) Stress-induced premature senescence in BJ and hTERT-BJ1 human foreskin fibroblasts. FEBS Lett 523(1):157–162
Demaria M, Ohtani N, Youssef Sameh A, Rodier F, Toussaint W, Mitchell James R et al (2014) An essential role for senescent cells in optimal wound healing through secretion of PDGF-AA. Dev Cell 31(6):722–733
Deschênes-Simard X, Gaumont-Leclerc M-F, Bourdeau V, Lessard F, Moiseeva O, Forest V et al (2013) Tumor suppressor activity of the ERK/MAPK pathway by promoting selective protein degradation. Genes Dev 27(8):900–915
Dou Z, Ghosh K, Vizioli MG, Zhu J, Sen P, Wangensteen KJ et al (2017) Cytoplasmic chromatin triggers inflammation in senescence and cancer. Nature 550(7676):402–406
Efimova EV, Mauceri HJ, Golden DW, Labay E, Bindokas VP, Darga TE et al (2010) Poly(ADP-ribose) polymerase inhibitor induces accelerated senescence in irradiated breast cancer cells and tumors. Cancer Res 70(15):6277–6282
el-Deiry WS, Harper JW, O'Connor PM, Velculescu VE, Canman CE, Jackman J et al (1994) WAF1/CIP1 is induced in p53-mediated G1 arrest and apoptosis. Cancer Res 54(5):1169–1174
Fang C, Wei X, Wei Y (2016) Mitochondrial DNA in the regulation of innate immune responses. Protein Cell 7(1):11–16
Farr JN, Xu M, Weivoda MM, Monroe DG, Fraser DG, Onken JL et al (2017) Targeting cellular senescence prevents age-related bone loss in mice. Nat Med 23(9):1072–1079
Forsyth NR, Evans AP, Shay JW, Wright WE (2003) Developmental differences in the immortalization of lung fibroblasts by telomerase. Aging Cell 2(5):235–243
Fouquerel E, Barnes RP, Uttam S, Watkins SC, Bruchez MP, Opresko PL (2019) Targeted and persistent 8-oxoguanine base damage at telomeres promotes telomere loss and crisis. Mol Cell 75(1):117–130.e6
Fumagalli M, Rossiello F, Clerici M, Barozzi S, Cittaro D, Kaplunov JM et al (2012) Telomeric DNA damage is irreparable and causes persistent DNA-damage-response activation. Nat Cell Biol 14(4):355–365
Gasser S, Orsulic S, Brown EJ, Raulet DH (2005) The DNA damage pathway regulates innate immune system ligands of the NKG2D receptor. Nature 436(7054):1186–1190
Gaston D, Tsaousis AD, Roger AJ (2009) Chapter 2 predicting proteomes of mitochondria and related organelles from genomic and expressed sequence tag data. In: Methods in Enzymology, vol 457. Academic Press, pp 21–47
Geng Y, Eaton EN, Picon M, Roberts JM, Lundberg AS, Gifford A et al (1996) Regulation of cyclin E transcription by E2Fs and retinoblastoma protein. Oncogene 12(6):1173–1180
Gluck S, Guey B, Gulen MF, Wolter K, Kang TW, Schmacke NA et al (2017) Innate immune sensing of cytosolic chromatin fragments through cGAS promotes senescence. Nat Cell Biol 19(9):1061–1070
Grabowska W, Sikora E, Bielak-Zmijewska A (2017) Sirtuins, a promising target in slowing down the ageing process. Biogerontology 18(4):447–476
Grollman AP, Moriya M (1993) Mutagenesis by 8-oxoguanine: an enemy within. Trends Genet 9(7):246–249
Groslambert M, Py BF (2018) Spotlight on the NLRP3 inflammasome pathway. J Inflamm Res 11:359–374
Haigis MC, Sinclair DA (2010) Mammalian sirtuins: biological insights and disease relevance. Annu Rev Pathol Mech Dis, 253–295
Harman D (1972) The biologic clock: the mitochondria? J Am Geriatr Soc 20(4):145–147
Harrison DE, Strong R, Sharp ZD, Nelson JF, Astle CM, Flurkey K et al (2009) Rapamycin fed late in life extends lifespan in genetically heterogeneous mice. Nature 460(7253):392–395

Hayflick L (1965) The limited in vitro lifetime of human diploid cell strains. Exp Cell Res 37:614–636

Hayflick L, Moorhead P (1961) The serial cultivation of human diploid cell strains. Exp Cell Res 25:585–621

Helman A, Klochendler A, Azazmeh N, Gabai Y, Horwitz E, Anzi S et al (2016) $p16^{Ink4a}$-induced senescence of pancreatic beta cells enhances insulin secretion. Nat Med 22:412

Henson SM, Lanna A, Riddell NE, Franzese O, Macaulay R, Griffiths SJ et al (2014) p38 signaling inhibits mTORC1-independent autophagy in senescent human $CD8^{+}$ T cells. J Clin Investig 124(9):4004–4016

Herranz N, Gallage S, Mellone M, Wuestefeld T, Klotz S, Hanley CJ et al (2015) mTOR regulates MAPKAPK2 translation to control the senescence-associated secretory phenotype. Nat Cell Biol 17(9):1205–1217

Hewitt G, Jurk D, Marques FD, Correia-Melo C, Hardy T, Gackowska A et al (2012) Telomeres are favoured targets of a persistent DNA damage response in ageing and stress-induced senescence. Nat Commun 3:708

Hogan KA, Chini CCS, Chini EN (2019) The multi-faceted ecto-enzyme CD38: roles in immunomodulation, cancer, aging, and metabolic diseases. Front Immunol 10:1187

Huang X, Wullschleger S, Shpiro N, McGuire VA, Sakamoto K, Woods YL et al (2008) Important role of the LKB1-AMPK pathway in suppressing tumorigenesis in PTEN-deficient mice. Biochem J 412(2):211–221

Hubackova S, Davidova E, Rohlenova K, Stursa J, Werner L, Andera L et al (2019) Selective elimination of senescent cells by mitochondrial targeting is regulated by ANT2. Cell Death Differ 26(2):276–290

Humbert N, Navaratnam N, Augert A, Da Costa M, Martien S, Wang J et al (2010) Regulation of ploidy and senescence by the AMPK-related kinase NUAK1. EMBO J 29(2):376–386

Ichim G, Lopez J, Ahmed Shafiq U, Muthalagu N, Giampazolias E, Delgado ME et al (2015) Limited mitochondrial permeabilization causes DNA damage and genomic instability in the absence of cell death. Mol Cell 57(5):860–872

Imai S, Guarente L (2014) NAD+ and sirtuins in aging and disease. Trends Cell Biol 24(8):464–471

Ivanov A, Pawlikowski J, Manoharan I, van Tuyn J, Nelson DM, Rai TS et al (2013) Lysosome-mediated processing of chromatin in senescence. J Cell Biol 202(1):129–143

Jeon OH, Kim C, Laberge RM, Demaria M, Rathod S, Vasserot AP et al (2017) Local clearance of senescent cells attenuates the development of post-traumatic osteoarthritis and creates a pro-regenerative environment. Nat Med 23(6):775–781

Jiang S, Park DW, Stigler WS, Creighton J, Ravi S, Darley-Usmar V et al (2013) Mitochondria and AMP-activated protein kinase-dependent mechanism of efferocytosis. J Biol Chem 288(36):26013–26026

Jones RG, Plas DR, Kubek S, Buzzai M, Mu J, Xu Y et al (2005) AMP-activated protein kinase induces a p53-dependent metabolic checkpoint. Mol Cell 18(3):283–293

Jurk D, Wilson C, Passos JF, Oakley F, Correia-Melo C, Greaves L et al (2014) Chronic inflammation induces telomere dysfunction and accelerates ageing in mice. Nat Commun 2:4172

Justice JN, Nambiar AM, Tchkonia T, LeBrasseur NK, Pascual R, Hashmi SK et al (2019) Senolytics in idiopathic pulmonary fibrosis: Results from a first-in-human, open-label, pilot study. EBioMedicine 40:554–563

Kang HT, Lee HI, Hwang ES (2006) Nicotinamide extends replicative lifespan of human cells. Aging Cell 5(5):423–436

Kaplon J, Zheng L, Meissl K, Chaneton B, Selivanov VA, Mackay G et al (2013) A key role for mitochondrial gatekeeper pyruvate dehydrogenase in oncogene-induced senescence. Nature 498(7452):109–112

Kim SJ, Mehta HH, Wan J, Kuehnemann C, Chen J, Hu JF et al (2018) Mitochondrial peptides modulate mitochondrial function during cellular senescence. Aging (Albany NY) 10(6):1239–1256

Kirkwood T, Cremer T (1982) Cytogerontology since 1881: a reappraisal of August Weissmann and a review of modern progress. Hum Genet 60:101–121
Krizhanovsky V, Yon M, Dickins R, Hearn S, Simon J, Miething C et al (2008) Senescence of activated stellate cells limits liver fibrosis. Cell 134(4):657–667
Krtolica A, Parrinello S, Lockett S, Desprez PY, Campisi J (2001) Senescent fibroblasts promote epithelial cell growth and tumorigenesis: a link between cancer and aging. Proc Natl Acad Sci USA 98(21):12072–12077
Laberge R-M, Sun Y, Orjalo AV, Patil CK, Freund A, Zhou L et al (2015) MTOR regulates the pro-tumorigenic senescence-associated secretory phenotype by promoting IL1A translation. Nat Cell Biol 17(8):1049–1061
Lagouge M, Larsson NG (2013) The role of mitochondrial DNA mutations and free radicals in disease and ageing. J Intern Med 273(6):529–543
Lee AC, Fenster BE, Ito H, Takeda K, Bae NS, Hirai T et al (1999) Ras proteins induce senescence by altering the intracellular levels of reactive oxygen species. J Biol Chem 274(12):7936–7940
Lee CW, Wong LL, Tse EY, Liu HF, Leong VY, Lee JM et al (2012) AMPK promotes p53 acetylation via phosphorylation and inactivation of SIRT1 in liver cancer cells. Cancer Res 72(17):4394–4404
Levy MZ, Allsopp RC, Futcher AB, Greider CW, Harley CB (1992) Telomere end-replication problem and cell aging. J Mol Biol 225(4):951–960
Li Y, Xu S, Mihaylova MM, Zheng B, Hou X, Jiang B et al (2011) AMPK phosphorylates and inhibits SREBP activity to attenuate hepatic steatosis and atherosclerosis in diet-induced insulin-resistant mice. Cell Metab 13(4):376–388
Lin S-J, Defossez P-A, Guarente L (2000) Requirement of NAD and SIR2 for life-span extension by calorie restriction in saccharomyces cerevisiae. Science 289(5487):2126–2128
Longo VD, Kennedy BK (2006) Sirtuins in aging and age-related disease. Cell 126(2):257–268
Macip S, Igarashi M, Fang L, Chen A, Pan Z, Lee S et al (2002) Inhibition of p21-mediated ROS accumulation can rescue p21-induced senescence. EMBO J 21(9):2180–2188
Macip S, Igarashi M, Berggren P, Yu J, Lee SW, Aaronson SA (2003) Influence of induced reactive oxygen species in p53-mediated cell fate decisions. Mol Cell Biol 23(23):8576–8585
Mai S, Klinkenberg M, Auburger G, Bereiter-Hahn J, Jendrach M (2010) Decreased expression of Drp1 and Fis1 mediates mitochondrial elongation in senescent cells and enhances resistance to oxidative stress through PINK1. J Cell Sci 123(Pt 6):917–926
Mandal P, Pritchard MT, Nagy LE, Mandal P, Pritchard MT, Nagy LE (2010) Anti-inflammatory pathways and alcoholic liver disease: role of an adiponectin/interleukin-10/heme oxygenase-1 pathway. World J Gastroenterol 16:1330–1336
Manzella N, Santin Y, Maggiorani D, Martini H, Douin-Echinard V, Passos JF et al (2018) Monoamine oxidase-A is a novel driver of stress-induced premature senescence through inhibition of parkin-mediated mitophagy. Aging Cell 17(5):e12811
McBride HM, Neuspiel M, Wasiak S (2006) Mitochondria: more than just a powerhouse. Curr Biol 16(14):R551–R560
McCool KW, Miyamoto S (2012) DNA damage-dependent NF-κB activation: NEMO turns nuclear signaling inside out. Immunol Rev 246(1):311–326
Miceli MV, Jazwinski SM (2005) Common and cell type-specific responses of human cells to mitochondrial dysfunction. Exp Cell Res 302(2):270–280
Michikawa Y, Mazzucchelli F, Bresolin N, Scarlato G, Attardi G (1999) Aging-dependent large accumulation of point mutations in the human mtDNA control region for replication. Science 286(5440):774–779
Mihaylova MM, Shaw RJ (2011) The AMPK signalling pathway coordinates cell growth, autophagy and metabolism. Nat Cell Biol 13(9):1016–1023
Mirzayans R, Andrais B, Scott A, Paterson MC, Murray D (2010) Single-cell analysis of p16^{INK4a} and p21^{WAF1} expression suggests distinct mechanisms of senescence in normal human and Li-Fraumeni Syndrome fibroblasts. J Cell Physiol 223(1):57–67
Mitchell P, Moyle J (1969) Estimation of membrane potential and pH difference across the cristae membrane of rat liver mitochondria. Eur J Biochem 7(4):471–484

Moiseeva O, Bourdeau V, Roux A, Deschênes-Simard X, Ferbeyre G (2009) Mitochondrial dysfunction contributes to oncogene-induced senescence. Mol Cell Biol 29(16):4495–4507
Mootha VK, Bunkenborg J, Olsen JV, Hjerrild M, Wisniewski JR, Stahl E et al (2003) Integrated analysis of protein composition, tissue diversity, and gene regulation in mouse mitochondria. Cell 115(5):629–640
Morgan MJ, Liu ZG (2011) Crosstalk of reactive oxygen species and NF-κB signaling. Cell Res 21(1):103–115
Motta MC, Divecha N, Lemieux M, Kamel C, Chen D, Gu W et al (2004) Mammalian SIRT1 represses forkhead transcription factors. Cell 116(4):551–563
Munoz-Espin D, Canamero M, Maraver A, Gomez-Lopez G, Contreras J, Murillo-Cuesta S et al (2013) Programmed cell senescence during mammalian embryonic development. Cell 155(5):1104–1118
Murphy Michael P (2009) How mitochondria produce reactive oxygen species. Biochem J 417(1):1–13
Musi N, Valentine JM, Sickora KR, Baeuerle E, Thompson CS, Shen Q et al (2018) Tau protein aggregation is associated with cellular senescence in the brain. Aging Cell 17(6):e12840
Nelson G, Wordsworth J, Wang C, Jurk D, Lawless C, Martin-Ruiz C et al (2012) A senescent cell bystander effect: senescence-induced senescence. Aging Cell 11(2):345–349
Neupert W, Herrmann JM (2007) Translocation of proteins into mitochondria. Annu Rev Biochem 76:723–749
Nicholls DG (1974) The influence of respiration and ATP hydrolysis on the proton-electrochemical gradient across the inner membrane of rat-liver mitochondria as determined by ion distribution. Eur J Biochem 50(1):305–315
Nunnari J, Suomalainen A (2012) Mitochondria: in sickness and in health. Cell 148(6):1145–1159
Ogrodnik M, Miwa S, Tchkonia T, Tiniakos D, Wilson CL, Lahat A et al (2017) Cellular senescence drives age-dependent hepatic steatosis. Nat Commun 8:15691
Ogrodnik M, Zhu Y, Langhi LGP, Tchkonia T, Kruger P, Fielder E et al (2019) Obesity-induced cellular senescence drives anxiety and impairs neurogenesis. Cell Metab 29(5):1061–1077.e8
Ohtani K, DeGregori J, Nevins JR (1995) Regulation of the cyclin E gene by transcription factor E2F1. Proc Natl Acad Sci U S A. 92(26):12146–12150
Opresko PL, Fan J, Danzy S, Wilson DM 3rd, Bohr VA (2005) Oxidative damage in telomeric DNA disrupts recognition by TRF1 and TRF2. Nucleic Acids Res 33(4):1230–1239
Owen MR, Doran E, Halestrap AP (2000) Evidence that metformin exerts its anti-diabetic effects through inhibition of complex 1 of the mitochondrial respiratory chain. Biochem J 348(Pt 3):607–614
Owusu-Ansah E, Yavari A, Mandal S, Banerjee U (2008) Distinct mitochondrial retrograde signals control the G1-S cell cycle checkpoint. Nat Genet 40(3):356–361
Pagliarini DJ, Calvo SE, Chang B, Sheth SA, Vafai SB, Ong S-E et al (2008) A mitochondrial protein compendium elucidates complex I disease biology. Cell 134(1):112–123
Palmer AK, Xu M, Zhu Y, Pirtskhalava T, Weivoda MM, Hachfeld CM et al (2019) Targeting senescent cells alleviates obesity-induced metabolic dysfunction. Aging Cell 18(3):e12950
Passos JF, Saretzki G, von Zglinicki T (2007a) DNA damage in telomeres and mitochondria during cellular senescence: is there a connection? Nucleic Acids Res 35(22):7505–7513
Passos J, Saretzki G, Ahmed S, Nelson G, Richter T, Peters H et al (2007) Mitochondrial dysfunction accounts for the stochastic heterogeneity in telomere-dependent senescence. PLoS Biol 5(5)
Passos J, Nelson G, Wang C, Richter T, Simillion C, Proctor C et al (2010) Feedback between p21 and reactive oxygen production is necessary for cell senescence. Mol Syst Biol 6:347
Petersen S, Saretzki G, von Zglinicki T (1998) Preferential accumulation of single-stranded regions in telomeres of human fibroblasts. Exp Cell Res 239(1):152–160
Peyton KJ, Liu XM, Yu Y, Yates B, Durante W (2012) Activation of AMP-activated protein kinase inhibits the proliferation of human endothelial cells. J Pharmacol Exp Ther 342(3):827–834

Rattan R, Giri S, Singh AK, Singh I (2005) 5-Aminoimidazole-4-carboxamide-1-beta-D-ribofuranoside inhibits cancer cell proliferation in vitro and in vivo via AMP-activated protein kinase. J Biol Chem 280(47):39582–39593

Richter T, Zglinicki TV (2007) A continuous correlation between oxidative stress and telomere shortening in fibroblasts. Exp Gerontol 42(11):1039–1042

Riley JS, Quarato G, Cloix C, Lopez J, O'Prey J, Pearson M et al (2018) Mitochondrial inner membrane permeabilisation enables mtDNA release during apoptosis. EMBO J 37(17)

Rodier F, Coppe J-P, Patil CK, Hoeijmakers WAM, Munoz DP, Raza SR et al (2009) Persistent DNA damage signalling triggers senescence-associated inflammatory cytokine secretion. Nat Cell Biol 11(8):973–979

Roos CM, Zhang B, Palmer AK, Ogrodnik MB, Pirtskhalava T, Thalji NM et al (2016) Chronic senolytic treatment alleviates established vasomotor dysfunction in aged or atherosclerotic mice. Aging Cell 15(5):973–977

Ryan MT, Hoogenraad NJ (2007) Mitochondrial-nuclear communications. Annu Rev Biochem 76(1):701–722

Sasaki T, Maier B, Bartke A, Scrable H (2006) Progressive loss of SIRT1 with cell cycle withdrawal. Aging Cell 5(5):413–422

Schafer MJ, Miller JD, LeBrasseur NK (2017) Cellular senescence: implications for metabolic disease. Mol Cell Endocrinol 455:93–102

Schmidt O, Pfanner N, Meisinger C (2010) Mitochondrial protein import: from proteomics to functional mechanisms. Nat Rev Mol Cell Biol 11:655

Seluanov A, Gorbunova V, Falcovitz A, Sigal A, Milyavsky M, Zurer I et al (2001) Change of the death pathway in senescent human fibroblasts in response to DNA damage is caused by an inability to stabilize p53. Mol Cell Biol 21(5):1552–1564

Serra V, von Zglinicki T, Lorenz M, Saretzki G (2003) Extracellular superoxide dismutase is a major antioxidant in human fibroblasts and slows telomere shortening. J Biol Chem 278(9):6824–6830

Serrano M, Hannon GJ, Beach D (1993) A new regulatory motif in cell-cycle control causing specific inhibition of cyclin D/CDK4. Nature 366(6456):704–707

Serrano M, Lin AW, McCurrach ME, Beach D, Lowe SW (1997) Oncogenic ras provokes premature cell senescence associated with accumulation of p53 and p16(INK4a). Cell 88(5):593–602

Shackelford DB, Shaw RJ (2009) The LKB1-AMPK pathway: metabolism and growth control in tumour suppression. Nat Rev Cancer 9(8):563–575

Sickmann A, Reinders J, Wagner Y, Joppich C, Zahedi R, Meyer HE et al (2003) The proteome of Saccharomyces cerevisiae mitochondria. Proc Natl Acad Sci 100(23):13207–13212

Signes A, Fernandez-Vizarra E (2018) Assembly of mammalian oxidative phosphorylation complexes I–V and supercomplexes. Essays Biochem 62(3):255–270

Stockl P, Hutter E, Zwerschke W, Jansen-Durr P (2006) Sustained inhibition of oxidative phosphorylation impairs cell proliferation and induces premature senescence in human fibroblasts. Exp Gerontol 41(7):674–682

Storer M, Mas A, Robert-Moreno A, Pecoraro M, Ortells MC, Di Giacomo V et al (2013) Senescence is a developmental mechanism that contributes to embryonic growth and patterning. Cell 155(5):1119–1130

Takahashi A, Ohtani N, Yamakoshi K, Iida S, Tahara H, Nakayama K et al (2006) Mitogenic signalling and the p16(INK4a)-Rb pathway cooperate to enforce irreversible cellular senescence. Nat Cell Biol 8(11):1291–1297

Tarrago MG, Chini CCS, Kanamori KS, Warner GM, Caride A, de Oliveira GC et al (2018) A potent and specific CD38 inhibitor ameliorates age-related metabolic dysfunction by reversing tissue NAD(+) decline. Cell Metab 27(5):1081–1095.e10

Thompson PJ, Shah A, Ntranos V, Van Gool F, Atkinson M, Bhushan A (2019) Targeted elimination of senescent beta cells prevents type 1 diabetes. Cell Metab 29(5):1045–1060.e10

van der Veer E, Ho C, O'Neil C, Barbosa N, Scott R, Cregan SP et al (2007) Extension of human cell lifespan by nicotinamide phosphoribosyltransferase. J Biol Chem 282(15):10841–10845

Victorelli S, Passos JF (2017) Telomeres and cell senescence-size matters not. EBioMedicine 21:14–20
von Zglinicki T (2001) Telomeres and replicative senescence: is it only length that counts? Cancer Lett 168(2):111–116
von Zglinicki T (2002) Oxidative stress shortens telomeres. Trends Biochem Sci 27(7):339–344
von Zglinicki T, Saretzki G, Döcke W, Lotze C (1995) Mild hyperoxia shortens telomeres and inhibits proliferation of fibroblasts: a model for senescence? Exp Cell Res 220(1):186–193
von Zglinicki T, Pilger R, Sitte N (2000) Accumulation of single-strand breaks is the major cause of telomere shortening in human fibroblasts. Free Radic Biol Med 28(1):64–74
Walaszczyk A, Dookun E, Redgrave R, Tual-Chalot S, Victorelli S, Spyridopoulos I et al (2019) Pharmacological clearance of senescent cells improves survival and recovery in aged mice following acute myocardial infarction. Aging Cell 18(3):e12945
Wan M, Hua X, Su J, Thiagarajan D, Frostegard AG, Haeggstrom JZ et al (2014) Oxidized but not native cardiolipin has pro-inflammatory effects, which are inhibited by Annexin A5. Atherosclerosis 235(2):592–598
Wang E, Lee MJ, Pandey S (1994) Control of fibroblast senescence and activation of programmed cell death. J Cell Biochem 54(4):432–439
Wang W, Chen JX, Liao R, Deng Q, Zhou JJ, Huang S et al (2002) Sequential activation of the MEK-extracellular signal-regulated kinase and MKK3/6-p38 mitogen-activated protein kinase pathways mediates oncogenic ras-induced premature senescence. Mol Cell Biol 22(10):3389–3403
Wang W, Yang X, López de Silanes I, Carling D, Gorospe M (2003) Increased AMP: ATP ratio and AMP-activated protein kinase activity during cellular senescence linked to reduced HuR function. J Biol Chem 278(29):27016–27023
Wang H, Wang M, Wang H, Böcker W, Iliakis G (2005) Complex H2AX phosphorylation patterns by multiple kinases including ATM and DNA-PK in human cells exposed to ionizing radiation and treated with kinase inhibitors. J Cell Physiol 202(2):492–502
Watanabe S, Kawamoto S, Ohtani N, Hara E (2017) Impact of senescence-associated secretory phenotype and its potential as a therapeutic target for senescence-associated diseases. Cancer Sci 108(4):563–569
West AP, Khoury-Hanold W, Staron M, Tal MC, Pineda CM, Lang SM et al (2015) Mitochondrial DNA stress primes the antiviral innate immune response. Nature 520:553
Wiley CD, Velarde MC, Lecot P, Liu S, Sarnoski EA, Freund A et al (2016) Mitochondrial dysfunction induces senescence with a distinct secretory phenotype. Cell Metab 23(2):303–314
Wilkinson JE, Burmeister L, Brooks SV, Chan CC, Friedline S, Harrison DE et al (2012) Rapamycin slows aging in mice. Aging Cell 11(4):675–682
Xu Y, Tao X, Shen B, Horng T, Medzhitov R (2000) Structural basis for signal transduction by the Toll/interleukin-1 receptor domains. Nature 408:111
Xu M, Bradley EW, Weivoda MM, Hwang SM, Pirtskhalava T, Decklever T et al (2017) Transplanted senescent cells induce an osteoarthritis-like condition in mice. J Gerontol Ser A Biol Sci Med Sci 72(6):780–785
Xu M, Pirtskhalava T, Farr JN, Weigand BM, Palmer AK, Weivoda MM et al (2018) Senolytics improve physical function and increase lifespan in old age. Nat Med 24(8):1246–1256
Xue W, Zender L, Miething C, Dickins R, Hernando E, Krizhanovsky V et al (2007) Senescence and tumour clearance is triggered by p53 restoration in murine liver carcinomas. Nature 445(7128):656–660
Yakes FM, Van Houten B (1997) Mitochondrial DNA damage is more extensive and persists longer than nuclear DNA damage in human cells following oxidative stress. Proc Natl Acad Sci 94(2):514–519
Yosef R, Pilpel N, Tokarsky-Amiel R, Biran A, Ovadya Y, Cohen S et al (2016) Directed elimination of senescent cells by inhibition of BCL-W and BCL-XL. Nat Commun 7:11190
Yousefzadeh MJ, Zhu Y, McGowan SJ, Angelini L, Fuhrmann-Stroissnigg H, Xu M et al (2018) Fisetin is a senotherapeutic that extends health and lifespan. EBioMedicine 36:18–28

Zhou J, Huang W, Tao R, Ibaragi S, Lan F, Ido Y et al (2009) Inactivation of AMPK alters gene expression and promotes growth of prostate cancer cells. Oncogene 28(18):1993–2002

Zhu Y, Tchkonia T, Pirtskhalava T, Gower AC, Ding H, Giorgadze N et al (2015) The Achilles' heel of senescent cells: from transcriptome to senolytic drugs. Aging Cell 14(4):644–658

Zhu Y, Tchkonia T, Fuhrmann-Stroissnigg H, Dai HM, Ling YY, Stout MB et al (2016) Identification of a novel senolytic agent, navitoclax, targeting the Bcl-2 family of anti-apoptotic factors. Aging Cell 15(3):428–435

Zorova LD, Popkov VA, Plotnikov EY, Silachev DN, Pevzner IB, Jankauskas SS et al (2018) Mitochondrial membrane potential. Anal Biochem 552:50–59

Zwerschke W, Mazurek S, Stockl P, Hutter E, Eigenbrodt E, Jansen-Durr P (2003) Metabolic analysis of senescent human fibroblasts reveals a role for AMP in cellular senescence. Biochem J 376(Pt 2):403–411

The manufacturer's authorised representative in the EU is Springer Nature Customer Service Centre GmbH, Europaplatz 3, 69115 Heidelberg, Germany. If you have any concerns regarding our products, please contact ProductSafety@springernature.com

Printed and bound by CPI Group (UK) Ltd, Croydon, CR0 4YY
30/09/2026
02225648-0005